Current Research in Abiotic and Biotic Stress in Plants

Current Research in Abiotic and Biotic Stress in Plants

Editor: Thomas Flynn

STATES
ACADEMIC PRESS
www.statesacademicpress.com

States Academic Press,
109 South 5th Street,
Brooklyn, NY 11249, USA

Visit us on the World Wide Web at:
www.statesacademicpress.com

ISBN: 978-1-63989-134-4 (Hardback)

Cataloging-in-Publication Data

Current research in abiotic and biotic stress in plants / edited by Thomas Flynn.
p. cm.
Includes bibliographical references and index.
ISBN 978-1-63989-134-4
1. Plants--Effect of stress on. 2. Plant physiology. 3. Crops--Effect of stress on.
4. Crops--Physiology. I. Flynn, Thomas.
QK754 .C87 2022
581.7--dc23

Table of Contents

Preface

The purpose of the book is to provide a glimpse into the dynamics and to present opinions and studies of some of the scientists engaged in the development of new ideas in the field from very different standpoints. This book will prove useful to students and researchers owing to its high content quality.

Stresses faced by plants due to changing external environment are classified into abiotic and biotic stress. Biotic stresses occur due to the infestation of plants by various organisms such as fungi, bacteria, viruses, herbivores and nematodes. These agents acts as parasites and deprive plants of their nutrients. Abiotic stresses include salination of water, drought, floods, desertification and extreme weather conditions. To combat these situations, plants undergo various biological changes such as altered generation, pathogen-associated molecular patterns, etc. Across the world, these stresses are responsible for loss of harvest and decreased crop yield. Increasing scarcity of water and changing weather patterns are predicted to increase stresses in plants. This book brings forth some of the most innovative concepts and elucidates the unexplored aspects of abiotic and biotic stress in plants. From theories to research to practical applications, case studies related to all contemporary topics of relevance to this field have been included in it. This book is a vital tool for all researching or studying abiotic and biotic stresses in plants as it gives incredible insights into emerging trends and concepts.

At the end, I would like to appreciate all the efforts made by the authors in completing their chapters professionally. I express my deepest gratitude to all of them for contributing to this book by sharing their valuable works. A special thanks to my family and friends for their constant support in this journey.

Editor

1

Secondary Metabolism in *Amaranthus* spp. — A Genomic Approach to Understand its Diversity and Responsiveness to Stress in Marginally Studied Crops with High Agronomic Potential

Paula Andrea Castrillón-Arbeláez and John Paul Délano Frier

Abstract

Cultivated amaranths are crops with an unrealized agronomical potential despite their high nutritional value and nutraceutic properties of their seeds and/ or leaves. They tolerate growing conditions unsuitable for cereals, and are tolerant to biotic aggressors. Several *Amaranthus* species are abundant of sources of secondary metabolites, mostly phenylpropanoids, predominantly in seeds and leaves, many of which may confer health benefits associated with their antioxidant properties. They could also act as defensive compounds against predators or pathogens. Recent biochemical and molecular approaches partly defined the mechanisms responsible for grain amaranth´s tolerance against biotic stress. However, the role played by secondary metabolites in (a)biotic stress amelioration in amaranth is practically unknown. Our group has identified several genes coding for enzymes involved in secondary metabolism pathways in A. *hypochondriacus,* in addition to related regulatory transcription factors. More than 50% of these genes involve the phenylpropanoid pathway. In this chapter, the role played by this pathway in (a)biotic stress amelioration in plants will be briefly reviewed, followed by an examination of its involvement in the conferral of nutraceutic properties to amaranth plants. A description of the progress obtained so far regarding the characterization of phenylpropanoid genes in grain amaranth will close this chapter.

Keywords: (a)biotic stress, grain amaranth, phenylpropanoids

1. Introduction

Species belonging to the Amaranthaceae compose a diverse and interesting family of plants. They can develop in highly contrasting habitats, from arid and semi-arid zones, where they can survive in sandy alkaline and/ or serpentine soils, to disturbed tropical forests. A minority are found in aquatic, semi-aquatic or marine environments. Their high affinity to saline conditions stems, in part, from to the weedy nature of most of the species that constitute the *Alternanthera, Amaranthus, Celosia, Chamissoa, Froelichia, Gomphrena, Guillemi-nea,* and *Iresine* genera. These plants are widely distributed in tropical and subtropical zones due to their ability to colonize or invade diverse habitats [1].

The *Amaranthus* genus is highly diverse, including approximately 70 species. A fraction of these may grow in saline soils; in this regard, *A. greggii* is considered to be a marker of saliferous soils. This genus includes grain-producing species, the most important being *A. cruentus, A. caudatus,* and *A. hypochondriacus*. Grain amaranths are valued for the high protein content and multiple nutraceutic properties of their seeds [2,3]. Moreover, they possess an inherent tolerance to high temperatures and drought, traits which have been associated with their C4 physiology, indeterminate flowering habit and superior water use efficiency due to their ability to grow long tap roots and develop an extensive lateral root system [4–6]. In concordance with related species within the Amaranthaceae, grain amaranths can tolerate poor and saline soils conditions and erratic rains unsuitable for the cultivation of other grain crops. Besides, they are not particularly susceptible to major diseases or insect pests [5–8]. Their high tolerance to severe defoliation, produced by mechanical means [9,10] or by insect folivory [6,11], is believed to contribute to this trait. In addition, they can readily respond to chemical elicitors of defense responses, such as jasmonic acid (JA) [12-14] or benzothiadiazole (BTH) [15], to increase their resistance against highly damaging phloem-feeding insect pests, such as the tarnished bug *Lygus lineolaris*, or against potentially lethal pathogenic bacteria. However, the contribution of secondary metabolites to (a)biotic stress tolerance in grain amaranth has been barely explored.

In this chapter, we shall first review the role played by phenylpropanoid pathway in the amelioration of both biotic and abiotic stress in plants. Then, we shall proceed to de-scribe the state of the art with regard to the phenylpropanoid pathway in *Amaranthus*, mostly in relation to stress, but also in relation to the their role in the biosynthesis of compounds with proposed nutraceutic properties. Finally, a thorough description of the progress we have achieved so far with respect to the genomic characterization of this particular secondary metabolic pathway in grain amaranth, mostly in terms of stress tolerance and/ or resistance will be made. The information gathered as a result of these efforts will expand the knowledge, perhaps into unknown territory, about the chemical diversity in plants, particularly in grain amaranth and related species, many of which can thrive in extreme habitats.

2. Reactive Oxygen Species (ROS)

2.1. Brief description of their ubiquitous and malignant role in plant stress and the antioxidant defense mechanisms induced for their control — What is known regarding amelioration of ROS-related damage in amaranth plants during stress?

The abundance of reactive oxygen species (ROS) tends to increase in the tissues of plants exposed to diverse environmental challenges including contact with heavy metal contaminants in soil, water or salinity stress, which is often accompanied by high light and temperature, and nutrient deficiency [16]. Most stress conditions in plants cause an accumulation of ROS, such as superoxide ion, hydrogen peroxide, oxygen-containing radicals, and others. These chemical entities can produce extensive oxidative damage in the apoplastic compartment and may also harm cellular membranes by lipid peroxidation. Additionally, they can have an impact on ion homeostasis mechanisms, which are crucial for many stress tolerance mechanisms, by interfering with ion fluxes [17]. ROS detoxification frequently involves the combined action of both antioxidant metabolites such as ascorbate, glutathione, tocopherols, and ROS-detoxifying enzymes such as superoxide dismutase (SOD), ascorbate peroxidase (APX), and catalase (CAT) [18,19]. In this sense, overproduction of antioxidants in response to drought-induced oxidative stress has been frequently found to be associated with the drought stress tolerance of different plant species [20,21]. Also, enhanced drought tolerance has been generated in several different transgenic plants transformed with genes encoding diverse types of antioxidant-related enzymes or metabolites (e.g., SOD, APX, monodehydroascorbate reductase, and tocopherol cyclase, a key enzyme of tocopherol biosynthesis, [22]). Regarding grain amaranth, a series of proteomic studies performed in plants exposed to drought and saline stress detected the accumulation of various antioxidant enzymes, similar to those mentioned above [23,24]. In addition, the participation of antioxidant genes, such as *AhCAT*, *AhAPX*, and *AhSOD*, was implied by findings showing their induced-up-regulation in grain amaranth plants primed to resist infection by pathogenic bacteria [15]. However, a further characterization of the additional six *CAT*, four *APX*, and three *SOD* genes identified in grain amaranth [12] remains to be performed.

Glycine betalaine (GB) is a quaternary ammonium compound that acts as an osmolite with protective functions in plants subjected to the osmotic stress normally produced under drought, high temperature and/ or excessive salinity conditions. GB accumulation in the cytoplasm reduces ion toxicity, and ameliorates the highly damaging effects caused by the usually simultaneous presence of dehydration, salinity, and extreme temperature stresses. The protective effect is proposed to be exerted by the stabilization of macromolecular structures, and/ or by the protection of chloroplasts, particularly the photosystem II complex. The latter is believed to involve the thermodynamic stabilization of the indirect interaction of extrinsic photosystem II complex proteins with membrane phosphatidylcholine moieties [25]. GB is synthesized by the two-step oxidation of choline. The first step is catalyzed by choline monooxygenase, followed by the action of betaine aldehyde dehydrogenase. Both genes have been described in different amaranth species [12,26,27], whereas the presence of GB has been reported in all Amaranthaceae species examined so far, with the exception of *Chenopodium quinoa* and *Noaea mucronata*. However, the GB concentration needed to exert a protective effect in plants of this family has been found to vary widely in a species-specific manner [25]. On the other hand, GB-accumulating transgenic plants usually show a significant improvement in their tolerance to different abiotic stress conditions [28].

2.2. *Betacyanins* in amaranth: More than pigments?

Betalains are water-soluble, nitrogen-containing pigments that are found only in one group of angiosperms, the Caryophyllales. For reasons that remain unresolved, they have never been found jointly with anthocyanins in the same plant [29,30]. This particular trait has been employed for chemo-taxonomical purposes. These pigments can be divided into the red-violet betacyanins and the yellow betaxanthins. Both are immonium conjugates of betalamic acid covalently bonded with cyclo-dihydroxyphenylalanine (cDOPA) glucosides, which can undergo further acylations [31–33]. These pigments are possibly needed for the optical attraction of pollinators and seed dispersers. Regarding stress, a protective role against accumulating ROS has been inferred from a number of studies [34–36] whereas the betacyanin amaranthine has also been proposed to exert protective effects against photo-oxidative damage in *A. tricolor* [36].

Recently, an analysis of key genes/ enzymes of the betacyanin biosynthetic pathway in *A. hypochondriacus* was performed. The study included genes coding for cyclo-DOPA glycosyl-transferase (*AhcDOPA5-GT*), two 4, 5-DOPA-extradiol-dioxygenase isoforms (*AhDODA-1* and *AhDODA-2*, respectively), a betanidin 5-glycosyl transferase (*AhB5-GT*), and an ortholog of the cytochrome P-450 R gene (*CYP76AD1*). The expression pattern of these genes, together with DOPA oxidase tyrosinase assays, was determined in both green and red tissues. The results obtained suggested that two apparently independent glycosylation pathways leading to betanins could be operating in *A. hypochondriacus* to synthesize amaranthine. In addition, these genes/ enzymes were shown to be induced differentially in a tissue-specific and genotype-specific manner in response to different stimuli, including water- and salt-stress and insect herbivory. The results obtained from the abiotic stress assays suggested that genes other than those examined in this study were probably contributing significantly to pigment content in tissues of stressed *A. hypochondriacus* plants. They also offered the possibility that these genes, due to their high induction by insect herbivory, particularly in acyanic plants, could function in defense responses against chewing insect pests by still undetermined mechanisms [37].

In addition to pigments, diverse phytochemical studies have shown that amaranth plants are capable of synthetizing a notable diversity of secondary metabolites [3,38,39]. Although many of these compounds are not considered to be essential for the primary needs of the plant, they are certainly required for survival in and/ or adaptation to challenging environmental conditions. Many of these compounds may be employed in amaranth and other plants as signaling compounds, in defense and/ or for communication with other organisms, such as pollinators [40–44].

The most commonly found secondary metabolite families found in amaranth and related species are phenylpropanoids, including flavonoids, phenolic acids, and their related amides, followed by alkaloids and terpenoids. From an anthropocentric perspective, some of these chemicals, including betacyanins, flavonoids polyphenols, and phenolic acids, are responsible for conferring amaranth and quinoa tissues with the bioactive antioxidant activity associated with their well-documented health benefit effects [3,38,45–49].

3. Phenylpropanoid secondary metabolites

Plants have accumulated a great diversity of phenolic compounds as a result of their long process of evolutionary adaptation. Approximately 40% of these compounds are derived from the highly diverse phenylpropanoid metabolism. The phenylpropanoid compounds constitute a highly diverse assortment of phenylalanine-derived secondary metabolites. These include flavonoids, which are generally sub classified into the anthocyanins, proanthocyanidins, flavonols, isoflavonoids, phlobaphenes, flavanones, and flavones subgroups, which are found in the majority of higher plants, in addition to the aurone subgroup, which is widespread, but not ubiquitous. Also included are monolignols, lignans, coumarins, phenolic acids, quinines, stilbenoids, and xanthones. Other phenolics include alkylmethoxyphenols, alkylphenols, curcuminoids, furacoumarins, hydroxybenzaldehydes, hydroxybenzoketones, hydroxycinnamaldehydes, hydroxycoumarins, hydroxyphenylpropenes, methoxyphenols, naphthoquinones, phenolic terpenes, and tyrosols.

Most of the flowers and fruits pigments employed for pollinator attraction and seed dispersal are water-soluble anthocyanins [50]. Proanthocyanidins, or condensed tannins, are colorless flavonoid polymers produced by the condensation of flavan-3-ol units [51]. Similarly colorless are the abundant flavonols which are usually found in the form of mono-, di-, or triglycosides [52]. Isoflavonoids and phlobaphenes are groups of flavonoids characterized by being predominantly found in the Papilionoideae family [53] or by their red pigmentation which results from the polymerization of flavan-4-ols [54], respectively. Based on their carbon skeleton, the ubiquitous phenolic acids can belong to the hydroxycinnamic acid type (chlorogenic, ferulic, rosmarinic, and sinapic acids) or to the hydroxybenzoic acid type (p-hydroxybenzoic, vanillic, and protocatechuic acids). Finally, the stilbenes represent a small family of phenylpropanoid metabolites dispersed in over 70 unrelated plant species [55–57]. Interestingly, the latter compounds are induced in response to several biotic and abiotic stimuli or by functionally related elicitors, such as methyl jasmonate (MeJA), and ethylene. Flavonoids are also involved in the regulation of auxin transport [58–60] and participate in the chemical dialog established between the plant roots and nitrogen-fixing bacteria and in signaling pathways designed to modulate ROS levels in plant tissues [61,62]. These compounds are also determining factors of male fertility and precursors for the synthesis of lignin [63–65]. The latter is an aromatic heteropolymer that confers mechanical strength to the cell wall and rigidity to plant stems, whose synthesis involves the assembly of p-coumaryl, coniferyl, and sinapyl alcohol monolignols. Lignin is also a waterproof insulator for cell walls and, as such, facilitates the transport and assimilation of water through the vascular system. It also provides protection against wounding, UV light, and pathogen attack [66].

As noted, phenylpropanoids have notable structural and biological function diversity. In terms of defense in plants, phenylpropanoids can be classified into three broad categories according to their function. Those having signaling activity, those known as phytoanticipins, which are part of the basal defensive arsenal of the plant and constitutively accumulate in certain plants tissues, and those whose *de novo* accumulation in plants, as phytoalexins, is induced in response to a biotic aggressor [67,68]. Phenylpropanoids also contribute to human diet and health. Their recognized bioactive properties have acquired medicinal and nutritional importance, mostly due to their antioxidant, antibacterial, antiviral and other reported activities or as effective therapeutic agents against certain types of cancer, cardiovascular pathologies, diabetes, osteoporosis, and neurodegenerative illnesses associated with oxidative stress [43,69,70].

3.1. Phenylpropanoid biosynthesis: A profusely branched pathway

Phenylpropanoids contain at least one aromatic ring with one or more hydroxyl groups, and are synthesized via the shikimate pathway alone or in combination with the mevalonate pathway. The first three steps in the synthesis of phenylpropanoid-derived compounds are catalyzed by phenylalanine ammonia lyase (PAL), cinnamate 4-hydroxylase (C4H), and 4-coumarate coenzyme A ligase (4CL), collectively referred to as the general phenylpropanoid pathway (GPP). GPP products then serve as precursors for phenylpropanoid-derived compounds [71,72].

The deamination of phenylalanine to cinnamic acid catalyzed by phenylalanine ammonia lyase (PAL, EC 4-3.1-5) is the initial step shared by all phenylpropanoid secondary metabolites. PAL is a conserved homotetrameric protein that is a key enzyme in the phenylpropanoid pathway of higher plants [73–76]. PAL enzymes are grouped as families having many isoforms that are responsive to different developmental and environmental stimuli [77,78].

Cinnamate is the basic structure from which simple phenylpropanoids with the basic C6–C3 carbon skeleton of phenylalanine are produced, via a series of hydroxylation, methylation, and dehydration reactions. This group includes compounds such as p-coumaric, caffeic, ferulic, and sinapic acids and simple coumarins, which rarely accumulate as free acids inside plant cells, being usually conjugated to sugars, cell wall carbohydrates, or organic acids. Salicylic, benzoic, and other acids are uncharacteristic phenylpropanoids that lack the three-carbon side chain, even though they originate from cinnamate and p-coumarate, whereas a large number of stress-induced phenylpropanoids are derived from the C15 flavonoid skeleton, which is synthesized via the chalcone synthase (CHS)-catalyzed condensation of p-coumaroyl-coenzyme A (COA) and three molecules of malonyl-COA. The tetrahydroxychalcone product resulting from the CHS-catalyzed step in most plants is further converted to other flavonoids, such as flavones, flavanones, flavanols, anthocyanins, and 3-deoxyanthocyanidins (Figure 1) [67,68]. Lignin and suberin represent an increased level of complexity, since they are large polymers, constructed from monolignol phenylpropanoid precursors, whose composition varies from species to species.

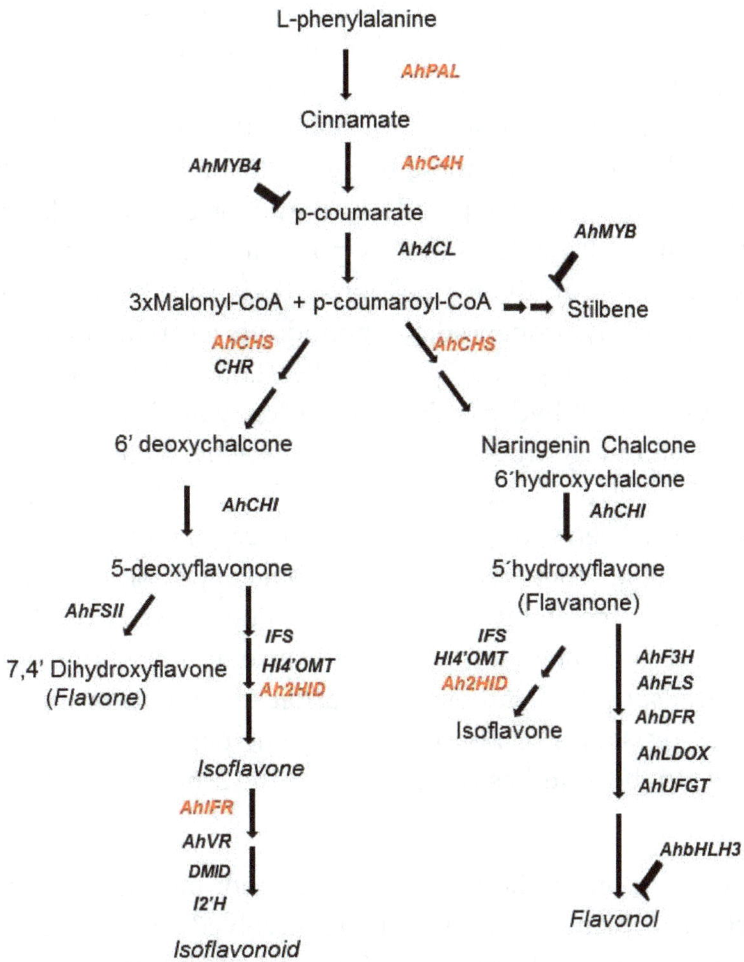

Figure 1. Schematic representation of phenylpropanoid biosynthesis. The image is modified from the original version in [68]. C4H, cinnamate 4-hydroxylase; CHI, chalcone isomerase; CHR, chalcone reductase; CHS, chalcone synthase; 4CL, 4-coumarate:CoA ligase; DMID, 7,2′-dihydroxy-4′-methoxy-isoflavonol dehydratase; FSII, flavone synthase II; 2HID, 2-hydroxyisoflavanone dehydratase; HI4′OMT, 2-hydroxyisoflavanone 4′-*O*-methyltransferase; IFR, isoflavone reductase; IFS, isoflavone synthase; I2′H, isoflavone 2′-hydroxylase; PAL, L-phenylalanine ammonia-lyase; VR, vestitone reductase; F3H, flavanone 3-hydroxylase; UFGT, UDP-glucose flavonol 3-O-glucosyl transferase; FLS, flavonol synthase; DFR, dihydroflavonol 4-reductase; LDOX, leucoanthocyanidin dioxygenase. The genes in red text represent those found to be induced by stress conditions in the grain amaranth transcriptomic analysis [14].

3.2. Regulation of phenylpropanoid biosynthesis: A complex scenario

The biosynthesis and accumulation of secondary metabolites, including phenylpropanoids, are usually tissue- and developmental-stage-specific. As mentioned above, phenylpropanoids can be present as pigments in leaves, flowers, fruits, and seeds or participate in the establishment of mutualistic or detrimental interactions either with beneficial fungi or bacteria or with pathogenic oomycetes [79]. They can participate in the synthesis of lignins and related fibrous materials associated with changes in the cell wall occurring concomitantly with development, in response to stress [66], or in the determination of pollen function. The latter involves the conjugation of polyamines with hydroxycinnamic acid [63–65,80–82]. Numerous factors mediate the expression of phenylpropanoid genes, including sugar levels, transcription factor

(TF) regulation, and diverse types of stress. Sucrose, for instance, has a dual function, first by providing carbon for phenylpropanoid metabolism, and second, by modulating transcriptional and post-translational regulation of many pigment-related genes [83,84]. Recently, a sugar-related regulatory loop was described in which the induction, by sucrose, of AN1, a MYB TF that activates the phenylpropanoid biosynthetic pathway, was self-regulated by the increased sucrolytic activity induced in parallel by the action of AN1 [85]. Structural variability in secondary metabolism is also determined by post-translational chemical modifications of the primary chemical structure by diverse reactions. This is a mechanism that profoundly alters the biological activity of phenylpropanoid compounds via its ability to modify various critical biochemical parameters, including stability, solubility, and/or localization within the cell. For instance, the glycosylation of hydroxycinnamic acids was found to have an important participation during N-limiting stress conditions in *Populus*, which is a plant known for the great diversity of its phenylpropanoid compounds [86]. Similarly, the glycosylation of diverse secondary metabolites has been shown to be needed to regulate oxidative stress responses in various plant species [87,88]. Moreover, methylation and acylation, for example, can favor the volatilization of secondary metabolites by the generation of esters or ethers. A well-known example is salicylic acid (SA), a phenolic acid which is temporarily transformed to its methyl ester to facilitate its transport to distal tissues in order to regulate systemic signaling during the systemic acquired resistance response to biotrophic pathogens [89,90]. Volatile phenylpropanoid esters can also be found as components of the aroma of fruits such as banana and strawberries [91].

Controlled transcription of biosynthetic genes is one major mechanism regulating secondary metabolite production in plant cells. Several TFs involved in the regulation of metabolic pathway genes have been isolated and studied. Synthesis of more than one class of phenylpropanoid-derived compounds is predominantly under the control of V-myb myeloblastosis viral oncogene homolog (MYB) proteins of the R2R3-MYB class that can act both as transcriptional activators and repressors [92]. The participation of these TFs in many phenylpropanoid-related processes has been extensively recorded in various plant species. In Arabidopsis, several R2R3-MYB members have been implicated as positive regulators of lignin synthesis. For instance, the secondary cell wall-associated AtSND1 protein, in association with related proteins, starts a cascade of events that regulate secondary cell wall formation by inducing the expression of the *AtMYB83* and *AtMYB46* R2R3-MYB genes that then trigger the expression of *AtMYB58*, *AtMYB63*, and *AtMYB85*. These, in turn, upregulate various lignin synthesis genes by interacting with their promoter AC elements [93-96]. Conversely, other R2R3-MYB TFs have been shown to act as repressors of the monolignol pathway in both mono and dicot plant species, thereby leading to the suppression of lignin biosynthesis [97–102]. Anthocyanin pigment synthesis is regulated predominantly by a transcriptional complex consisting of three proteins: (i) R2R3-MYB, (ii) basic-helix-loop-helix (bHLH), and (iii) WD-repeat (WDR), which is known as the MBW complex. The first report describing the formation of a MYB bHLH complex for the activation of anthocyanin biosynthesis was made in maize [103]. Subsequent studies recognized the multiple dependence between MYB TFs (i.e., transparent testa2, TT2), WD40, and bHLH (i.e., TT8) for the regulation of the *BANYULS* gene coding for an anthocyanidin reductase required for proanthocyanidin biosynthesis [104].

Complex formation, initiated by the activation of R2R3-MYB genes can be induced by environmental stress conditions or in response to developmental cues. Known targets of this complex are genes encoding dihydroflavonol 4-reductase (DFR), bHLH2, and, curiously, two MYB repressors whose activation leads to a self-regulatory feedback repression loop [105]. The AtDOF4-2, BrMYB4, and AtMYB4 TFs have been reported also as negative regulators of flavonoid and lignan biosynthesis [80,106,107]. In Arabidopsis, anthocyanin biosynthesis via the MBW complex has been demonstrated to be stimulated in response to light, sucrose, nitrogen, and JA [83,108–110].

In addition, anthocyanin patterning and spatial localization are mainly determined by R2R3-three subgroups of MYB activators, many of which have been identified in plants [92]. The MYB, bHLH, and WDR transcription factors have also been shown to be prevalent in the regulation of proanthocyanidin genes [105,111]. Conversely, the regulation of the flavonol pathway is species-specific, and diverges from the above regulatory mechanisms by its diversity, which may require the action of either a single MYB TF, the formation of an MYB-bHLH dimer or an MBW complex. Additionally, augmented flavonol content in Arabidopsis has been found to result from the association of members of the plant-specific teosinte branched1, cycloidea, and proliferating cell nuclear antigen factor, or TCP TF family that interact with AtMYB12 and AtMYB111 [112]. Additionally, the expression of *AtMYB12* in response to both visible and UV-B light is regulated by the basic leucine zipper transcription factor HY5 [113]. Experimental evidence gathered to date clearly indicates that MYB regulation pattern is dependent on cell and tissue type, developmental stage, and environmental conditions. However, information regarding the mechanistic basis of MYB protein responses to biotic and abiotic stimuli remains limited.

3.3. Phenylpropanoid and other less abundant secondary metabolites in amaranth: Nutraceutical properties and suggested defensive roles

Several chemical analyses of diverse tissues of *Amaranthus* species indicate that they differentially accumulate diverse types of secondary metabolites (Table 1). A particular interest in the study of phenolic acids, flavonoids, and other polyphenolics in diverse amaranth species consumed as vegetables and grain has been generated by their high antioxidant activity. As mentioned above, this is mostly because this property has been associated epidemiologically with a decreased risk of diseases associated with oxidative stress, such as cancer and cardiovascular disorder [114]. Consequently, several phenolic acids, flavonoids, and their glycosides have been identified in various *Amaranthus* species (Table 2). These compounds have been isolated using various solvent combinations in various fresh plant tissues obtained from plants at different developmental stages and/ or grown under diverse ambient conditions. They have also been isolated from tissues subjected to diverse processing, from milling to cooking. The concentration of many of these compounds has been observed to vary widely between species and varieties within a species, tissue type, processing and/ or growing conditions, including exposure to (a)biotic stress [3]. A number of selected examples will be described next to illustrate this point.

Metabolic group	[1]cru	*hyp*	*hyb*	*cau*	*pan*	*spi*	*ret*	*liv*	*gan*	*tri*
Flavonoids	[2]+	+	+	+		+				
Betacyanins	+	+	+							+
Phenolic acids	+	+	+	+		+				
Tannins	+	+	+	+		+				
Steroids and triterpenoids	+	+	+	+		+				
Saponins	+	+	+	+		+				
Coumarins					+					
Alkaloids		+				+				
Amarantholidosides				+			+			
Polyhydroxylated nerolidols							+			
Carotenoids	+			+				+	+	
Chlorophylls	+									+
Phytate	+	+		+						
Resinols		+								
Tocopherols and tocotrienols								+		

[1]Amaranthus species included are: **cru**= *cruentus;***hyp**= *hypochondriacus;***hyb**= *hybridus;***cau**= *caudatus;***pan** = *paniculatus;* **spi** = *spinosus;* **ret** = *retroflexus;* **liv** = *lividus;* **gan** = *gangeticus,* and **tri** = *tricolor.*

[2]The + sign represents species in which these compounds have been detected.

Table 1. The diversity of secondary metabolites in amaranth.

Compound	[1]hyp	cru	cau	hyb	man	spi	pan	ret	tri	gan
Total phenolic acids	[2]S, F	S,V	S							
Gallic acid	F, L	S, Sp		L			S			
Vanilic acid	F, S, L	S		L	S					
Cinnamic acid		S								
Syringic acid	Fl, S	Sp								
p-Coumaric acid	S, L	S, Sp, L	S	L	S		S			
Ferulic acid	Fl, L	L, S, Sp	S, IF	L			S			
Protocatechuic acid	S	S	S, Sp		S		S			
p-Hydroxybenzoic acid	Fl, S, L	S	S	L	S		S			
Caffeic acid	L	L, S	S	L			S			
Chlorogenic acid	L			L						

Compound	[1]hyp	cru	cau	hyb	man	spi	pan	ret	tri	gan
Sinapic acid	L			L			S			
Salicylic acid	L		S	L			S			
Caffeoylquinic acids						St				
Cumaroylquinic acids:						St				
Feruoylquinic acids						St				
Hydroxycinnamates (quinic acid)	L			L		St, L				
Flavonoids	Fl, S	L		L						
Rutin	S, L, St, Fl	S, L, St, Fl, Sp	S, L, St, Fl	S, L, St, Fl		WP, St		S, L, St, Fl	S, L, St, Fl	
Isoquercetin	S, L								L	
Quercetin: released from rutin	S, L, St, Fl	S, L, St, Fl	S, L, St, Fl	S, L, St, Fl		L		S, L, St, Fl		L
Quercetin diglycoside						St				
Quercetin-3-O-glucoside						St				St
Nicotiflorin	S	S			S					
Vitexin		S								
Isovitexin		S				St				
Orientin		Sp								
Morin		Sp								
neohesperidin and hesperidin		Sp								
Kaempferol diglycoside					L	St				
Tannins	S, L	S, L								
Polyphenolics	S, L	S, L		S, L						
Anthocyanins	S									
Betains, betacyanins	L	L			L				L	
Phenolic Amides	L				L					L

[1]Amaranthus species included are: **hyp** = hypochondriacus; **cru** = cruentus; **cau** = caudatus; **hyb** = hybridus; **man** = mantegazzianus; **spi** = spinosus; **pan** = paniculatus; **ret** = retroflexus; **tri** = tricolor, and **gan** = gangeticus.

[2]Tissues or plant developmental stage examined include: **F** = flower; **Fl** = seed flour; **L** = leaf; **S** = seed; **Sp** = sprout; **St** = stem, and **WP** = whole plant.

Table 2. Diversity of phenylpropanoid metabolites reported in different *Amaranthus* species (modified and amended from [3])

A recent study reported a significantly variable content of bioactive substances and phenolic contents in leaves of various cultivars of *A. tricolor* and *A. hypochondriacus* [49]. For instance, leaf color attributes and betacyanins varied widely among the cultivars. Also, the hyperoside flavonoid was found only in one *A. hypochondriacus* cultivar, in contrast to isoqercetin and rutin, which were abundantly found in all amaranth cultivars examined. SA, syringic, gallic, vanilic, ferulic, p-coumaric, and sinapic acids were also common phenolic acids detected in all amaranth cultivars, whereas significant amounts of ellagic and sinapic acids were only detected in *A. hypochondriacus* cultivars. Lastly, total phenol content was found to be strikingly greater than total phenol index in *A. tricolor*. The free phenolic acid profile in seed ethanol extracts isolated from *A. caudatus* and *A. paniculatus* was also found to be significantly different [115]. However, the differences observed had only a slight influence on their antioxidant activity. A subsequent study performed with seeds of *A. cruentus* showed that processing (i.e., popping or flaking) and cultivation area had significant effects on their total phenolic acid content, whereas differences in individual phenolic acids (e.g., ferulic acid in processed seeds) were highly variable and were found not to have statistical significance [116]. A similar influence of climatic and agro-technical factors on the polyphenol content of amaranth seeds was described prior to this report [117].

In this study, the levels of 11 different polyphenols, including three flavonoids, i.e. rutin, isoquercitrin, and nicotiflorin, and 8 phenolic acids, i.e., protocatechuic, vanillic, 4-hydroxybenzoic, p-coumaric, syringic, caffeic, ferulic, and salicylic acids, were analyzed in mature seeds of 18 *Amaranthus* genotypes, including *A. cruentus, A. hypochondriacus, A. mantegazzianus,* and one grain amaranth hybrid. All were cultivated in three different countries in two continents and in two different locations within the same country. Interestingly, the results derived from principal component analysis (PCA) showed that varying environmental conditions had a contrasting effect on the rutin and nicotiflorin flavonoid levels in the seeds, which largely affected the former. In contrast to the above study, individual phenolic acids in seed samples, such as p-coumaric and protocatechuic acids, were found to be descriptors of climatic and other variations between the different locations studied. Besides, genotype-dependent effects were also observed, since polyphenols content in *A. hypochondriacus* displayed the lowest variation between cultivation sites and the highest content of flavonoids. Comparable results were obtained from the analysis of the aerial tissues of *A. mantegazzianus* and the grain amaranth hybrid plants grown in the same locations. However, this study included the determination of additional compounds such as hydroxycinnamyl amides (N-trans-feruloyltyramine and N-transferuloyl- 4-O-methyldopamine) and betaines (glycinebetaine and trigonelline) [118].

Once more, PCA clearly identified that samples from one location (i.e., in Argentina) differed from all other experimental sites by having a higher content of most compounds analyzed. Phenolic acids were, once again, a key group of compounds since their analysis permitted the separation of the different experimental groups, while separation of both amaranth genotypes could be performed primarily by the higher contents of trigonelline and the two hydroxycinnamyl amides present in *A. mantegazzianus*. Additionally, the contents of polyphenols and betaines in the aerial parts of grain amaranth were found to be very dependent on growing conditions. Further analysis revealed that trigonelline and the two hydroxycinnamyl amides could be tentatively used for chemo-taxonomic classification. Mention must be made of the presence of these and other cinnamoylphenethylamines (i.e., caffeoyltyramine, feruloyldopamine, sinapoyltyramine, p-coumaroyltyramine, and feruloyl-4-O-methyldopamine), which had been just previously reported in the Amaranthaceae for the first time [119]. Feruloyldopamine was different from the other cinnamoylphenethylamines detected by its seemingly ubiquitous presence in the genus *Amaranthus* and to its moderate antifungal activity. In addition to these effects, cinnamoylphenethylamines were previously associated with various other biological activities, such as the potentiation of antibiotics and inhibition of prostaglandin biosynthesis. In the same context, a recent study also determined that leaves and flowers of *A. hybridus*, as well as their extracts, possessed higher antioxidant activities compared to stems and seeds. An on-line HPLC- 2, 2-diphenyl-1-picrylhydrazyl radical assay determined that rutin was the main radical scavenger in these amaranth tissues/extracts, compared to other phenolic compounds detected, such as nicotiflorin, isoquercitrin, 4-hydroxybenzoic, and p-coumaric acids [38].

The influence of other experimental effects on polyphenol levels, such as tissue type, ripeness, or time of harvest was demonstrated by results obtained from a two-year field study performed with various grain and foliar amaranth species, in addition to two amaranth hybrids [120]. The tissue-type effect was clearly demonstrated by results that showed a more than 300-fold difference in rutin content between seeds and leaves. This study also showed that rutin was predominantly found in mature amaranth leaves, in accordance with previous reports describing a progressive accumulation of rutin in maturing amaranth and other rutin-accumulating plants, such as common buckwheat. Genotype-dependent effects were again observed, since noticeable variations between the species and even between the varieties belonging to the same species were detected. For instance, the highest rutin contents found in *A. retroflexus* leaves examined just before harvest contrasted with those measured in *A. tricolor*, which had approximately 12-fold and 67-fold lower rutin contents in leaves and flowers of plants sampled at the same developmental stage, respectively.

Other environmental factors, such as light, or the lack of it, have been also found to selectively influence the level of certain phenolic compounds. Thus, growth of *A. cruentus* sprouts in plain daylight had no effect on gallic acid content but increased the amount of rutin, whereas growth in darkness led to the accumulation of vitexin and isovitexin [121]. In related studies, the antioxidant activity and related color parameters were analyzed in *A. tricolor* and other pigmented leafy vegetables when grown under different photoperiods and light intensities. Betacyanin, chlorophyll, total polyphenol, and antioxidant activity peaked in leaves of *A. tricolor* plants maintained under a 12 h photoperiod but were severely reduced when exposed to constant light for 24 h. The quality and intensity of the light were also found to be significant factors, since shading by blue polyethylene sheets increased betacyanins, polyphenols, and antioxidant activity [122–124]. Sample preparation and the subsequent extraction and measurement protocols are factors that are also known to significantly affect the analyses of polyphenols content in amaranth species. Several studies, for example, have reported a wide range of tannin content in seeds and leaves of various grain and vegetable amaranth species [125,126].

Aside from expected genotype- and tissue-related variations, the differences observed were also suggested to be caused by the analytical methods employed for tannin content determination. Posterior processing procedures (e.g., roasting of seeds, cooking, or blanching of leaves, etc.) were also found to be significant factors affecting tannin content in amaranth seeds or leaves. Subsequently, a pertinent study assessed that the nutraceutical value of leaves of *A. cruentus* and *A. hybridus*, measured in terms of *in vitro* antioxidant (provided mostly by the presence of polyphenols, tannins, flavonoids, and betalains) and xanthine oxidase inhibitory activities, was highly dependent on the type of solvents used for extraction [47]. Similar variations were obtained in another study in which the combined effect of the extraction method and type of solvent on the antioxidant capacity and total phenolic content of extracts from seeds or leaves of *A. hypochondriacus* was evaluated [127]. In this respect, the extraction of phenolic compounds from seeds is believed to be affected by the complex interaction existing between phenolic acids and their cell wall constituents. For example, an investigation of the association of ferulic acid, an alkali-extractable phenolic acid, with the dietary insoluble fiber and non-starch polysaccharides of seeds of *A. caudatus* led to the identification of three complex compounds: O-(6-O-trans-feruloyl-β-D-galactopyranosyl)-(1 → 4)-D-galactopyranose, O-(2-O-transferuloyl-α-L-arabinofurano-syl)-(1 → 5)-L-arabinofuranose, and O-α-L-arabinofuranosyl-(1 → 3)-O-(2-O-trans-feruloyl-α-Larabinofuranosyl)-(1 → 5)- L arabinofuranose. This study also demonstrated that ferulic acid in amaranth seed cell walls is predominantly bound to pectic arabinans and galactans [128]. Additionally, the presence of ferulate associations with polysaccharides of dietary fibers could be considered to have taxonomic potential based on a previous study in which ferulate cross-links in the cell walls of dietary fibers were found to be restricted to species belonging to families of the Caryophyllales [129].

Two recent reports focused on direct or indirect changes in polyphenolic content in grain amaranth plants exposed to different stress conditions. The first one determined changes in the abundance of 3 flavonoid glucosides (rutin, nicotiflorin, and isoquercitin), 9 phenolic compounds (coumaric, vanillic, caffeic, syringic, ferulic, sinapic, protocatechuic, salicylic, and 4-hydroxybenzoic acid) and 3 betalains (amaranthine, iso-amaranthine, and betanin) in leaves of five varieties of three grain amaranth species subjected to insect folivory, in a one-year field trial [44]. Multivariate regression analysis revealed significant and predictable differences in the chemical composition of the leaves between grain amaranth genotypes. A similar analytical approach indicated that 8 of the 15 compounds analyzed in the plants, including all 3 flavonoid glucosides, 2 betalains, and 3 phenolic acids, had significant linear relationships with insect herbivory in the field. However, the experiment was not designed to determine biological relevance of the herbivory-induced accumulation of some of these metabolites in amaranth leaves. Thus, the possibility that phenolics could have been acting as feeding deterrents, phagostimulants, digestion inhibitors, digestion stimulants, toxins, toxicity reducers, signal inhibitors, and/or signal transducers in damaged grain amaranth remained unanswered [130]. In potato (*Solanum tuberosum*), for instance, rutin is known to accumulate to high levels only in varieties which are resistant to *Pectobacterium atrosepticum*, a very destructive necrotrophic bacterial pathogen. On the other hand, rutin is considered to be a susceptibility factor with respect to *Phytophthora infestans* infections [131].

In a related proteomic study, the upregulation of transcription factors (i.e., DOF and MIF) was found to be coupled with the downregulation of caffeic acid O-methyltransferase, an isoflavone reductase-like protein, and two different S-adenosylmethionine synthetases, which are enzymes related to secondary metabolism associated with flavonoid and lignin synthesis [23]. Based on these results, the authors suggested that repressed root growth in grain amaranth plants subjected to severe drought is an adaptive response occurring in response to decreased root lignification. This proposal is in accordance to other reports showing that roots of plants exposed to different stresses may change their lignin content and composition [132]. One possible advantage derived from reduced lignification of the roots, particularly in the elongation zone, is that it may facilitate growth recovery once drought stress has been alleviated [133].

3.4. Secondary metabolism biosynthetic pathways and related genes in amaranth: Wandering into unknown territory?

The information provided above indicates that a potentially high nutritive and medicinal benefit may be derived from the consumption of amaranth seeds and foliage, which are high in antioxidant phenolic compounds, among other health-enhancing constituents. Until recently, information about the biosynthesis of these bioactive compounds and of the genes coding for the respective biosynthetic enzymes was practically null in amaranth. However, a recent transcriptomic study of grain amaranth leaves subjected to various stress treatments [14] revealed the presence of several genes involved in secondary metabolism, mostly in the phenylpropanoid biosynthetic pathway. The rest of this chapter will concentrate on the description of their characteristics and possible functions.

The transcriptomic study permitted the identification of 95 genes that code for the enzymes that most probably form part of the secondary metabolite biosynthetic flow in *A. hypochondriacus*. These are shown in Table 3. The transcriptomic data also uncovered the presence of several TFs that could be involved in the regulation of distinct branches of the secondary metabolism biosynthetic processes in grain amaranth, including the phenylpropanoid pathway.

Metabolic Group	Putative gene	[4]No	Function
ID[1]	Indole 3-glycerol phosphate lyase (*AhIGL*)	1	Emission of volatile compounds related to defense [169].
ID[1]	Anthranilate synthase (*AhANS*)	2	Production of tryptophan pathway metabolites as part of an Arabidopsis defense response (e.g., camalexin) [170, 171].
ID[1]	*AhWRKY33*, transcription factor	1	Pathogen-induced transcription factor; camalexin biosynthesis [172].
A[2]	Berberine bridge enzymes (*AhBBE*)	3	Upregulation of these genes was observed at specific stages of development and in response to osmotic stress and pathogens attack [173,174].
A[2]	Strictosidine synthase like (*AhSSL*)		The four SSL-coding genes in Arabidopsis are regulated individually, suggesting specific roles in

Metabolic Group	Putative gene	[4]No	Function
			basal (SSL4) and inducible (SSL5-7) plant defense mechanisms [175].
A[2]	Tropinone reductase (*AhTR*)	4	The *TR-I* gene expression was found to be upregulated on exposure to signal molecules (MeJA and SA) and in response to mechanical injury [176].
A[2]	Hyoscyamine 6-hydroxylase (*AhH6H*)	2	The overexpression of *H6H* increases scopolamine production [177].
T[3]	Acetoacetyl-CoA thiolase (*AhAACT*)	2	The AACT is a regulatory enzyme in isoprenoid biosynthesis involved in abiotic stress adaptation [178].
T[3]	3-Hydroxy-3-methyl-glutaryl-CoA reductase (*AhHMGR*)	3	HMGR activity induced by wounding, elicitor, or pathogen challenge is correlated with increased *HMGR* mRNA. The mutant *hmgr1* shows early senescence and sterility, as well as a dwarf phenotype [179, 180].
T[3]	1-Deoxy-xylulose 5-phosphate synthase (*AhDXS*)	1	*DXS* genes type II been proposed to be primarily involved in defense responses and secondary metabolism [181, 182].
T[3]	Sesquiterpene synthase (*AhSS*)	2	Involved in indirect defense of maize against herbivore attack [183].
T[3]	Squalene synthase (*AhSqS*)	1	The downregulation of *SqS* reduced stomatal conductance and conserved more moisture; it also led to increased grain yield and improved drought tolerance [184].
T[3]	Squalene epoxidase (*AhSqE*) y β-amyrin synthase (*AhβAS*)	4	The synthesis of saponins and expression of respective genes (*squalene synthase, squalene epoxidase,* and *β-amyrin synthase*) is induced in response to elicitors or MeJA treatment [185,186].
T[3]	*AhWRKY1*, transcription factor	1	The orthologous genes, *GaWRK1*, regulates the expression of δ-cadinene synthase needed for the synthesis for phytoalexin gossypol. This gene is also part of the defensive responses to pathogens [187,188].

[1] ID: Indole derivates

[2] A: alkaloids

[3] T: terpenes

[4] Number of genes detected in the grain amaranth transcriptome [14]

Table 3. Secondary metabolism genes identified in grain amaranth

In this respect, the phenylpropanoid pathway is the best represented, with more than 69.5% of the above 95 biosynthesis-related genes coding for enzymes related to their biosynthesis [14]. Their position in the intricate phenylpropanoid biosynthetic reaction pathway is shown in Figure 1. Many of these genes have been amply characterized in other plant species. However, those described below code for enzymes that should be highly active, considering that, as mentioned in previous sections of this chapter, amaranth plants have been frequently found to be unusually rich in these compounds (Table 4). Thus, these particular grain amaranth enzymes and/or genes could have attractive biochemical and/or regulatory properties that could offer potentially important biotechnological applications. In addition, four TFs similar to those described above as key regulators of this biosynthetic pathway are described.

Flavonoids (μg/mg)	*Zea mays*	*Sorghum sp.*	*Amaranthus sp.*
Catequin	0.15-0.23	0.07-0.22	-
Rutin	0.38-0.47	0.17-0.30	4.0-10 (4798-15531)*
Quercetin/isoquercetin	-	0.1-0.65	0.3-0.5 (27.8-279.5)*
Nicotiflorin	-	-	4.8-7.2 (141-1281)*

*Numbers in parentheses indicate the quantity of these compounds detected in leaves

Table 4. Flavonoid content in amaranth, compared to other grain-producing C4 species

As already mentioned, PAL catalyzes the first committed step of the phenylpropanoid pathway, which is shared by all compounds produced by downstream ramifications of the pathway (Figure 1). It also represents a bifurcation point between primary and secondary metabolism. Diverse environmental stimuli and developmental programs regulate PAL. It is induced by lignin demands for cell wall fortification and by both biotic (e.g., pathogen and insect damage) and abiotic stresses (e.g., UV irradiation, low temperatures, and nutrient deficiency) [134,135]. PAL activity has been found in all the higher plants analyzed so far, and in some fungi and a few bacteria, but not in animals. In all species studied, *PAL* is part of a multigenic family. For instance, four *PAL* genes have been described in Arabidopsis, five in *Populus tricocharpa* and tomato (*Solanum lycopersicum*) and sixteen in potato (*S. tuberosum*). Two genes have been identified so far in *A. hypochondriacus* [14]. In Arabidopsis, the induction of *PAL* genes is closely related to a defense-related accumulation of flavonoids [136]. Similarly, the induced expression of *AhPAL1* in *A. hypochondriacus* and *A. cruentus* plantlets pre-treated with the BTH, a priming agent, and subsequently infected by *Clavibacter michiganensis* or various *Pseudomonas syringae* pathovars, all of which are opportunistic but highly damaging aggressors of grain amaranths, could contribute the increased resistance observed [15]. However, the exact relationship of this gene with defense-related accumulation of phenylpropanoid compounds and/ or cell wall fortification via lignin incrustations has not been evaluated.

A number of R2R3 MYB transcription factors are known to be able to transactivate *PAL* promoters to control the tissue-specific expression of this gene. An *in silico* analysis of the 64

MYB TF identified in grain amaranth revealed, however, that only five of them are candidates to participate in the phenylpropanoid pathway. In common with orthologs identified in other plants species (Figure 2), these amaranth TFs may also activate the phenylpropanoid pathway via the activation of *PAL* expression [135]. Available transcriptomic data support their role in the regulation of stress responses. Thus, AhMYB5 (Figure 2A) was found to be induced by bacterial infection, which coincided with the participation of AtMYB44, its homolog in Arabidopsis, in the regulation of resistance responses against pathogens and/or water and salt stress [14,137,138]. However, nothing is known in amaranth regarding the interaction of these MYB TFs with other regulatory components, such as certain bHLH and WD40 proteins. Further *in silico* analysis of amaranth transcriptomic data revealed the presence of 50 genes belonging to either of the two gene families. However, only AhbHLH3, whose nomenclature is indicative of its similitude with a similarly named ortholog TF present in *Beta vulgaris* (Figure 2B), was found to have the potential to participate in the regulation of the phenylpropanoid pathway in addition to a possible role in the control of the JA-dependent wound response, via its potential interaction with JAZ proteins [139].

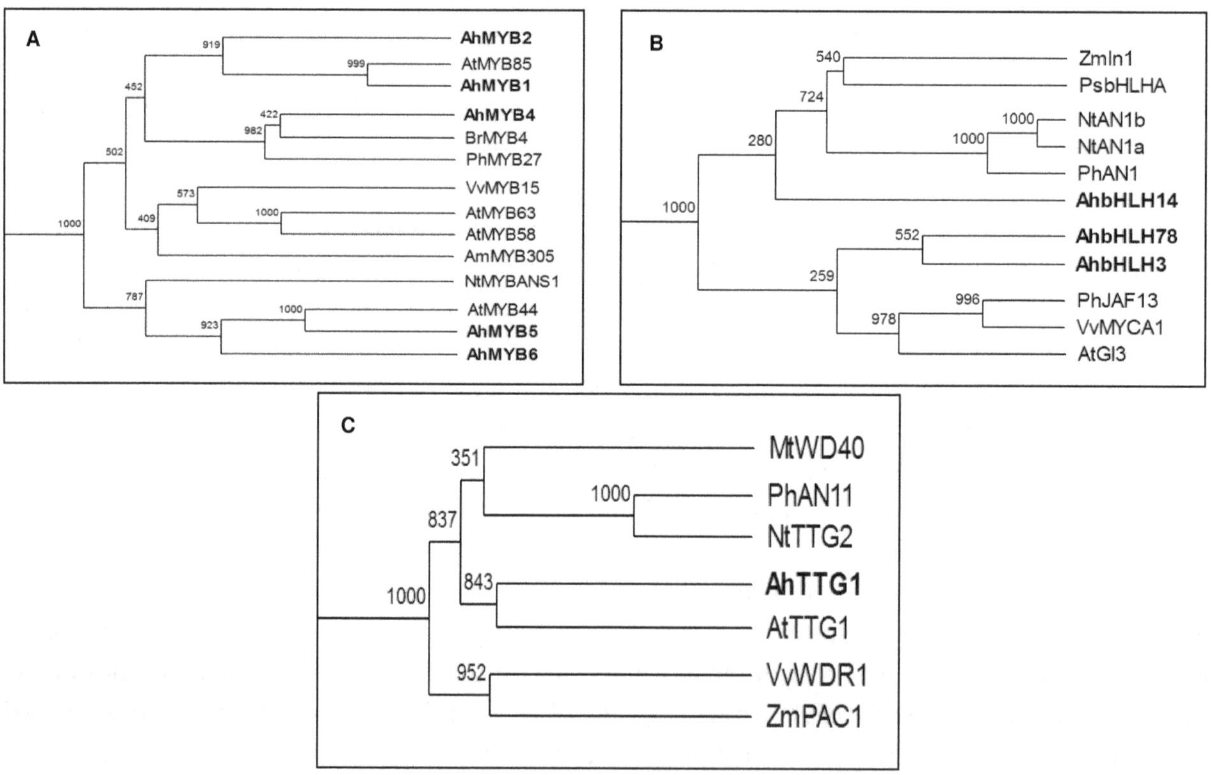

Figure 2. Phylogenetic trees of transcription factors from amaranth and other species: MYB (A), bHLH (B), and WD40 (C). *Amaranthus hypochondriacus*: AhMyb1, AhMyb2, AhMyb4, AhMyb5, and AhMyb6; AhbHLH3, AhbHLH14, and AhbHLH78, and AhTTG1. *Arabidopsis thaliana*: AtMYB85 = AEE84639, AtMYB63 = NP_178039.1.1, AtMYB58 = EFH69169.1, AtMYB44 = NP_201531, AtGL3: NP_680372, AtTTG1: AED93321.1; *Brassica rapa*: BrMYB4 = ADZ98868.1; *Petunia hybrida*: PhMYB27 = AHX24372.1, PhAN11: AAC18914, PhAn1: AAG25927, PhJAF13: AAC39455; *Vitis vinifera*: VvMYB15 = AHA83524.1, VvMYCA1 = ABM92332, VvWDR1 = ABF66625; *Zea mays*: ZmIn1 = AAB03841, ZmPAC1: AAM76742; *Pisum sativum*: PsbHLHA = ADO13282; *Nicotiana tabacum*: NtAN1a = AEE99257; NtAN1b = AEE99258, NtTTG2 = ACN87316; *Medicago truncatula*: MtWD40: ABW08112

Correspondingly, the *AhTTG1* gene, which codifies for a protein having a conserved WD40 domain, shared 77% homology with the Arabidopsis *AtTTG1* gene (Figure 2C). Based on the observed similarity, it could be hypothesized that AhTTG1 could participate in flavonol (quercetin)-dependent developmental processes, such as the control of root growth under abiotic stress conditions [140].

Cinnamate 4-hydroxylase (C4H) is the second key enzyme in the phenylpropanoid pathway and catalyzes the hydroxylation of *trans*-cinnamic acid to *p*-coumaric acid, which is the biosynthetic precursor of flavonoids, phytoalexins, lignin, pigments, and many other defense-related compounds. C4H belongs to the large CYP73 family of cytochrome P450 monooxygenases (P450s). In addition to its central role in the phenylpropanoid biosynthetic pathway, it is also involved in the biosynthesis of various other compounds (e.g., fatty acids, alkaloids, and terpenoids) and may participate in the detoxification of herbicides and pesticides. In accordance with many key biosynthetic genes, *C4H* genes form extensive multi-gene families in various plant species, including *Populus*, orange, and pea [141-143]. Several studies have shown that *C4H* genes have a tissue-specific expression pattern, and that similar to *PAL* genes, they may be also induced by wounding, pathogen infection, and nutrient deficiency [144–147]. Four genes, *AhC4H1-4,* have been identified in grain amaranth. Based on transcriptomic and other data, only the *AhC4H2* is induced by (a)biotic stress [14,134]. However, nothing is known regarding its participation in the synthesis of phenylpropanoids, and perhaps other compounds, in grain amaranth.

The enzyme 4-coumaric acid CoA ligase (4CL) plays an important role in the biosynthesis of lignin precursors such as hydroxycinnamate-CoA thioesters. The *4CL* gene is differentially expressed during development in various tissues and presents multiple isoforms with different substrate specificities. In Arabidopsis, four 4CL isozymes have been identified. Of these, 4CL1 and 4CL2 are known to be involved in the lignin biosynthesis, while 4CL3 participates in flavonoid and other non-lignin biosynthesis pathways. In *P. trichocarpa*, 17 genes showing sequence similarity with known *4CLs* were identified, whereas a similar 4CL1 enzyme was detected in developing xylem tissues of *P. tremuloides*. In addition, *Ptr4CL2* was proposed be involved in flavonoids biosynthesis. Their role in lignin biosynthesis was demonstrated by reports showing that the downregulation of 4CL1 in Arabidopsis, 4CL1 in poplar, and 4CL3 in rice resulted in reduced lignin content [134]. The *4CL* gene family is divided into two main groups. Group II contains those genes which are associated with flavonoid biosynthesis, including two that were identified in grain amaranth: *Ah4CL2-3* (Figure 3). Conversely, the *Ah4CL1* gene could be part of group I and, thereby, participate in lignin biosynthesis (Figure 3). Numerous studies have reported the induction of *4CL* genes by wounding, UV radiation, and pathogen infection in potato, soybean, Arabidopsis, and rice [148–153].

Chalcone synthase (CHS) is another key enzyme of the flavonoid/ isoflavonoid biosynthesis pathway. Besides being part of the plant developmental program, *CHS* gene expression is induced in plants exposed to diverse stress conditions such as UV light, excess salinity, insect herbivory, and bacterial or fungal infection. *CHS* expression leads to the accumulation of flavonoid and isoflavonoid phytoalexins and is involved in the SA defense pathway. Multiple

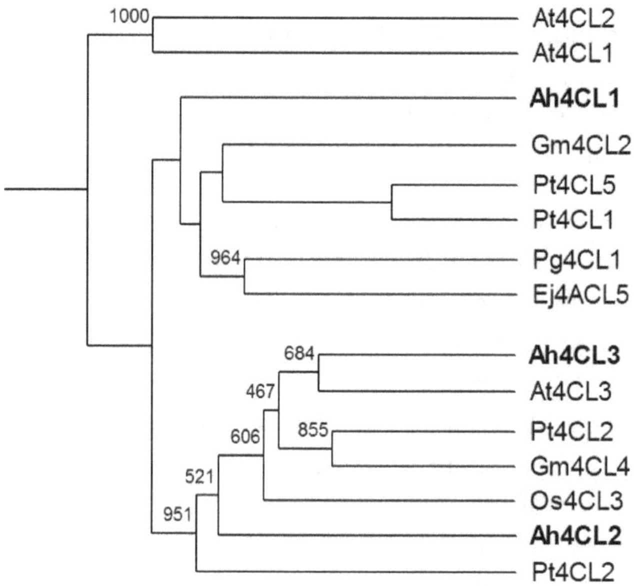

Figure 3. Phylogenetic tree of genes encoding 4-coumarate CoA ligase from amaranth and other species. *Amaranthus hypochondriacus*: AhCL1, AhCL2, and AhCL3. *Arabidopsis thaliana*: At4CL1 = U18675, At4CL2 = B1GUZ3, At4CL3 = AY376730; *Glycine max:* Gm4CL2 = X69954, Gm4CL4 = X69955; *Populus tremuloides:* Pt4CL1 = AF041049, Pt4CL2 = AF041050, *Populus* trichocarpa: Pt4CL5 = EU603299; *Populus generosa*: Pg4CL1 = AF008184; Ej4ACL5 = EF685345; *Oryza sativa:* Os4CL3 = L43362; *Pinus taeda*: Pt4CL2 = U12013.

copies of the CHS gene have been detected in several plants including *Gerbera hybrida, Petunia hybrida, Ipomoea sp., Cannabis sativa,* and *Pisum sativum*. In contrast to grain amaranth, in which two CHS genes (*AhCHS1-2*) have been identified, only single copies of the *CHS* gene have been found in Arabidopsis, parsley, and snapdragon. The structural and catalytic domains present in the hypothetical AhCHS1-2 enzymes are shown in Figure 4. Chalcone isomerase (EC 5.5.1.6), CHI, is one of the most important intermediate enzymes in the flavonoid pathway whose activity involves the modification of substrates previously synthesized by CHS. Although these modifications can occur spontaneously, the efficiency of the reactions are 10^7-fold higher if catalyzed by CHI. A consequence of CHI activity is the typical lack of chalcones and naringenin chalcone in plants, due to their rapid isomerization to naringenin by this enzyme. However, other reactions such as the isomerization of 6-deoxychalcone to 5-deoxyflavanone are rather slow because of the intramolecular hydrogen bond in the substrate molecule. CHIs are classified into two types, and their distribution is highly family-specific. CHIs generally found in non-legumes exclusively catalyze the isomerization 6-hydroxychalcone to 5-hydroxyflavanone. CHIs with this catalytic function are referred to as type I CHIs. On the other hand, most, if not all, of the CHIs found so far in leguminous plants (referred to as type II CHIs) recognize both 6-deoxychalcone and 6-hydroxychalcone as substrates, yielding 5-deoxyflavanone and 5-hydroxyflavanone, respectively. Although more additional information is needed, available data indicate that each of the two *AhCHI* genes (i.e., *AhCHI1-2*) identified in grain amaranth may belong to the different types described, as shown in Figure 5. This characteristic could allow the synthesis of an increased diversity of phenylpropanoid metabolites in amaranth and in related species [134].

```
AhCHS1    LKFLDWVKKLLLRPSRSGANLSQKITHVIMCTTSGVDMPGADYQLTKLLGLRPSVRRFML
AhCHS2    --------------------------------------PGADYQLTKLLGLRPSVRRFML
MsCHS     ------------------------ITHLIVCTTS-VDMPGADYQLTKLLGLRPYVKRYMM

AhCHS1    YQQGCFAGGTVLRLAKDLAENNRGARVLVVCSEITAICFRGPTETHLDSMVGQALFGDGA
AhCHS2    YQQGCFAGGTVLRLAKDLAENNRGARVLVVCSEITAICFRGPTETHLDSMVGQALFGDGA
MsCHS     YQQGCFAGGTVLRLAKDLAENNKGARVLVVCSEVTAVTFRGPSDTHLDSLVGQALFGDGA

AhCHS1    GALIVGADPDELIERPLFKMVWTAQTILPDSEGAIDGHLREVGLTFHLLKDVPGLISKNI
AhCHS2    GALIVGADPNEPIERPLFKMVWTAQTIIPDSEGAIDGHLREVGLTFHLLKDVPGLISKNI
MsCHS     AALIVGSDPVPEIEKPIFEMVWTAQTIAPDSEGAIDGHLREAGLTFHLLKDVPGIVSKNI

AhCHS1    NKALEDAFLPLGINDWNSIFWIAHPGGPAILDQVEQKLGLNEEKLTATRNVLRDYGNMSS
AhCHS2    NKALEDAFNPFGINDWNSIFWIAHPGGPAILDQVEEKLGLDKEKLKATRNVLRDYGNMSS
MsCHS     NKALVEAFEPLGISDYNSIFWIAHPGGPAILDQVEQKLALKPEKMKATREVLSEYGNMSS

AhCHS1    ACVLFILDEMRKKSSEEGKATTGEGLDWGVLFGFGPGLTVETVVLHSVPLNY-
AhCHS2    ACVLFILDEMRKKSCEDGKSTTGEGLHWGVLFGFGPGLTVETVVLQSVPIAET
MsCHS     ACVLFILDEMRKKSAQDGLKTTGEGLEWGVLFGFGPGLTIETVVLRSVTI---
```

Figure 4. Alignment of chalcone synthase enzymes identified in grain amaranth (AhCHS1-2) and *Medicago sativa* (MsCHS). The amino acid residues found in the catalytic site are shown in red, whereas those that compose the structural domains are highlighted in yellow [203].

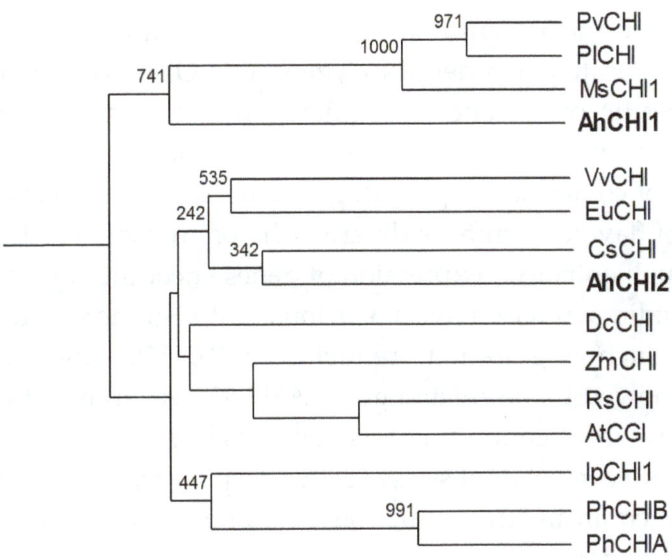

Figure 5. Phylogenetic tree of genes encoding chalcone isomerase from amaranth and other species. *Amaranthus hypochondriacus*: AhCHI1 and AhCHI2. *Zea mays:* ZmCHI = Z22760.1; *Medicago sativa*: MsCHI1 = M31079.1; *Pueraria lobata:* PlCHI = D63577.1; *Phaseolus vulgaris*: PvCHI = XM_00712628.1; *Arabidopsis thaliana*: AtCGI = M86358; *Raphanus sativus*: RsCHI-AF031921.1; *Petunia* hybrida: PhCHIA = AF233637.1, PhCHIB = X14590.1; *Ipomoea purpurea*: IpCHI1 = AF028238.1; *Dianthus* caryophyllus: DcCHI = Z67989.1; *Elaeagnus umbellata*: EuCHI = AF061808.1; *Citrus sinensis* CsCHI = AB011794.1; *Vitis vinífera:* VvCHI = X75963.1.

Flavanone 3-hydroxylase (EC 1.14.11.9), F3H, is a key enzyme in the flavonoid biosynthetic pathway, catalyzing the 3-hydroxylation of (2S)-flavanones, such as naringenin to dihydroflavonols. In soybean seeds the downregulation of *F3H* is accompanied by increased accumulation of isoflavonoids [154]. Seven copies of the F3H gene have been identified in wheat, barley, and rye. F3H activity has been detected in young flower petals and its expression is associated with disease resistance in plants [155]. The antisense repression of the *F3H* gene in carnation flowers (*Dianthus caryophyllus*) and strawberries reduced anthocyanin levels while compounds such as methyl benzoate and 2-hydroxymethyl benzoate (responsible for the flower's fragrance) or polyphenols, including p-coumaroyl-glucose and pcoumaroyl-1-acetate among many others, were found to accumulate [156,157].

Tolerance to UV radiation and severe water deprivation in the extremophyte *Reaumuria soongorica* was associated with increased expression and enzymatic activity of F3H. This increase was also correlated with flavonoid accumulation in consequent antioxidant activity [155]. In grain amaranth, the *AhF3H1* gene, sharing a 77% of homology with the *F3H* gene in *R. soongorica*, was identified. This finding suggests that this gene could be an important factor in the proposed role played by the phenylpropanoid pathway in the tolerance mechanisms used by extremophytes, many of which belong the Caryophyllales order, to thrive in the highly adverse environmental they inhabit [1,158-160]. The use of this gene, and perhaps others related to the phenylpropanoid biosynthetic pathway, offers the attractive possibility of their application for biotechnological purposes in commercial crops.

Flavonol synthase (EC 1.14.11.23), FLS, is another relevant enzyme due to its crucial participation in the conversion of several precursors needed for different branches of the flavonoid biosynthesis. For instance, the biosynthesis of flavonols from dihydroflavonols is catalyzed by FLS, a soluble 2-oxoglutarate-dependent dioxygenase (2-ODD) (Figure 6). The expression is tissue- and organ-specific organ and is regulated by various light intensities, pathogen infection, and herbivore attack [134,161–163]. Silencing of the *FLS* gene in tobacco led to suppression of flavonol content through a decrease in transcript level of flavonol synthase. Moreover, silencing of flavonol synthase diverted the pathway toward catechin and epicatechin production through enhanced expression of genes encoding dihydroflavonol reductase (DFR) and anthocyanidin synthase (ANS). Curiously, it also increased the activity of ROS-detoxifying enzymes such as glutatione reductase (GR), APX, and CAT. Grain amaranth transcriptomic data showed that a related gene, *AhFLS2*, was induced by both drought and salinity stress. The above information suggests that this gene is a promising candidate for crop improvement, acting as a controller of stress-related responses, perhaps by its participation in the biosynthesis of cathequin and the regulation of other protective biochemical processes that could help ameliorate oxidative stress in plants [164].

Dihydroflavonol 4-reductase (DFR, EC 1.1.1.219), DFR, is a pivotal enzyme in the flavonoid biosynthetic pathway that plays a crucial role in producing simple and condensed anthocyanins. This enzyme catalyzes the production of flavan-3, 4-diols (leucoanthocyanidins) via the reduction of three colorless dihydroflavonols, i.e., dihydrokaempferol, dihydroquercetin, and dihydromyricetin. These compounds are also intermediates of flavonol biosynthesis, occurring through the flavonol synthase reaction. These leucoanthocyanidins are subsequently

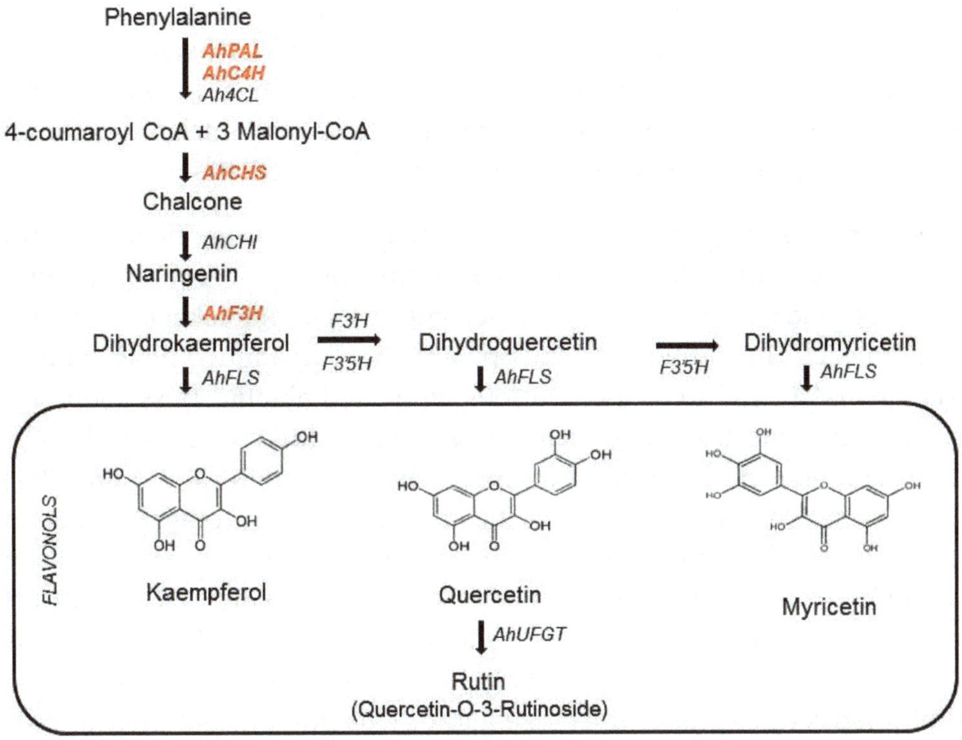

Figure 6. Flavonol biosynthesis in plants. The figure is modified from the original version in [136]. PAL, phenylalanine ammonia lyase; C4H, cinnamate 4-hydroxylase; 4CL, 4-coumarate CoA ligase; CHS, chalcone synthase; CHI, chalcone isomerase; F3H, flavone 3-hydroxylase; F3′H, flavonoid 3′-hydroxylase; F3′5′H, flavonoid 3′5′-hydroxylase; FLS, flavonol synthase. The genes in red text represent those found to be induced by stress conditions in the grain amaranth transcriptomic analysis [14].

converted to pelargonidin, cyanidin, and delphinidin, respectively. DFR can accept wide range of substrates, although substrate specificity of DFR has been shown to vary depending on the specific types of anthocyanins that accumulate in a given plant species. In plants, DFR can be either present as a single gene or as a multicopy gene family. Single DFR genes have been found in *Arabidopsis*, grape, tomato, rice, snapdragon, rose, barley, buckwheat, and grain amaranth (*AhDFR*) while multicopy DFR genes have been found in *Vitis vinifera*, *Ipomoea purpurea*, *Populus*, lotus, and *M. truncatula*. Increased flower pigmentation has been observed by transformation of petunia with a heterologous DFR. Also, the expression of *DFR* has been shown to be spatially and developmentally regulated, may be organ-specific, and its induction leads to the accumulation of anthocyanins in different plant tissues. Furthermore, tobacco plants overexpressing *CsDFR* showed early flowering and significantly higher seed yields, in addition to increased resistance to insect herbivory and antioxidant potential [165]. Other external factors can modulate the expression of *DFR*, including light and UV radiation, exogenous sucrose or JA, and cold or freezing stress [134,166].

Leucoanthocyanidin dioxygenase (LDOX: 1.14.11.19), also called 2-oxoglutarate iron-dependent dioxygenase (2-ODD) or anthocyanidin synthase (ANS), is also involved in anthocyanin biosynthesis and catalyzes the conversion of colorless to colored leucoanthocyanidin. Expression of *LDOX* has been detected in different organs of Shiraz grapevine, such as leaves, roots, seeds, flowers, berry skin, and flesh. An *LDOX* cDNA has been cloned from *Arabidopsis* and the *transparent testa18* and *transparent testa19* mutants were subsequently shown to be *ldox* mutants. In developing *V. vinifera* grapes, the expression of *LDOX* was detected both before and after the ripening stage [134]. The detection of two *LDOX* genes in grain amaranth, one of which (*AhLDOX1*) was responsive to bacterial infection, was intriguing, considering the taxonomic restriction of anthocyanin synthesis in the Caryophyllales.

As mentioned, lignin is the generic term for a large group of aromatic polymers that result from the oxidative combinatorial coupling of 4-hydroxyphenylpropanoids (Figure 7). These polymers are deposited predominantly in the secondary walls of thickened cells, making them rigid and impermeable. In addition to developmentally programmed deposition of lignin, its biosynthesis can also be induced upon various biotic and abiotic stress conditions, such as wounding, pathogen infection, metabolic stress, and perturbations in cell wall structure. Activation of the monolignol precursor biosynthesis in the apoplast requires the combined activity of enzymes such as peroxidases (POX), laccases (LAC), or other polyphenol oxidases that transfer electrons from monolignols to electron receptors. These apoplastic enzymes interact with ROS such as hydrogen peroxide or superoxide, which act as electron receptors or modulators of POX and LAC enzymes through their signaling function. Once oxidized, monolignol radicals can bind to other similarly formed radicals to form the three-dimensionally cross-linked structures that characterize lignin. This polymerization process constitutes the final step of lignin biosynthesis.

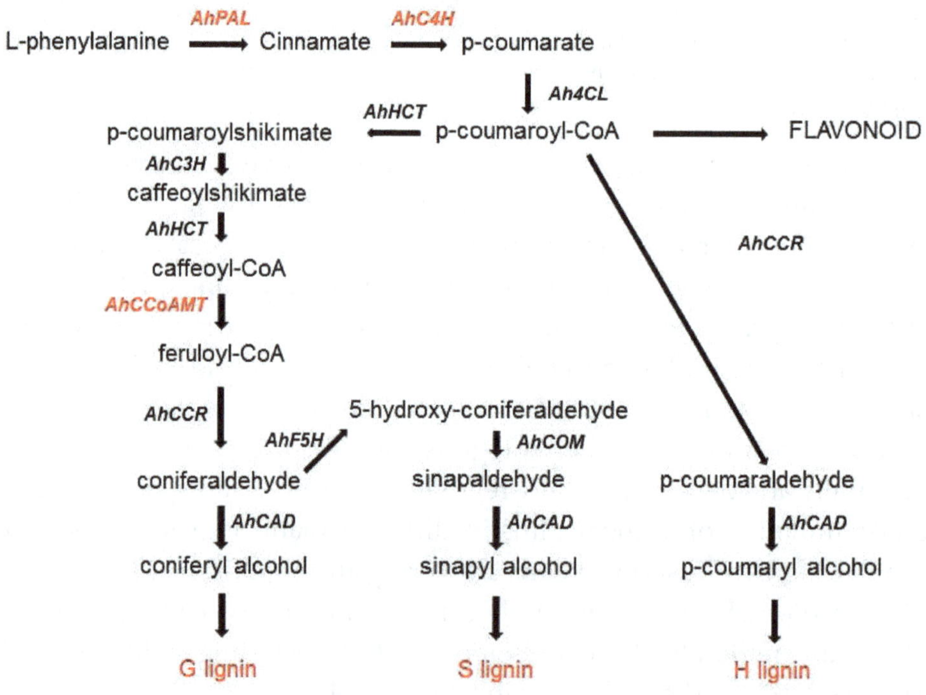

Figure 7. The main biosynthetic route toward the monolignols p-coumaryl, coniferyl, and sinapyl alcohol (the figure is modified from the original version [66]. PAL, phenylalanine ammonia-lyase; C4H, cinnamate 4-hydroxylase; 4CL, 4-coumarate:CoA ligase; C3H, p-coumarate- 3-hydroxylase; HCT, p-hydroxycinnamoyl-CoA: quinate/shikimate p-hydroxycinnamoyltransferase; CCoAOMT, caffeoyl-CoA o-methyltransferase; CCR, cinnamoyl-CoA reductase; F5H, ferulate 5-hydroxylase; COMT, caffeic acid O ethyltransferase; CAD, cinnamyl alcohol dehydrogenase. The genes in red text represent those found to be induced by stress conditions in the grain amaranth transcriptomic analysis [14].

Contrary to lignin, lignans represent a structurally diverse class of plant-specialized metabolites that are ubiquitously distributed in all land plants. They are presumed to have a predominantly defensive role *in planta,* considering several reports that have described their antibacterial and/ or antifungal activities. As such, lignans can act as both phytoanticipins and phytoalexins. In many tree species, the constitutive deposition of lignans in the heartwood is believed to reinforce durability, longevity, and resistance to many wood-rotting fungi. On the other hand, *de novo* formation of lignans in response to fungal attack has been reported in both woody and non-woody species [68]. Our transcriptomic-derived data indicate the possible presence of lignans biosynthetic genes in grain amaranth (Figure 8). This finding is in accordance with a recent report describing the deposition of these metabolites in seeds of *Amaranthus* species. The presence of lignans in amaranth seeds has been linked with some of its medicinal properties, considering the presumed antioxidant, antiviral, antitumoral, antibacterial, antifungic, insecticidal, estrogenic, and anti-estrogenic properties assigned to these particular secondary metabolites [3]. The flux of carbon to lignin biosynthesis can be seriously affected by changes in the expression levels of diverse other genes. Examples of a number of these genes, together with their respective homologs identified in grain amaranth, is described in Table 5.

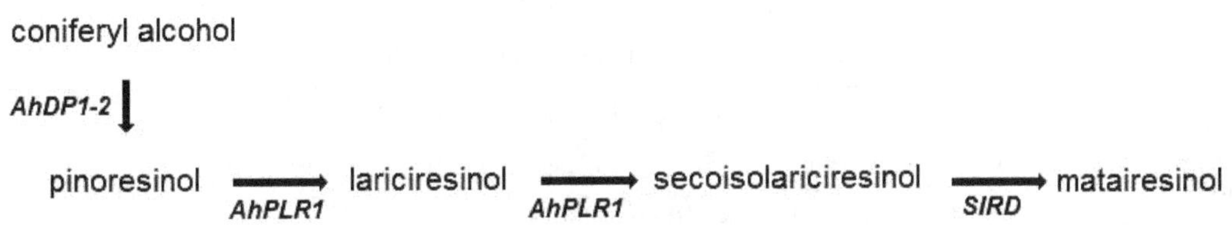

Figure 8. Lignan biosynthetic pathway. DP, dirigent protein; PLR, pinoresinol/lariciresinol reductase; SIRD, secoisolariciresinol dehydrogenase. Two lignan-biosynthesis-related genes were found in grain amaranth; two of them code for *DP* and one for PLR.

Coumarins may be subclassified as simple coumarins (benzo-α-pyrones syn 1, 2-benzopyrone), 7-oxygenated coumarins (furanocoumarins syn. furobenzo-α-pyrones or furocoumarins) or pyranocoumarins (benzodipyran-2-ones). Simple coumarins, furanocoumarins, and pyranocoumarins share the same biosynthetic pathway, whereas the most common phenylcoumarins (i.e., coumestans) originate from isoflavone. Coumarin is characterized for its pleasant vanilla-like odor. The presence of this metabolite has been reported in a diversity of plants, including members of the Fabaceae, Lauraceae, Lamiacea, Apiaceae, Asteracea, Rutaceae, and Amaranthaceae (*Spinoside,* isoflavone). However, the transcriptomic data yielded no gene related to the biosynthesis of these interesting compounds, probably because they were not inducible under the stress conditions employed. Nevertheless, search for these genes in amaranth should continue due to the great interest invested in coumarins due to their potent physiological, bacteriostatic, and anti-tumor activity [167, 168]. Moreover, it has been proposed that coumarins may be engineered to have novel therapeutic properties by further derivatization of their backbone structure.

Gene	[1]No. Seq.	Function in stress	Reference(s)
Cinnamoyl-CoA-reductase (*AhCCR*)	1	In wheat, this gene is involved in pathogen defense signaling. In tomato and other plants, the repression of this gene produced severe development-related phenotypes.	[189,190]
Cinnamyl alcohol dehydrogenase (*AhCAD*)	16	The silencing *CAD* led to changes in the lignin structure, fiber quality, and/or improved digestibility. It also led to development related phenotypes during vegetative and reproductive stages. Pathogen resistance was also modified. Phylogenetic analysis suggests that the *AhCAD2* gene could be involved in stress tolerance responses.	[189,191,192]
Caffeoyl CoA 3-O-methyltransferase (*AhCCoAOMT*)	3	This gene can be highly induced by drought and cold stress, suggesting a possible role in plant abiotic stress resistance. In maize and wheat, *CCoAOMT* is involved in the mechanism of resistance/susceptibility against fusariosis and powdery mildew, respectively.	[193–195]
4-Coumaroyl shikimate 3-hydroxylase/ p-coumarate 3-hydroxylase (*AhC3H*)	1	C3H downregulation led to a dramatic accumulation of several glucosides of *p*-coumaric acid, including phenylglucosides (populoside, grandidentatin, and trichocarposide). This modification may be an acceptable strategy to increase the secondary metabolite diversity against pathogens.	[196]
Caffeic acid O-methyltransferase (*AhCOMT*)	3	In wheat, silencing of this gene is linked to increased susceptibility to fungal pathogens, such as *Blumeria graminis*. Conversely, it led to an augmented resistance to *Hyaloperonospora arabidopsidis* in Arabidopsis.	[193,197]
Ferulic acid hydroxylase (*AhF5H*)	2	*F5H* transcripts accumulated in the leaves in response to mechanical wounding or the application of related elicitors, such as ethylene, ABA, and hydrogen peroxide (H_2O_2). *f5h* mutants were unable to develop normally under UV light, probably because of a lack of UV-radiation protective synapate esters. This gene is essential for expression of anthocyanin biosynthesis-associated genes and anthocyanin accumulation under photo-oxidative stress.	[198–201]
Hydroxycinnamoyl CoA: shikimate hydroxycinnamoyl transferase (*AhHCT*).	2	Downregulation of the HCT gene in alfalfa plants exhibited constitutive activation of defense responses, triggered by release of bioactive cell wall fragments and production of hydrogen peroxide generated as a result of impaired secondary cell wall integrity.	[202]

[1] Number of genes detected in the grain amaranth transcriptome [14].

Table 5. Lignin biosynthetic pathway genes identified in grain amaranth

Finally, many stress-induced phenylpropanoids are classified as phytoalexins. These are antimicrobial compounds synthesized in response to pathogen attack. They include pterocarpans (e.g., glyceollin), isoflavans, prenylated isoflavonoids (e.g., kievitone), stilbenes, psoralens, coumarins, 3-deoxyanthocyanidins, flavonols (e.g., quercetin, kaempferol), and aurones. Most of these compounds have trivial names, such as the coumarins umbelliferone (7-hydroxycoumarin), esculetin (6, 7-dihydroxycoumarin), scopoletin (6-methoxy-7-hydroxycoumarin), and others. No phenylpropanoid phytoalexins have been reported in amaranth. Findings from a recent report [15] suggested, however, that they could be participating as part of the induced defense responses against bacterial pathogens produced by the application of defense-related inductors, such as BTH, JA, or the exposure to avirulent pathogens. This proposal is supported by the induced expression of *AhPAL* in addition to several other known defensive genes, including those coding for ROS-detoxifying enzymes and pathogenesis-related proteins.

4. Conclusion

The several hypotheses raised by the discovery of the numerous stress-related phenylpropanoid genes in grain amaranth represents a strong incentive for the initiation and subsequent deepening of secondary metabolite studies in *Amaranthus* plants, which may yield promising results in various areas of interest, including food science and nutrition, medicine and stress plant physiology, among others.

Acknowledgements

Paula A. Castrillón-Arbeláez was supported financially by a one-year Post-Doctoral Fellowship awarded by the Autonomous National University of México (UNAM).

Author details

Paula Andrea Castrillón-Arbeláez[1] and John Paul Délano Frier[2*]

*Address all correspondence to: jdelano@ira.cinvestav.mx

1 Escuela Nacional de Estudios Superiores (ENES)-UNAM, Unidad León, Predio El Saucillo y El Potrero. C.P.. León, Gto., México

2 Centro de Investigación y de Estudios Avanzados del I. P. N. (Cinvestav), Unidad Irapuato, Gto., México

References

[1] Sánchez-del Pino I, Flóres-Olvera H, Valdés J. La familia Amaranthaceae en la flora halófita y gipsófila de México. Anales del Instituto de Biología Universidad Autónoma de México 1999;70:29–135.

[2] Caselato-Sousa VM, Amaya-Farfan J. State of knowledge on amaranth grain: a comprehensive review. J Food Sci 2012;77:R93–104.

[3] Venskutonis PR, Kraujalis P. Nutritional components of amaranth seeds and vegetables: a review on composition, properties, and uses. Compr Rev Food Sci Food Saf 2013;12:381–412.

[4] Kauffman CS, Weber LE. Grain amaranth. In: Janick J, Simon JE. (Eds.) Advances in New Crops, Timber Press, Portland, OR; 1990. pp. 127–139.

[5] Johnson BL, Henderson TL. Water use patterns of grain amaranth in the northern Great Plains. Agron J 2002;94:1437–43.

[6] Omami EN, Hammes PS, Robbertse PJ. Differences in salinity tolerance for growth and water-use efficiency in some amaranth (*Amaranthus spp.*) genotypes. New Zea J Crop Horticult Sci. 2006;34:11–22.

[7] Espitia-Rangel E, Ed. Amaranto: Ciencia y Tecnología, vol 2. Celaya, Guanajuato, México; 2012. 354 p.

[8] Brenner D, Baltensperger D, Kulakow P, Lehmann J, Myers R, Slabbert M, Sleugh B. Genetic resources and breeding of *Amaranthus*. Plant Breed Rev 2000;19:227–85.

[9] Castrillon-Arbelaez PA, Martinez-Gallardo N, Arnaut HA, Tiessen A, Delano-Frier JP. Metabolic and enzymatic changes associated with carbon mobilization, utilization and replenishment triggered in grain amaranth (*Amaranthus cruentus*) in response to partial defoliation by mechanical injury or insect herbivory. BMC PlantBiol 2012;12:163.

[10] Vargas-Ortiz E, Espitia-Rangel E, Tiessen A, Delano-Frier JP. Grain amaranths are defoliation tolerant crop species capable of utilizing stem and root carbohydrate reserves to sustain vegetative and reproductive growth after leaf loss. PLoS ONE 2013;8:e67879.

[11] Niveyro S, Salvo A. Taxonomic and functional structure of phytophagous insect communities associated with grain amaranth. Neotrop Entomol 2014;43:532–40.

[12] Délano-Frier JP, Martínez-Gallardo NA, Martínez-de la Verga O, Salas-Araiza MD, Barbosa-Jaramillo ER, Torres A, Vargas P, Borodanenko A. The effect of exogenous jasmonic acid on induced resistance and productivity in amaranth (*Amaranthus hypochondriacus*) is influenced by environmental conditions. J Chem Ecol 2004;30:1001–34.

[13] Sánchez-Hernández C, Martínez-Gallardo N, Guerrero-Rangel A, Valdés-Rodríguez S, Délano-Frier J. Trypsin and a-amylase inhibitors are differentially induced in

leaves of amaranth (*Amaranthus hypochondriacus*) in response to biotic and abiotic stress. Physiol Plant 2004;122:254–64.

[14] Délano-Frier JP, Avilés-Arnaut H, Casarrubias-Castillo K, Casique-Arroyo G, Castrillón-Arbeláez PA, Herrera-Estrella L, Massange-Sánchez J, Martínez-Gallardo NA, Parra-Cota FI, Vargas-Ortiz E, Estrada-Hernández MG. Transcriptomic analysis of grain amaranth (*Amaranthus hypochondriacus*) using 454 pyrosequencing: comparison with *A. tuberculatus*, expression profiling in stems and in response to biotic and abiotic stress. BMC Genomics 2011;12:363.

[15] Casarrubias-Castillo K, Martinez-Gallardo NA, Delano-Frier JP. Treatment of *Amaranthus cruentus* with chemical and biological inducers of resistance has contrasting effects on fitness and protection against compatible Gram positive and Gram negative bacterial pathogens. J Plant Physiol 2014;171:927–39.

[16] Gill SS, Tuteja N. Reactive oxygen species and antioxidant machinery in abiotic stress tolerance in crop plants. Plant Physiol Biochem 2010;48:909–30.

[17] Baier M, Kandlbinder A, Golldack D, Dietz KJ. Oxidative stress and ozone: perception, signalling and response. Plant Cell Environ 2005;28:1012–20.

[18] Mittler R. Oxidative stress, antioxidants and stress tolerance. Trend Plant Sci 2002;7:405–10.

[19] Neill S, Desikan R, Hancock J. Hydrogen peroxide signalling. Curr Opin Plant Biol 2002;5:388–95.

[20] Pastori GM, Foyer CH. Common components, networks, and pathways of cross-tolerance to stress. The central role of "redox" and abscisic acid-mediated controls. Plant Physiol 2002;129:460–8.

[21] Sunkar R, Kapoor A, Zhu JK. Posttranscriptional induction of two Cu/Zn superoxide dismutase genes in Arabidopsis is mediated by downregulation of miR398 and important for oxidative stress tolerance. Plant Cell 2006;18:2051–65.

[22] Noreen S, Ashraf M. Modulation of salt (NaCl)-induced effects on oil composition and fatty acid profile of sunflower (*Helianthus annuus* L.) by exogenous application of salicylic acid. J Sci Food Agric 2010;90:2608–16.

[23] Huerta-Ocampo JA, Leon-Galvan MF, Ortega-Cruz LB, Barrera-Pacheco A, De Leon-Rodriguez A, Mendoza-Hernandez G, de la Rosa AP. Water stress induces up-regulation of DOF1 and MIF1 transcription factors and down-regulation of proteins involved in secondary metabolism in amaranth roots (*Amaranthus hypochondriacus* L.). Plant Biol 2011;13:472–82.

[24] Huerta-Ocampo JA, Barrera-Pacheco A, Mendoza-Hernández CS, Espitia-Rangel E, Mock HP, Barba de la Rosa AP. Salt stress-induced alterations in the root proteome of *Amaranthus cruentus* L. J Proteome Res 2014;13:3607–27.

[25] Slama I, Abdelly C, Bouchereau A, Flowers T, Savoure A. Diversity, distribution and roles of osmoprotective compounds accumulated in halophytes under abiotic stress. Ann Bot 2015;115:433–47.

[26] Meng YL, Wang YM, Zhang B, Nii N. Isolation of a choline monooxygenase cDNA clone from *Amaranthus tricolor* and its expression under stress conditions. Cell Res 2001;11:187–93.

[27] Bhuiyan NH, Hamada A, Yamada N, Rai V, Hibino T, Takabe T. Regulation of betaine synthesis by precursor supply and choline monooxygenase expression in *Amaranthus tricolor*. J Exp Bot 2007;58:4203–12.

[28] Chen TH, Murata N. Glycinebetaine protects plants against abiotic stress: mechanisms and biotechnological applications. Plant Cell Environ 2011;34:1–20.

[29] Brockington SF, Walker RH, Glover BJ, Soltis PS, Soltis DE. Complex pigment evolution in the Caryophyllales. New Phytol 2011;190:854–64.

[30] Gandia-Herrero F, Garcia-Carmona F. Biosynthesis of betalains: yellow and violet plant pigments. Trend Plant Sci 2013;18:334–43.

[31] Cai YZ, Sun M, Corke H. Identification and distribution of simple and acylated betacyanins in the Amaranthaceae. J Agric Food Chem2001;49:1971–8.

[32] Strack D, Vogt T, Schliemann W. Recent advances in betalain research. Phytochemistry 2003;62:247–69.

[33] Tanaka Y, Sasaki N, Ohmiya A. Biosynthesis of plant pigments: anthocyanins, betalains and carotenoids. Plant J 2008;54:733–49.

[34] Sepulveda-Jimenez G, Rueda-Benitez P, Porta H, Rocha-Sosa M. A red beet (*Beta vulgaris*) UDP-glucosyltransferase gene induced by wounding, bacterial infiltration and oxidative stress. J Exp Bot 2005;56:605–11.

[35] Chang-Quan W, Min C, Ji-Qiang Z, Bao-Shan W. Betacyanin accumulation in the leaves of C3 halophyte *Suaeda salsa* L. is induced by watering roots with H_2O_2. Plant Sci 2007;172:1–7.

[36] Shao L, Li Y, Wu X, Peng C. Comparison on antioxidative capability in leaves of red and green amaranth (*Amaranthus tricolor* L.) under high temperature stress. Plant Physiol Commun 2008;44:923–6.

[37] Casique-Arroyo G, Martinez-Gallardo N, Gonzalez de la Vara L, Delano-Frier JP. Betacyanin biosynthetic genes and enzymes are differentially induced by (a)biotic stress in *Amaranthus hypochondriacus*. PLoS ONE 2014;9:e99012.

[38] Kraujalis P, Venskutonis PR, Kraujaliene V, Pukalskas A. Antioxidant properties and preliminary evaluation of phytochemical composition of different anatomical parts of amaranth. Plant Food Human Nutr 2013;68:322–8.

[39] Viswa Preeth GP, Paul Das M. Comparative study on phytochemical parameters of *Amaranthus caudatus* and *Amaranthus hybridus*. J Chem Pharm Res 2014;6:1462–5.

[40] Simmonds MS. Flavonoid-insect interactions: recent advances in our knowledge. Phytochemistry 2003;64:21–30.

[41] Lattanzio V, Lattanzio VM, Cardinali A. Role of phenolics in the resistance mechanisms of plants against fungal pathogens and insects. Phytochem Adv Res 2006;661:23–67.

[42] Carlsen S, Fomsgaard I. Biologically active secondary metabolites in white clover (*Trifolium repens* L.)-a review focusing on contents in the plant, plant-pest interactions and transformation. Chemoecology 2008;18:129–70.

[43] Falcone Ferreyra ML, Rius SP, Casati P. Flavonoids: biosynthesis, biological functions, and biotechnological applications. Front Plant Sci 2012;3:222.

[44] Niveyro SL, Mortensen AG, Fomsgaard IS, Salvo A. Differences among five amaranth varieties (*Amaranthus* spp.) regarding secondary metabolites and foliar herbivory by chewing insects in the field. Arthropod-Plant Interact 2013;7:235–45.

[45] Alvarez-Jubete L, Arendt EK, Gallagher E. Nutritive value of pseudocereals and their increasing use as functional gluten-free ingredients. Trend Food Sci Technol 2010;21:106–13.

[46] Arif T, Mandal TK, Dabur R. Natural products: anti-fungal agents derived from plants. J Asian Nat Prod Res 2011;11:621–38.

[47] Nana FW, Hilou A, Millogo JF, Nacoulma OG. Phytochemical composition, antioxidant and xanthine oxidase inhibitory activities of *Amaranthus cruentus* L. and *Amaranthus hybridus* L. extracts. Pharmaceuticals 2012;5:613–28.

[48] Pannu J, Thalwal S, Gupta A. Comparison of antimicrobial activity and phytochemical constituents of *in vivo* and *in vitro* grown *Amaranthus spinosus* plant. Int J Pharm Pharm Sci 2013;5:703–7.

[49] Khanam UKS, Oba S. Bioactive substances in leaves of two amaranth species, *Amaranthus tricolor* and *A. hypochondriacus*. Can J Agricult Sci 2013;93:47–58.

[50] Zhang Y, Butelli E, Martin C. Engineering anthocyanin biosynthesis in plants. Curr Opin Plant Biol 2014;19:81–90.

[51] Xie DY, Sharma SB, Paiva NL, Ferreira D, Dixon RA. Role of anthocyanidin reductase, encoded by BANYULS in plant flavonoid biosynthesis. Science 2003;299:396–9.

[52] Stracke R, Ishihara H, Huep G, Barsch A, Mehrtens F, Niehaus K, Weisshaar B. Differential regulation of closely related R2R3-MYB transcription factors controls flavonol accumulation in different parts of the Arabidopsis thaliana seedling. Plant J 2007;50:660–77.

[53] Yi J, Derynck MR, Chen L, Dhaubhadel S. Differential expression of CHS7 and CHS8 genes in soybean. Planta 2010;231:741–53.

[54] Grotewold E, Drummond BJ, Bowen B, Peterson T. The myb-homologous P gene controls phlobaphene pigmentation in maize floral organs by directly activating a flavonoid biosynthetic gene subset. Cell 1994;76:543–53.

[55] Sotheeswaran S, Pasupathy V. Distribution of resveratrol oligomers in plants. Phytochemistry 1993;32:1083–92.

[56] Yu CK, Springob K, Schmidt J, Nicholson RL, Chu IK, Yip WK, Lo C. A stilbene synthase gene (*SbSTS1*) is involved in host and nonhost defense responses in sorghum. Plant Physiol 2005;138:393–401.

[57] Counet C, Callemien D, Collin S. Chocolate and cocoa: new sources of trans-resveratrol and trans-piceid. Food Chem 2006;98:649–57.

[58] Brown DE, Rashotte AM, Murphy AS, Normanly J, Tague BW, Peer WA, Taiz L, Muday GK. Flavonoids act as negative regulators of auxin transport *in vivo* in Arabidopsis. Plant Physiol 2001;126:524–35.

[59] Wasson AP, Pellerone FI, Mathesius U. Silencing the flavonoid pathway in *Medicago truncatula* inhibits root nodule formation and prevents auxin transport regulation by rhizobia. Plant Cell 2006;18:1617–29.

[60] Peer WA, Murphy AS. Flavonoids and auxin transport: modulators or regulators? Trend Plant Sci 2007;12:556–63.

[61] Taylor LP, Grotewold E. Flavonoids as developmental regulators. Curr Opin Plant Biol 2005;8:317–23.

[62] Agati G, Azzarello E, Pollastri S, Tattini M. Flavonoids as antioxidants in plants: location and functional significance. Plant Sci 2012;196:67–76.

[63] Winkel-Shirley B. Flavonoid biosynthesis. A colorful model for genetics, biochemistry, cell biology, and biotechnology. Plant Physiol 2001;126:485–93.

[64] Al-Ghazi Y, Bourot S, Arioli T, Dennis ES, Llewellyn DJ. Transcript profiling during fiber development identifies pathways in secondary metabolism and cell wall structure that may contribute to cotton fiber quality. Plant Cell Physiol 2009;50:1364–81.

[65] Schilmiller AL, Stout J, Weng JK, Humphreys J, Ruegger MO, Chapple C. Mutations in the cinnamate 4-hydroxylase gene impact metabolism, growth and development in Arabidopsis. Plant J 2009;60:771–82.

[66] Boerjan W, Ralph J, Baucher M. Lignin biosynthesis. Annu Rev Plant Biol 2003;54:519–46.

[67] Dixon RA, Paiva NL. Stress-induced phenyl-propanoid metabolism. Plant Cell 1995;7:1085–97.

[68] Naoumkina MA, Zhao Q, Gallego-Giraldo L, Dai X, Zhao PX, Dixon RA. Genome-wide analysis of phenylpropanoid defence pathways. Mol Plant Pathol 2010;11:829–46.

[69] Mattila P, Hellstrom J. Phenolic acids in potatoes, vegetables, and some of their products. J Food Comp Anal 2007;20:152–60.

[70] Habauzit V, Morand C. Evidence for a protective effect of polyphenols-containing foods on cardiovascular health: an update for clinicians. Ther Adv Chronic Dis 2012;3:87–106.

[71] Winkel BS. Metabolic channeling in plants. Annu Rev Plant Biol 2004;55:85–107.

[72] Vogt T. Phenylpropanoid biosynthesis. Mol Plant 2010;3:2–20.

[73] Bowles DJ. Defense-related proteins in higher plants. Annu Rev Biochem 1990;59:873–907.

[74] Bate NJ, Orr J, Ni W, Meromi A, Nadler-Hassar T, Doerner PW, Dixon RA, Lamb CJ, Elkind Y. Quantitative relationship between phenylalanine ammonia-lyase levels and phenylpropanoid accumulation in transgenic tobacco identifies a rate-determining step in natural product synthesis. Proc Natl Acad Sci USA 1994;91:7608–12.

[75] Reichert AI, He XZ, Dixon RA. Phenylalanine ammonia-lyase (PAL) from tobacco (*Nicotiana tabacum*): characterization of the four tobacco *PAL* genes and active heterotetrameric enzymes. Biochem J 2009;424:233–42.

[76] Rawal HC, Singh NK, Sharma TR. Conservation, divergence, and genome-wide distribution of *PAL* and *POX A* gene families in plants. Int J Genom 2013;2013:678969.

[77] Shang QM, Li L, Dong CJ. Multiple tandem duplication of the phenylalanine ammonia-lyase genes in *Cucumis sativus* L. Planta 2012;236:1093–105.

[78] Hemmati S. Phenylalanine ammonia-lyase through evolution: a bioinformatic approach. Trend Pharm Sci 2015;1:10–14.

[79] Badri DV, Weir TL, van der Lelie D, Vivanco JM. Rhizosphere chemical dialogues: plant-microbe interactions. Curr Opin Biotechnol 2009;20:642–50.

[80] Skirycz A, Jozefczuk S, Stobiecki M, Muth D, Zanor MI, Witt I, Mueller-Roeber B. Transcription factor AtDOF4;2 affects phenylpropanoid metabolism in *Arabidopsis thaliana*. New Phytol 2007;175:425–38.

[81] Hassan S, Mathesius U. The role of flavonoids in root-rhizosphere signalling: opportunities and challenges for improving plant-microbe interactions. J Exp Bot 2012;63:3429–44.

[82] Sinha R, Rajam MV. RNAi silencing of three homologues of S-adenosylmethionine decarboxylase gene in tapetal tissue of tomato results in male sterility. Plant Mol Biol 2013;82:169–80.

[83] Teng S, Keurentjes J, Bentsink L, Koornneef M, Smeekens S. Sucrose-specific induction of anthocyanin biosynthesis in Arabidopsis requires the *MYB75/PAP1* gene. Plant Physiol 2005;139:1840–52.

[84] Yan J, Wang B, Jiang Y, Cheng L, Wu T. *GmFNSII*-controlled soybean flavone metabolism responds to abiotic stresses and regulates plant salt tolerance. Plant Cell Physiol 2014;55:74–86.

[85] Payyavula RS, Singh RK, Navarre DA. Transcription factors, sucrose, and sucrose metabolic genes interact to regulate potato phenylpropanoid metabolism. J Exp Bot 2013;64:5115–31.

[86] Babst BA, Chen HY, Wang HQ, Payyavula RS, Thomas TP, Harding SA, Tsai CJ. Stress-responsive hydroxycinnamate glycosyltransferase modulates phenylpropanoid metabolism in *Populus*. J Exp Bot 2014;65:4191–200.

[87] Kylli P, Nousiainen P, Biely P, Sipila J, Tenkanen M, Heinonen M. Antioxidant potential of hydroxycinnamic acid glycoside esters. J Agric Food Chem 2008;56:4797–805.

[88] Meissner D, Albert A, Bottcher C, Strack D, Milkowski C. The role of UDP-glucose:hydroxycinnamate glucosyltransferases in phenylpropanoid metabolism and the response to UV-B radiation in *Arabidopsis thaliana*. Planta 2008;228:663–74.

[89] Rivas-San Vicente M, Plasencia J. Salicylic acid beyond defence: its role in plant growth and development. J Exp Bot 2011;62:3321–38.

[90] Zheng XY, Zhou M, Yoo H, Pruneda-Paz JL, Spivey NW, Kay SA, Dong X. Spatial and temporal regulation of biosynthesis of the plant immune signal salicylic acid. Proc Natl Acad Sci USA 2015;112:9166–73.

[91] Schwab W. Metabolome diversity: too few genes, too many metabolites? Phytochemistry 2003;62:837–49.

[92] Liu J, Osbourn A, Ma P. MYB Transcription factors as regulators of phenylpropanoid metabolism in plants. Mol Plant 2015;8:689–708.

[93] Zhong R, Richardson EA, Ye ZH. The MYB46 transcription factor is a direct target of SND1 and regulates secondary wall biosynthesis in Arabidopsis. Plant Cell 2007;19:2776–92.

[94] Zhong R, Lee C, Zhou J, McCarthy RL., Ye ZH. A battery of transcription factors involved in the regulation of secondary cell wall biosynthesis in Arabidopsis. Plant Cell 2008;20:2763–82.

[95] McCarthy RL, Zhong R, Ye ZH. MYB83 is a direct target of SND1 and acts redundantly with MYB46 in the regulation of secondary cell wall biosynthesis in Arabidopsis. Plant Cell Physiol 2009;50:1950–64.

[96] Zhou J, Lee C, Zhong R, Ye ZH. MYB58 and MYB63 are transcriptional activators of the lignin biosynthetic pathway during secondary cell wall formation in Arabidopsis. Plant Cell 2009;21:248–66.

[97] Legay S, Lacombe E, Goicoechea M, Briere C, Seguin A, Mackay J, Grima-Pettenati J. Molecular characterization of EgMYB1, a putative transcriptional repressor of the lignin biosynthetic pathway. Plant Sci 2007;173:542–9.

[98] Fornale S, Shi X, Chai C, Encina A, Irar S, Capellades M, Fuguet E, Torres JL, Rovira P, Puigdomenech P, Rigau J, Grotewold E, Gray J, Caparros-Ruiz D. ZmMYB31 directly represses maize lignin genes and redirects the phenylpropanoid metabolic flux. Plant J 2010;64:633–44.

[99] Legay S, Sivadon P, Blervacq A S, Pavy N, Baghdady A, Tremblay L, Levasseur C, Ladouce N, Lapierre C, Seguin A, Hawkins S, Mackay J, Grima-Pettenati J. *EgMYB1*, an R2R3 MYB transcription factor from eucalyptus negatively regulates secondary cell wall formation in Arabidopsis and poplar. New Phytol 2010;188:774–86.

[100] Shen H, He X, Poovaiah CR, Wuddineh WA, Ma J, Mann DG, Wang H, Jackson L, Tang Y, Stewart CN, Chen F, Dixon RA. Functional characterization of the switchgrass *(Panicum virgatum)* R2R3-MYB transcription factor PvMYB4 for improvement of lignocellulosic feedstocks. New Phytol 2012;193:121–36.

[101] Omer S, Kumar S, Khan BM. Over-expression of a subgroup 4 R2R3 type MYB transcription factor gene from *Leucaena leucocephala* reduces lignin content in transgenic tobacco. Plant Cell Rep 2013;32:161–71.

[102] Zhu L, Shan H, Chen S, Jiang J, Gu C, Zhou G, Chen Y, Song A, Chen F. The heterologous expression of the chrysanthemum R2R3-MYB transcription factor alters lignin composition and represses flavonoid synthesis in *Arabidopsis thaliana*. PLoS ONE 2013;8:e65680.

[103] Goff SA, Cone KC, Chandler VL. Functional analysis of the transcriptional activator encoded by the maize B gene: evidence for a direct functional interaction between two classes of regulatory proteins. Gene Develop 1992;6:864–75.

[104] Baudry A, Heim MA, Dubreucq B, Caboche M, Weisshaar B, Lepiniec L. TT2, TT8, and TTG1 synergistically specify the expression of *BANYULS* and proanthocyanidin biosynthesis in *Arabidopsis thaliana*. Plant J 2004;39:366–80.

[105] Albert NW, Davies KM, Lewis DH, Zhang H, Montefiori M, Brendolise C, Boase MR, Ngo H, Jameson PE, Schwinn KE. A conserved network of transcriptional activators and repressors regulates anthocyanin pigmentation in eudicots. Plant Cell 2014;26:962–80.

[106] Jin H, Cominelli E, Bailey P, Parr A, Mehrtens F, Jones J, Tonelli C, Weisshaar B, Martin C. Transcriptional repression by AtMYB4 controls production of UV-protecting sunscreens in Arabidopsis. EMBO J 2000;19:6150–61.

[107] Zhang L, Wang Y, Sun M, Wang J, Kawabata S, Li Y. *BrMYB4,* a suppressor of genes for phenylpropanoid and anthocyanin biosynthesis, is down-regulated by UV-B but not by pigment-inducing sunlight in turnip cv. Tsuda. Plant Cell Physiol 2014;55:2092–01.

[108] Qi T, Song S, Ren Q, Wu D, Huang H, Chen Y, Fan M, Peng W, Ren C, Xie D. The Jasmonate-ZIM-domain proteins interact with the WD-repeat/ bHLH/ MYB complexes to regulate jasmonate-mediated anthocyanin accumulation and trichome initiation in *Arabidopsis thaliana.* Plant Cell 2011;23:1795–814.

[109] Zhou LL, Shi MZ, Xie DY. Regulation of anthocyanin biosynthesis by nitrogen in TTG1-GL3/TT8-PAP1-programmed red cells of *Arabidopsis thaliana.* Planta 2012;236:825–37.

[110] Maier A, Schrader A, Kokkelink L, Falke C, Welter B, Iniesto E, Rubio V, Uhrig JF, Hulskamp M, Hoecker U. Light and the E3 ubiquitin ligase COP1/SPA control the protein stability of the MYB transcription factors PAP1 and PAP2 involved in anthocyanin accumulation in *Arabidopsis.* Plant J 2013;74:638–51.

[111] Schaart JG, Dubos C, Romero De La Fuente I, van Houwelingen AM, de Vos RC, Jonker HH, Xu W, Routaboul JM, Lepiniec L, Bovy AG. Identification and characterization of MYB-bHLH-WD40 regulatory complexes controlling proanthocyanidin biosynthesis in strawberry (*Fragaria x ananassa*) fruits. New Phytol 2013;197:454–67.

[112] Li S, Zachgo S. TCP3 interacts with R2R3-MYB proteins, promotes flavonoid biosynthesis and negatively regulates the auxin response in *Arabidopsis thaliana.* Plant J 2013;76:901–13.

[113] Stracke R, Favory JJ, Gruber H, Bartelniewoehner L, Bartels S, Binkert M, Funk M, Weisshaar B, Ulm R. The Arabidopsis bZIP transcription factor HY5 regulates expression of the PFG1/MYB12 gene in response to light and ultraviolet-B radiation. Plant Cell Environ 2010;33:88–103.

[114] Scalbert A, Manach C, Morand C, Remesy C, Jimenez L. Dietary polyphenols and the prevention of diseases. Crit Rev Food Sci Nut 2005;45:287–306.

[115] Klimczak I, Malecka M, Pacholek B. Antioxidant activity of ethanolic extracts of amaranth seeds. Nahrung/ Food 2002;46:184–6.

[116] Ogrodowska D, Czaplicki S, Zadernowski R, Mattila P, Hellstrom J, Naczk M. Phenolic acids in seeds and products obtained from *Amaranthus cruentus.* J Food Nutr Res 2012;51:96–101.

[117] Steffensen SK, Rinnan A, Mortensen AG, Laursen B, de Troiani RM, Noellemeyer EJ, Janovska D, Dusek K, Délano-Frier J, Taberner A, Christophersen C, Fomsgaard IS. Variations in the polyphenol content of seeds of field grown *Amaranthus* genotypes. Food Chem 2011;129:131–8.

[118] Steffensen SK, Pedersen HA, Labouriau R, Mortensen AG, Laursen B, de Troiani RM, Noellemeyer EJ, Janovska D, Stavelikova H, Taberner A, Christophersen C, Foms-

gaard IS. Variation of polyphenols and betaines in aerial parts of young, field-grown *Amaranthus* genotypes. J Agric Food Chem 2011;59:12073–82.

[119] Pedersen HA, Steffensen SK, Christophersen C, Mortensen AG, Jorgensen LN, Niveyro S, de Troiani RM, Rodriguez-Enriquez RJ, Barba-de la Rosa AP, Fomsgaard IS. Synthesis and quantitation of six phenolic amides in *Amaranthus* spp. J Agric Food-Chem 2010;58:6306–11.

[120] Kalinova J, Dadakova E. Rutin and total quercetin content in amaranth (*Amaranthus* spp.). Plant Food Human Nutr 2009;64:68–74.

[121] Pasko P, Sajewicz M, Gorinstein S, Zachwieja Z. Analysis of selected phenolic acids and flavonoids in *Amaranthus cruentus* and *Chenopodium quinoa* seeds and sprouts by HPLC. Acta Chromatograph 2008;20:661–72.

[122] Ali MB, Khandaker L, Oba S. Comparative study on functional components, antioxidant activity and color parameters of selected colored leafy vegetables as affected by photoperiods. J Food Agric Environ 2009;7:392–8.

[123] Ali MB, Khandaker L, Oba S. Changes in pigments, total polyphenol, antioxidant activity and color parameters of red and green edible amaranth leaves under different shade levels. J Food Agric Environ 2010;8:217–22.

[124] Khandaker L, Akond AS MGM, Ali MB, Oba S. Biomass yield and accumulations of bioactive compounds in red amaranth (*Amaranthus tricolor* L.) grown under different colored shade polyethylene in spring season. Sci Hort 2010;123:289–94.

[125] Gorinstein S, Lojek A, Číž M, Pawelzik E, Delgado-Licon E, Medina OJ, Moreno M, Salas IA, Goshev I. Comparison of composition and antioxidant capacity of some cereals and pseudocereals. Int J Food Sci Technol 2008;43:629–37.

[126] Kunyanga CN, Imungi JK, Okoth M, Momanyi C, Biesalski HK, Vadivel V. Antioxidant and antidiabetic properties of condensed tannins in acetonic extract of selected raw and processed indigenous food ingredients from Kenya. J Food Sci 2011;76:C560–7.

[127] López-Mejía OA, López-Malo A, Palou E. Antioxidant capacity of extracts from amaranth (*Amaranthus hypochondriacus* L.) seeds or leaves. Ind Crop Prod 2014;53:55–9.

[128] Bunzel M, Ralph J, Steinhart H. Association of non-starch polysaccharides and ferulic acid in grain amaranth (*Amaranthus caudatus* L.) dietary fiber. Mol Nutr Food Res 2005;49:551–9.

[129] Hartley RD, Harris PJ. Phenolic constituents of the cell walls of dicotyledons. Biochem Sys Ecol 1981;9:189–203.

[130] Johnson MTJ, Ives AR, Ahern J, Salminen J-P. Macroevolution of plant defenses against herbivores in the evening primroses. New Phytol 2014;203:267–79.

[131] Kroner A, Marnet N, Andrivon D, Val F. Nicotiflorin, rutin and chlorogenic acid: phenylpropanoids involved differently in quantitative resistance of potato tubers to biotrophic and necrotrophic pathogens. Plant Physiol Biochem 2012;57:23–31.

[132] Moura JC, Bonine CA, de Oliveira Fernandes Viana J, Dornelas MC, Mazzafera P. Abiotic and biotic stresses and changes in the lignin content and composition in plants. J Integ Plant Biol 2010;52:360–76.

[133] Vincent D, Lapierre C, Pollet B, Cornic G, Negroni L, Zivy M. Water deficits affect caffeate O-methyltransferase, lignification, and related enzymes in maize leaves. A proteomic investigation. Plant Physiol 2005;137:949–60.

[134] Ali MB. Secondary metabolites and environmental stress in plants: biosynthesis, regulation, and function. In: Ahmad P, Wani MR. (Eds.) Physiological Mechanisms and Adaptation Strategies in Plants Under Changing Environment, vol 2. Springer Science Business Media, New York; 2012. pp. 55–85.

[135] Zhang X, Liu CJ. Multifaceted regulations of gateway enzyme phenylalanine ammonia-lyase in the biosynthesis of phenylpropanoids. Mol Plant 2015;8:17–27.

[136] Kim DS, Hwang BK. An important role of the pepper phenylalanine ammonia-lyase gene (*PAL1*) in salicylic acid-dependent signalling of the defence response to microbial pathogens. J Exp Bot 2014;65:2295–306.

[137] Lu B, Sun W, Zhang S, Zhang C, Qian J, Wang X, Gao R, Dong H. HrpN Ea-induced deterrent effect on phloem feeding of the green peach aphid *Myzus persicae* requires *AtGSL5* and *AtMYB44* genes in *Arabidopsis thaliana.* J Biosci 2011;36:123–37.

[138] Persak H, Pitzschke A. Dominant repression by Arabidopsis transcription factor MYB44 causes oxidative damage and hypersensitivity to abiotic stress. Int J Mol Sci 2014;15:2517–37.

[139] Song S, Qi T, Fan M, Zhang X, Gao H, Huang H, Wu D, Guo H, Xie D. The bHLH subgroup IIId factors negatively regulate jasmonate-mediated plant defense and development. PLoS Genet 2013;9:e1003653.

[140] Nguyen HN, Kim JH, Hyun WY, Nguyen NT, Hong SW, Lee H. TTG1-mediated flavonols biosynthesis alleviates root growth inhibition in response to ABA. Plant Cell Rep 2013;32:503–14.

[141] Whitbred JM, Schuler MA. Molecular characterization of *CYP73A9* and *CYP82A1* P450 genes involved in plant defense in pea. Plant Physiol 2000;124:47–58.

[142] Betz C, McCollum TG, Mayer RT. Differential expression of two cinnamate 4-hydroxylases in 'Valencia' orange (*Citrus sinensis* Osbeck). Plant Mol Biol. 2001; 46: 741-748.

[143] Lu S, Zhou Y, Li L, Chiang VL. Distinct roles of cinnamate 4-hydroxylasegenes in *Populus*. Plant Cell Physiol 2006;47:905–14.

[144] Bell-Lelong DA, Cusumano JC, Meyer K, Chapple C. Cinnamate-4-hydroxylase expression in Arabidopsis. Regulation in response to development and the environment. Plant Physiol 1997;113:729–38.

[145] Raes J, Rohde A, Christensen JH, Van de Peer Y, Boerjan W. Genome-wide characterization of the lignification toolbox in Arabidopsis. Plant Physiol 2003;133:1051–71.

[146] Lillo C, Lea US, Ruoff P. Nutrient depletion as a key factor for manipulating gene expression and product formation in different branches of the flavonoid pathway. Plant Cell Environ 2008;31:587–601.

[147] Sadeghi M, Dehghan S, Fischer R, Wenzel U, Vilcinskas A, Kavousi HR, Rahnamaeian M. Isolation and characterization of isochorismate synthase and cinnamate 4-hydroxylase during salinity stress, wounding, and salicylic acid treatment in *Carthamus tinctorius*. Plant Signal Behav 2013;8:e27335.

[148] Ehlting J, Büttner D, Wang Q, Douglas CJ, Somssich IE, Kombrink E. Three 4-coumarate: coenzyme A ligases in *Arabidopsis thaliana* represent two evolutionarily divergent classes in angiosperms. Plant J 1999;19:9–20.

[149] Lindermayr C, Möllers B, Fliegmann J, Uhlmann A, Lottspeich F, Meimberg H, Ebel J. Divergent members of a soybean (*Glycine max* L.) 4-coumarate: coenzyme A ligase gene family. Eur J Biochem 2002;269:1304–15.

[150] Lindermayr C, Fliegmann J, and Ebel J. Deletion of a single amino acid residue from different 4-coumarate: CoA ligases from soybean results in the generation of new substrate specificities. J Biol Chem 2003;278:2781–6.

[151] Hectors K, Van Oevelen S, Guisez Y, Prinsen E, Jansen MAK. The phytohormone auxin is a component of the regulatory system that controls UV-mediated accumulation of flavonoids and UV-induced morphogenesis. Physiol Plant 2012;145:594–603.

[152] Sun C, Huang H, Xu C, Li X, Chen K. Biological activities of extracts from Chinese bayberry (*Myrica rubra* Sieb. et Zucc.): a review. Plant Food Human Nutr 2013;68:97–106.

[153] Xu Q, Yin XR, Zeng JK, Ge H, Song M, Xu CJ, Li X, Ferguson IB, Chen KS. Activator- and repressor-type MYB transcription factors are involved in chilling injury induced flesh lignification in loquat via their interactions with the phenylpropanoid pathway. J Exp Bot 2014;65:4349–59.

[154] Yu O, Shi J, Hession AO, Maxwell CA, McGonigle B, Odell JT. Metabolic engineering to increase isoflavone biosynthesis in soybean seed. Phytochemistry 2003;63:753–63.

[155] Liu M, Li X, Liu Y, Cao B. Regulation of flavanone 3-hydroxylase gene involved in the flavonoid biosynthesis pathway in response to UV-B radiation and drought stress in the desert plant, *Reaumuria soongorica*. Plant Physiol Biochem 2013;73:161–7.

[156] Zuker A, Tzfira T, Ben-Meir H, Ovadis M, Shklarman E, Itzhaki H, Forkmann G, Martens S, Neta-Sharir I, Weiss D, Vainstein A. Modification of flower color and fra-

grance by antisense suppression of the flavanone 3-hydroxylase gene. Mol Breed 2002;9:33–41.

[157] Jiang F, Wang JY, Jia HF, Jia W-S, Wang H-Q, Xiao M. RNAi-mediated silencing of the flavanone 3-hydroxylase gene and its effect on flavonoid biosynthesis in strawberry fruit. J Plant Growth Regul 2013;32:182–90.

[158] Oh DH, Dassanayake M, Bohnert HJ, Cheeseman JM. Life at the extreme: lessons from the genome. Genome Biol 2012;13:241.

[159] Buhmann A, Papenbrock J. An economic point of view, secondary compounds in halophytes. Funct Plant Biol 2013;40:952–67.

[160] Wang A-R, Song H-C, An H-M, Huang Q, Luo X, Dong J-Y. Secondary metabolites of plants from the genus *Chloranthus*: chemistry and biological activities. Chem Biodivers 2015;12:451–73.

[161] Mohanta TK, Occhipinti A, Atsbaha Zebelo S, Foti M, Fliegmann J, Bossi S, Maffei ME, Bertea CM. *Ginkgo biloba* responds to herbivory by activating early signaling and direct defenses. PLoS ONE 2012;7:e32822.

[162] Xu F, Li L, Zhang W, Sun N, Cheng S, Wang Y. Isolation, characterization, and function analysis of a flavonol synthase gene from *Ginkgo biloba*. Mol Biol Rep 2012;39:2285–96.

[163] Cheng S, Zhang W, Sun N, Xu F, Li L, Liao Y, Cheng H. Production of flavonoids and terpene lactones from optimized *Ginkgo biloba* tissue culture. Notulae Botanicae Horti Agrobot Cluj-Napoca 2014;42:88–93.

[164] Mahajan M, Joshi R, Gulati A, Yadav SK. Increase in flavan-3-ols by silencing flavonol synthase mRNA affects the transcript expression and activity levels of antioxidant enzymes in tobacco. Plant Biol 2012;14:725–33.

[165] Kumar V, Nadda G, Kumar S, Yadav SK. Transgenic tobacco overexpressing tea cDNA encoding dihydroflavonol 4-reductase and anthocyanidin reductase induces early flowering and provides biotic stress tolerance. PLoS ONE 2013;8:e65535.

[166] Ahmed NU, Park JI, Jung HJ, Yang TJ, Hur Y, Nou IS. Characterization of dihydroflavonol 4-reductase (DFR) genes and their association with cold and freezing stress in *Brassica rapa*. Gene 2014;550:46–55.

[167] Lacy A, O'Kennedy R. Studies on coumarins and coumarin-related compounds to determine their therapeutic role in the treatment of cancer. Curr Pharm Design 2004;10:3797–811.

[168] Bourgaud F, Hehn A, Larbat R, Doerper S, Gontier E, Kellner S, Matern U. Biosynthesis of coumarins in plants: a major pathway still to be unravelled for cytochrome P450 enzymes. Phytochem Rev 2006;5:293–308.

[169] Frey M, Stettner C, Pare PW, Schmelz EA, Tumlinson JH, Gierl A. An herbivore elicitor activates the gene for indole emission in maize. Proc Natl Acad Sci USA 2000;97:14801–6.

[170] Zook M. Biosynthesis of camalexin from tryptophan pathway intermediates in cell-suspension cultures of Arabidopsis. Plant Physiol 1998;118:1389–93.

[171] Niyogi KK, Fink GR. Two anthranilate synthase genes in Arabidopsis: defense-related regulation of the tryptophan pathway. Plant Cell 1992;4:721–33.

[172] Birkenbihl RP, Diezel C, Somssich IE. Arabidopsis WRKY33 is a key transcriptional regulator of hormonal and metabolic responses toward *Botrytis cinerea* infection. Plant Physiol 2012;159:266–85.

[173] Daniel B, Pavkov-Keller T, Steiner B, Dordic A, Gutmann A, Nidetzky B, Sensen C. W, van der Graaff E, Wallner S, Gruber K, Macheroux P. Oxidation of monolignols by members of the berberine bridge enzyme family suggests a role in cell wall metabolism. J Biol Chem 2015;290:18770–81.

[174] Gonzalez-Candelas L, Alamar S, Sanchez-Torres P, Zacarias L, Marcos JF. A transcriptomic approach highlights induction of secondary metabolism in citrus fruit in response to *Penicillium digitatum* infection. BMC Plant Biol 2010;10:194.

[175] Sohani MM, Schenk PM, Schultz CJ, Schmidt O. Phylogenetic and transcriptional analysis of a strictosidine synthase-like gene family in *Arabidopsis thaliana* reveals involvement in plant defence responses. Plant Biol 2009;11:105–17.

[176] Kushwaha AK, Sangwan NS, Trivedi PK, Negi AS, Misra L, Sangwan RS. Tropine forming tropinone reductase gene from *Withania somnifera* (Ashwagandha): biochemical characteristics of the recombinant enzyme and novel physiological overtones of tissue-wide gene expression patterns. PLoS ONE 2013;8:e74777.

[177] Palazón J, Moyano E, Cusidó RM, Bonfill M, Oksman-Caldentey KM, Piñol MT. Alkaloid production in *Duboisia* hybrid hairy roots and plants overexpressing the *h6h* gene. Plant Sci 2003;165:1289–95.

[178] Soto G, Stritzler M, Lisi C, Alleva K, Pagano ME, Ardila F, Mozzicafreddo M, Cuccioloni M, Angeletti M, Ayub ND. Acetoacetyl-CoA thiolase regulates the mevalonate pathway during abiotic stress adaptation. J Exp Bot 2011;62:5699–711.

[179] Yang Z, Park H, Lacy GH, Cramer CL. Differential activation of potato 3-hydroxy-3-methylglutaryl coenzyme A reductase genes by wounding and pathogen challenge. Plant Cell 1991;3:397–405.

[180] Suzuki M, Kamide Y, Nagata N, Seki H, Ohyama K, Kato H, Masuda K, Sato S, Kato T, Tabata S, Yoshida S, Muranaka T. Loss of function of *3-hydroxy-3-methylglutaryl coenzyme A reductase 1* (*HMG1*) in Arabidopsis leads to dwarfing, early senescence and male sterility, and reduced sterol levels. Plant J 2004;37:750–61.

[181] Phillips MA, Walter MH, Ralph SG, Dabrowska P, Luck K, Uros EM, Boland W, Strack D, Rodriguez-Concepcion M, Bohlmann J, Gershenzon J. Functional identification and differential expression of 1-deoxy-D-xylulose 5-phosphate synthase in induced terpenoid resin formation of Norway spruce (*Picea abies*). Plant Mol Biol 2007;65:243–57.

[182] Okada A, Shimizu T, Okada K, Kuzuyama T, Koga J, Shibuya N, Nojiri H, Yamane H. Elicitor induced activation of the methylerythritol phosphate pathway toward phytoalexins biosynthesis in rice. Plant Mol Biol 2007;65:177–87.

[183] Schnee C, Kollner TG, Held M, Turlings TC, Gershenzon J, Degenhardt J. The products of a single maize sesquiterpene synthase form a volatile defense signal that attracts natural enemies of maize herbivores. Proc Natl Acad Sci USA 2006;103:1129–34.

[184] Manavalan LP, Chen X, Clarke J, Salmeron J, Nguyen HT. RNAi-mediated disruption of squalene synthase improves drought tolerance and yield in rice. J Exp Bot 2012;63:163–75.

[185] Jenner H, Townsend B, Osbourn A. Unravelling triterpene glycoside synthesis in plants: phytochemistry and functional genomics join forces. Planta 2005;220:503–6.

[186] Suzuki H, Reddy MS, Naoumkina M, Aziz N, May GD, Huhman DV, Sumner LW, Blount JW, Mendes P, Dixon RA. Methyl jasmonate and yeast elicitor induce differential transcriptional and metabolic re-programming in cell suspension cultures of the model legume *Medicago truncatula*. Planta 2005;220:696–707.

[187] Tholl D. Terpene synthases and the regulation, diversity and biological roles of terpene metabolism. Curr Opin Plant Biol 2006;9:297–304.

[188] Pandey SP, Somssich IE. The role of WRKY transcription factors in plant immunity. Plant Physiol 2009;150:1648–55.

[189] Thevenin J., Pollet B., Letarnec B., Saulnier L., Gissot L., Maia-Grondard A., Lapierre C., Jouanin L. The simultaneous repression of CCR and CAD, two enzymes of the lignin biosynthetic pathway, results in sterility and dwarfism in *Arabidopsis thaliana*. Mol Plant 2011;4:70–82.

[190] Van der Rest B, Danoun S, Boudet AM, Rochange SF. Down-regulation of cinnamoyl-CoA reductase in tomato (*Solanum lycopersicum* L.) induces dramatic changes in soluble phenolic pools. J Exp Bot 2006;57:1399–411.

[191] Barakat A, Bagniewska-Zadworna A, Choi A, Plakkat U, DiLoreto DS, Yellanki P, Carlson JE. The cinnamyl alcohol dehydrogenase gene family in *Populus*: phylogeny, organization, and expression. BMC Plant Biol 2009;9:26.

[192] Preisner M, Kulma A, Zebrowski J, Dyminska L, Hanuza J, Arendt M, Starzycki M, Szopa J. Manipulating cinnamyl alcohol dehydrogenase (CAD) expression in flax affects fibre composition and properties. BMC Plant Biol 2014;14:50.

[193] Bhuiyan NH, Selvaraj G, Wei Y, King J. Gene expression profiling and silencing reveal that monolignol biosynthesis plays a critical role in penetration defence in wheat against powdery mildew invasion. J Exp Bot 2009;60:509–21.

[194] Tatout C, Grezes-Besset B, George P. Maize with enhanced tolerance to fungal pathogen. 2012; US Patent App. 12/447,023.

[195] Liu SJ, Huang YH, HE CJ, Fang C, Zhang YW. Cloning, bioinformatics and transcriptional analysis of caffeoyl-coenzyme A 3-*O*-methyltransferase in switchgrass under abiotic stress. J Integr Agric 2015; Doi:10.1016/S2095-3119(15)61038-3

[196] Coleman HD, Park JY, Nair R, Chapple C, Mansfield SD. RNAi-mediated suppression of *p*-coumaroyl-CoA 3'-hydroxylase in hybrid poplar impacts lignin deposition and soluble secondary metabolism. Proc Natl Acad Sci USA 2008;105:4501–6.

[197] Quentin M, Allasia V, Pegard A, Allais F, Ducrot PH, Favery B, Levis C, Martinet S, Masur C, Ponchet M, Roby D, Schlaich NL, Jouanin L, Keller H. Imbalanced lignin biosynthesis promotes the sexual reproduction of homothallic oomycete pathogens. PLoS Pathog 2009;5:e1000264.

[198] Landry LG, Chapple CC, Last RL. Arabidopsis mutants lacking phenolic sunscreens exhibit enhanced ultraviolet-B injury and oxidative damage. Plant Physiol 1995;109:1159–66.

[199] Kim YJ, Kim DG, Lee SH, Lee I. Wound-induced expression of the ferulate 5-hydroxylase gene in *Camptotheca acuminata*. Biochim Biophys Acta 2006;1760:182–90.

[200] Kim J, Choi B, Park YH, Cho BK, Lim HS, Natarajan S, Park SU, Bae H. Molecular characterization of ferulate 5-hydroxylase gene from kenaf (*Hibiscus cannabinus* L.). Sci World J 2013;2013:1-11.

[201] Maruta T, Noshi M, Nakamura M, Matsuda S, Tamoi M, Ishikawa T, Shigeoka S. Ferulic acid 5-hydroxylase 1 is essential for expression of anthocyanin biosynthesis-associated genes and anthocyanin accumulation under photooxidative stress in Arabidopsis. Plant Sci 2014;219–20:61–8.

[202] Gallego-Giraldo L, Jikumaru Y, Kamiya Y, Tang Y, Dixon RA. Selective lignin downregulation leads to constitutive defense response expression in alfalfa (*Medicago sativa* L.). New Phytol 2011;190:627–39.

[203] Ferrer JL, Jez JM, Bowman ME, Dixon RA, Noel JP. Structure of chalcone synthase and the molecular basis of plant polyketide biosynthesis. Nat Struct Biol 1999;6:775–84.

2

Histone Methylation - A Cornerstone for Plant Responses to Environmental Stresses?

Roberto Bobadilla and Alexandre Berr

Abstract

Plants are sessile organisms and, as such, their survival relies on their ability to respond quickly all along their life cycle to any kind of environmental stimuli, including abiotic and biotic stresses. In this respect, plants have developed efficient mechanisms of protection and/or adaptation to minimize deleterious effects of stress on their growth and development. In a stress type-dependent manner, external signals are firstly sensed. This step is then followed by the activation of particular signalling pathways, resulting ultimately in the rapid and specific modulation of the plant transcriptome. Currently, transcriptional regulation is considered as a central process in the build-up of plant responses to both abiotic and biotic stresses. Among mechanisms involved in transcriptional regulation, the combined effect of different histone tail post-translational modifications (PTMs; e.g. acetylation and methylation) through the activity of particular histone-modifying enzymes can lead to changes in the local chromatin structure environment and hence the underlying DNA accessibility.

By focusing on histone lysine methylation, in this chapter we highlight our current understanding of the transcriptional roles played by chromatin-remodelling mechanisms in regulating plant response/adaptation to different biotic and abiotic stresses. Based on recent advances, we further discuss the stability and transmission of such methylation marks to subsequent generations, with the underlying idea of an epigenetically based transcriptional memory of stresses in plants.

Keywords: Histone methylation and demethylation, histone methyltransferases and demethylases, biotic and abiotic stresses

1. Introduction

Stress, as we currently think of it, is a highly subjective phenomenon defined as a state of threatened homeostasis. Depending on their nature, external stresses are usually divided into biotic (i.e. herbivorous insects and pathogens such as fungi, bacteria and viruses) or abiotic (i.e. including, among others, high or low temperature, submergence or drought and salinity). During their lifetime, all living organisms inevitably and constantly face all sorts of environmental stresses that often occur suddenly and/or simultaneously. Classically, different strategies can be applied to minimize deleterious effects of stresses, such as resistance, tolerance, avoidance or escape. Being sessile, plants cannot escape and are therefore more prone to the deleterious effect of unfavourable environmental growth conditions. Because responses are critical to ensure their survival, plants have developed specific and efficient strategies that allow them to precisely perceive different environmental stresses and respond and/or adapt to them [1, 2]. In addition to preformed defence traits, plants have evolved inducible defence strategies. Indeed, upon perception, each stress will raise a complex and more or less specific repertoire of cellular and molecular responses implemented by the plant to minimize or prevent damage. Particularly, the stimulation of a given stress-signalling pathway after pathogen detection will be integrated into the plant cell nucleus through a set of regulatory transcription factor cascades, which prioritizes defence over growth-related cellular functions, while conserving enough valuable resources for survival and reproduction [3, 4]. Supporting the idea that the capacity of a plant to rapidly reprogramme its gene expression at the transcriptional level is an essential and common component of all plant response strategies to stress and disturbance; more than 1,000 transcription factors were found to be involved in stress responses [5, 6]. Because eukaryotic genes function in the context of chromatin, modifications and remodelling of the chromatin configuration from permissive for transcription to restrictive, and vice versa, may be an integral part of mechanisms involved in this vital transcriptional reprogramming. In this chapter, we review and discuss the current knowledge about the functional impact of chromatin changes on the transcriptional regulation of genes under different stress conditions, with particular emphasis on histone methylation/demethylation.

2. Chromatin structure and histone methylation/demethylation

In eukaryotes, genomic DNA in the cell nucleus is packaged in a complex and evolutionarily conserved structure named chromatin, with nucleosome as the basic unit. The nucleosome complex contains about 160–241 base pairs (bp) of DNA, a nucleosome core particle and the H1 linker histone. The nucleosome core particle is composed of an octamer of core histones, consisting of two H3–H4 dimers associated with two H2A–H2B dimers. About 146 bp of DNA is wrapped in ~1.65 negatively supercoiled circles around the histone octamer, while the linker DNA associated with H1 varies in length from 8 to 114 bp [7]. At first sight, the chromatin as it is described appears as a barrier, restricting the access of all kinds of enzymes that process the DNA. However, nucleosomes are not merely static but highly dynamic entities. Indeed, nucleosomes can be moved, stabilized/destabilized, disassembled/reassembled at particular genome locations in response to specific environmental signals or developmental cues [8]. This dynamic leads to a wide range of chromatin condensation states modulating the DNA accessibility, with euchromatin, being relaxed, and heterochromatin, being compacted. Therefore, in eukaryotic cells, an intimate connection exists between the structural organization of the genome and its functioning. For this reason, the level of chromatin condensation is directly related to all aspects of DNA metabolism, thus playing a major role in regulating transcription, DNA replication, DNA repair, recombination, transposition and chromosome segregation. In plants, changes in the chromatin structure were reported to affect various biological processes such as root growth, flowering, organogenesis, gametophyte or embryo formation [9–11].

In the nucleosome core particle, histones H2A, H2B, H3 and H4 possess two common regions, a histone-fold domain and a histone tail. The histone-fold domain is the most conserved region and the main element of histone dimerization [12]. The tail protrudes from the nucleosome core particle and is more variable and unstructured than the fold [13]. All four core histones have an N-terminal tail domain, but only histone H2A has an additional long C-terminal tail. Histone tails are extremely basic due to their particularly high content in basic amino acid residues, such as lysine and arginine [14]. Resulting positive charges allow them to closely associate with the negatively charged nucleosomal DNA through electrostatic interactions [15]. In addition, histone tails, especially N-terminal ones, may undergo diverse types of post-translational modifications such as acetylation or methylation. The great diversity of these modifications as well as the high number of amino acid residues that can be modified within histone tails, and the correlation between these modifications and various nuclear processes, lead to the hypothesis that the specific combination of histone modifications constitute a histone 'code' [16].

Technically, these histone marks can be localized by chromatin immunoprecipitation (ChIP) using specific antibodies against the modification [17]. Briefly, protein–DNA interactions are stabilized by cross-linking with formaldehyde; chromatin is sheared into small pieces to facilitate analysis and then immunoprecipitated using an antibody raised against a specific histone modification. Following enrichment, cross links are reversed to release DNA, which is then quantified by polymerase chain reaction (PCR) to measure the relative amount of the specific histone mark on selected plant genes. ChIP can also be combined with microarray hybridization (ChIP-chip) or high-throughput sequencing (ChIP-Seq), allowing the genome-wide discovery of DNA–histone modification interactions.

Methylation is the most abundant one compared with other histone PTMs. It can occur at both lysine (K) and arginine (R) residues of core histone tails. Further extending the indexing potential of this modification, mono-, di- and trimethylation of lysine and mono- and dimethylation (symmetric or asymmetric) of arginine are common at N-terminal tails of H2A, H2B, H3 and H4. Although histone acetylation is generally associated with active gene transcription, histone methylation can be associated with either active or silent gene expression, depending upon the histone, the methylated residue or the level (mono-, di- or tri-) of methylation. In *Arabidopsis*, genome-wide analyses revealed that trimethylations of H3K4 and H3K36

(H3K4me3 and H3K36me3) are generally enriched at actively transcribed genes, whereas H3K27me3 is associated with repressed genes and H3K9me2 and H4K20me1 are enriched at constitutive heterochromatin and silenced transposons [18]. For histone arginine methylation, a definitive role has not yet been clearly established. However, because the level of symmetric H3R2me2 and H4R3me2 was negatively correlated with the level of H3K4me3, a well-known mark reflecting active transcription, high levels of H3R2me2 and H4R3me2 are thought to cause transcriptional repression [19–21]. In contrast, asymmetric H4R3me2 was associated with gene activation [22, 23].

Histone methylation is relatively stable and can be established on lysine and arginine by two distinct families of enzymes, the histone lysine methyltransferases (HKMTs), all containing the evolutionary conserved catalytic SET domain in plants [24], and the protein arginine methyltransferases (PRMTs) [25], respectively. As a counterpart, methyl groups on histone can also be removed by at least two evolutionarily conserved classes of histone demethylases, the lysine-specific demethylase1 (LSD1) type and the Jumonji C (JmjC) domain-containing demethylases [26]. Histone methyltransferases and demethylases are well conserved in angiosperms and have been identified and classified on the basis of phylogenetic analyses and domain organization in several plants, including *Arabidopsis*, maize, tomato, rice, grapevine and *Brassica rapa*, [27–32]. However, cellular and molecular functions of many of these modifiers have not yet been addressed.

Although histone acetylation can directly modulate the chromatin structure, arginine and lysine methylation of histone tails can promote or prevent the docking of key transcriptional effector molecules, named readers, needed to 'translate' the code in order to determine the functional and structural outcome of the corresponding PTMs. Just as there are a large number of PTMs on histone tails, there are also numerous protein domains that recognize and bind to particular PTMs on these tails. For example, PTM-recognition domains such as plant homeodomain (PHD) fingers, chromodomains and Tudor domains all recognize methylated lysine residues [33].

3. Histone methylation changes associated with biotic stress conditions

Biotic stress is the result of the damage done to plants by insects or pathogens, such as bacteria or fungi. Plant pathogens are generally divided into two distinct categories: biotrophs, which colonize living plant tissue and obtain nutrients from living host cells, and necrotrophs, which depend on dead host tissue for nutrients and reproduction. To fend off pathogens with different infection strategies, plants have evolved complex defence mechanisms. Classically, the pathogen-sensing machinery induces signalling cascades that promote the accumulation of hormones such as salicylic acid (SA) or jasmonic acid (JA)/ethylene (ET) [34]. These hormones then orchestrate the overall plant defence reaction locally and systemically by inducing the transcriptional activation of defence genes through an intricate signalling network. In this part, we highlight recent examples illustrating how histone methylations condition major steps leading to immunity, ranging from initial pathogen perception to hormonal homeostasis changes for antimicrobial effector expression.

3.1. Histone methylation/demethylation in the defence against biotrophic pathogens

The phytohormone SA plays an important role in plant defence, from the induction of pathogen resistance (*PR*) genes against biotrophic bacteria (e.g. *Pseudomonas syringae*) to the establishment of systemic acquired resistance (SAR) [35]. Several studies suggested that the SA signalling pathway is notably controlled by histone methylation. Under normal growth conditions, *Arabidopsis* mutants for *SNI1* (*Suppressor of NPR1, Inducible*), a negative regulator of SAR required to dampen the basal expression of *PR* genes, presented an increased H3K4me2 on *PR1* [36]. Rather than being a constitutive mark of transcription, H3K4me2 was proposed to be involved in the fine-tuning of tissue-specific expression [37]. Using the functional SA-analogue S-methyl benzo [1,2,3] thiadiazole-7-carbothioate (BTH), an increased level of H3K4me2 on *PR1* was observed in wild-type plants 48 h after treatment and was not detected in mutants. Interestingly, when expressed in yeast, SNI1 also repressed transcription, suggesting a highly conserved mechanism of transcriptional repression. These results together with the structural similarity of SNI1 with armadillo repeat (ARM) proteins (i.e. a motif known to mediate protein–protein interactions) imply that SNI1 may form a scaffold for interaction with proteins that modulates the chromatin structure of *PR* genes, thus repressing their transcription. In addition, the presence of H3K4me2 detected on *PR1* before induction suggested that this mark is readily in place, providing the appropriate chromatin configuration for the efficient induction of *PR1* upon need. Using a similar approach, Alvarez-Venegas et al. [38] reported no significant changes in levels of H3K4me2 and H3K4me3 on *PR1* 24 h after the SA treatment [38]. This discrepancy may reflect differences in experimental conditions. Indeed, the action of the so-called 'SA-analog' BTH on gene transcription is significantly broader than the action of SA itself [39]. Moreover, samplings were performed 48 h versus 24 h after treatment. Together, because the H3K4 methylation increase does not occur immediately after the induction of *PR1*, this mark may not be directly related to the transcriptional induction itself, but later, for the maintenance/reinforcement of *PR1* expression.

The ARABIDOPSIS HOMOLOG OF TRITHORAX (ATX1) is a H3K4 trimethyltransferase providing basal resistance against *Pseudomonas syringae pv. tomato* (*Pst*); [40]. Despite being not induced by either *Pst* infection or SA, ATX1 positively and directly regulates the expression of the transcription factor WRKY70 through H3K4 trimethylation at the WRKY70 promoter. In addition, *atx1* mutant shows induced expression of the JA-inducible *THI1.2* gene and the reduced *PR1* expression without detectable changes in their chromatin, resulting in impaired resistance to *Pst* infection. Since the transcriptional factor WRKY70 was positioned at the convergence nod of the SA and JA signalling pathways, activating the SA-responsive *PR1* gene and repressing the JA inducible genes [41], ATX1 was proposed to indirectly regulate *PR1* and *THI1.2* through *WRKY70*. SET DOMAIN GROUP 8 (SDG8), another HKMT encoding the major *Arabidopsis* H3K36 di- and trimethyltransferase [42], was also involved in the plant-defence against *Pst*, but it was more upstream than ATX1 [43]. Indeed, SDG8 sustains the basal transcription of particular *R* genes (*RPM1* or *LAZ5*) by maintaining a basal level of H3K36me3, another histone mark tightly associated with active transcription. SDG8 is also required for the transcriptional induction of these *R* genes upon BTH treatment or *Pst* inoculation. However, this induction occurs without any detectable increase of H3K36me3. Therefore, in resting plants, SDG8 may establish a 'permissive' chromatin structure at some *R* genes by methylating H3K36, thus ensuring their basal expression and their transcriptional inducibility upon need. Similarly as *atx1* and *sdg8* mutants, loss-of-function mutants for the putative HKMT *SDG7* were also found to be more sensitive to *Pst* infection than wild-type plants [44]. The expression of other *R* genes seems to be under the control of histone methylation. Indeed, enhanced downy mildew 2 (EDM2) impacts disease resistance by controlling levels of H3K9me2 at an alternative polyadenylation site in the immune receptor gene *RPP7*, thus regulating the balance between full-length *RPP7* transcripts and prematurely polyadenylated transcripts, which do not encode the *RPP7* immune receptor [45, 46]. EDM2, as an epigenetic 'reader', contains two stretches of atypical PHD-finger motifs known to dock specifically several forms of methylated or unmethylated lysine residues on histones [47]. Besides this, EDM2 was also proposed to cooperate within a large protein complex with EMSY-like (AtEML) members, harbouring an Agenet domain related to the Tudor domain family of epigenetic 'readers' [48].

Apart from *Arabidopsis*, little is known about the regulatory role of histone methylation in the defence against pathogen attack in other plant species. In rice, the JmjC protein gene *JMJ705* encoding a histone lysine demethylase that specifically reverses H3K27me2/3 was found induced during infection with the bacterial pathogen *Xanthomonas oryzae*. JMJ705 was further involved in the dynamic removal of the basal H3K27me3 over defence-related genes, thereby increasing their basal expression and/or potentiating their higher expression upon biotic stress. Interestingly, the *JMJ705* overexpression resulted in an enhanced resistance to the bacterial pathogen, while its mutation reduces the plant resistance [49].

3.2. Histone methylation/demethylation in the defence against necrotrophic pathogens

While to combat biotrophic pathogens the plant activates mainly the SA signalling pathway, the activation of the JA/ET signalling pathway is prominent to mediate defences against necrotrophic pathogens and herbivorous insect attacks [50]. The involvement of histone methylation in the defence against necrotrophic pathogens is far less documented as compared with the defence against biotrophic pathogens. Besides being more susceptible to *Pst* [43], *sdg8* mutants were also reported to be more sensitive to necrotrophic fungal pathogens such as *Alternaria brassicicola* (*Alt*) and *Botrytis cinerea* [51]. This increased susceptibility was the consequence of the inefficient transcriptional induction of different genes along the JA/ET signalling pathway that was correlated with a stably weak level of H3K36me3 at these genes. Inversely, in wild-type plants, H3K36me3 together with gene expression were increased upon *Alt* infection or stimulation with exogenous MeJA. Under resting conditions, a similarly weak level of H3K36me3 was correlated with a reduced basal expression in *sdg8*. On that account, H3K36 methylation was proposed to act as a 'permissive' mark correlated with gene activity and readily in place at a subset of JA/ET signalling-related genes to raise their rapid and efficient transcriptional induction when required [52]. Interestingly, a stable and very low level of H3K27me3 was detected in defence effector genes. Because H3K27me3 is often associated with epigenetic silencing [53], this low H3K27me3 level may provide these genes with a reduced probability for undesired silencing, thus participating in the reactivity of plants to pathogen infections.

4. Histone methylation changes associated with abiotic stress conditions

Abiotic stresses such as heat, cold, drought, salinity and nutrient deficiency are inherent to every ecosystem and essentially unavoidable. Abiotic stresses are considered the most harmful factors in terms of growth and productivity of crops worldwide [54, 55], especially when they occur in combinations [56]. Here, we summarize and discuss various studies in order to clarify the functional involvement of different histone methylation marks in setting up plant responses to adverse environmental growth conditions.

4.1. Histone methylation/demethylation and the plant stress hormone ABA

The phytohormone abscisic acid (ABA) is a crucial signalling molecule playing versatile functions in regulating many developmental processes, including seed dormancy and germination [57, 58]. ABA also plays a pivotal role in adaptive stress processes, integrating both biotic and abiotic environmental constraints in a complex network of interacting pathways with crosstalks at different levels [59–61]. Currently, ABA is considered as a global regulator of stress responses that can dominantly control the switch in priority between the responses to biotic or abiotic stress, allowing plants to respond to the most severe threat [62].

The transition from heterotrophic to autotrophic development at the post-germinative stage (i.e. embryonic state) is highly vulnerable to osmotic stress [63]. During a period of osmotic stress, ABA promotes the expression of transcription factors such as ABI3 and ABI5, which in turn delay germination and lead to osmotolerance and survival [64]. In *Arabidopsis*, mutation in *PICKLE* (*PKL*), encoding a putative chromatin modifier, results in increased and abnormally sustained expression of *ABI3* and *ABI5* in response to exogenous ABA treatment. This sustained expression was correlated with reduced levels of H3K9me2 and H3K27me2, two methylation marks found in the chromatin of silent genes [65]. Based on these results, it was suggested that PKL might act on *ABI3* and *ABI5* to promote directly or indirectly the formation of a repressed chromatin state through a so-far-unknown mechanism.

In adult plants, the establishment of a response and tolerance to drought stress by ABA has been extensively studied and is well discussed in several outstanding reviews [58, 66]. Briefly, under drought conditions, water stress perception triggers ABA biosynthesis and increased tissue ABA accumulation, resulting in stomatal closure and reduced transpiration. Among major enzymes involved in the ABA biosynthesis pathway, NINE-CIS-EPOXYCAROTENOID DIOXYGENASE 3 (NCED3) is thought to be the rate-limiting enzyme [67]. In *Arabidopsis*, the increased transcription of *NCED3* upon dehydration was correlated with the binding of the HKMT ATX1 and the increased level of H3K4me3 at *NCED3* [68]. Therefore, the loss-of-function *atx1* mutant showed less tolerance to dehydration, notably because of the lower enrichment of RNA Polymerase II (RNAPII) and H3K4me3 at *NCED3* under stress. ATX1-modified H3K4me3 may thus have an important function in the transcriptional regulation of *NCED3*. However, it is still unclear whether this function is directly linked with the transcriptional induction or is more related to the reinforcement of the increased transcription upon stress perception.

4.2. Histone methylation/demethylation in response to water stresses

Water stresses including drought or submergence are major environmental factors limiting plant growth and crop productivity worldwide [69, 70]. Consequently, plants have evolved a variety of biochemical and physiological mechanisms to respond/adapt to these stresses [71, 72]. In the following section, we distinctly address the involvement of histone methylation in responses to drought and submergence.

4.2.1. Drought stress

Using *Arabidopsis*, the molecular response to water deficit was found to rely notably on the transcriptional regulation of stress-inducible genes with products thought to function in drought tolerance and response [73]. Using ChIP analyses in 15-day-old *Arabidopsis* seedlings, the level of the active mark H3K4me3 was found gradually enriched in response to dehydration stress, preferentially on the coding region of four drought-inducible genes (*RD29A*, *RD29B*, *RD20* and *RAP2.4*), and was correlated with their upregulation [74]. Consistent with this result, ATX1 was reported to be required for the efficient transcriptional induction of *RD29A* and *RD29B* during a dehydration stress response in an ABA-independent manner [68]. Also, for unknown reasons, it is worth noting that upon exogenous ABA treatment the transcriptional induction of *RD29A* and *RD29B* was stronger in *atx1* mutants than in wild-type plants (their basal transcript levels were lower in *atx1* than in wild-type, while their transcript levels upon ABA treatment were identical [68]. Next, for the ABA-dependent *RD29A* and ABA-independent *RAP2.4* genes, a time lag was observed between their transcriptional induction and the increase in H3K4me3 [74]. Based on these findings and the rapid saturation in RNAPII enrichment compared with H3K4me3 (i.e. already saturated 1 h after stress exposure for RNAPII, while H3K4me3 was still increasing up to 5 h), Kim et al. [74] concluded that the H3K4me3 enrichment may be established, gradually, in response to drought stress after full transcriptional activation of *RD29A* and *RAP2.4*. Because the timing of H3K4me3 enrichment followed subsequent to the RNAPII enrichment, H3K4me3 might be dispensable for the initiation of transcription. Finally, the gradual increase of H3K4me3 further indicates that the longer the stress lasts, the more H3K4me3 will be enriched, suggesting that the epigenetic responsiveness must depend on the intensity of a stress [75].

In a similar approach but using ChIP-Seq, van Dijk et al. [76] established the whole-genome distribution patterns of H3K4me1, H3K4me2 and H3K4me3 in 4-week-old rosette *Arabidopsis* leaves under dehydration stress conditions. They also observed a strong correlation between H3K4me3 abundance and transcripts levels from responding genes. Indeed, among the most strongly downregulated genes, an increase in H3K4me1 and a decrease in H3K4me3 were detected, suggesting the involvement of a histone demethylase in modulating the expression of a subset of stress-responsive genes. Supporting this finding, a putative PKDM7 subfamily-like H3K4 demethylase homologue and two putative demethylase enzymes containing a JmjC domain were found to be drought-inducible in two barley cultivars and in young peanut plants, respectively [77, 78]. Surprisingly, in contrast to the classical genome-wide H3K4me3 enrichment around the transcriptional start site observed in all other eukaryotes [79], H3K4me3 displayed a broader distribution on dehydration and ABA-inducible genes. Such an unusual profile may reflect a function not strictly related to transcription initiation, as will be discussed later in the stress memory section.

Through a genome-wide approach in rice seedling, Zong et al. [80] also uncovered a weak but positive correlation between H3K4me3 enrichment and the transcript level of some drought-responsive genes under drought stress. This correlation was extended to many genes involved in stress-related metabolite and hormone signalling pathways, further supporting the role played by H3K4me3 in the stress response [80]. However, because H3K4me3 is not the only histone mark for gene activation, this weak correlation may reflect that other active histone marks may also play important roles in regulating gene expression in response to stress in rice. Although these large data sets have provided much information on drought responses in rice, more detailed analyses will be required to elucidate whether the observed variations in H3K4 methylation are a cause or a consequence of the transcriptional changes triggered by water stress. Moreover, identifying key histone modification enzymes is indispensable to better understand the transcriptional regulatory network of the abiotic stress response.

4.2.2. Submergence

Submergence is a complex stress that encompasses many changes in environmental factors, including light intensity, pH and dissolved oxygen concentration. Alcoholic fermentation is important for the survival of plants especially under anaerobic environments [81]. In rice, *alcohol dehydrogenase 1* (*ADH1*) and *pyruvate decarboxylase 1* (*PDC1*) genes are involved in this anaerobic metabolism and their expression is reversibly induced (i.e. activated upon submergence and repressed upon re-aeration) [82]. Using these two genes as a model, Tsuji et al. [83] observed that the level of H3K4 methylation, specifically at both the 5′- and 3′-coding regions of *ADH1* and *PDC1*, was changed from a dimethylated state to a trimethylated state upon their transcriptional upregulation in response to submergence. This change was reverted back to its initial level following re-aeration, indicating that in this particular case, H3K4me3 does not serve as a memory mark of a prior transcriptional activity. Similarly to drought, these results highlight the dynamic and reversible change of histone H3K4 methylation at stress-related genes in response to the occurrence and disappearance of a stress.

4.3. Histone methylation/demethylation in response to salt stress

Salinity is also a serious factor affecting plants in several ways (i.e. water stress, ion toxicity, nutritional disorders, oxidative stress, alteration of metabolic processes, membrane disorganization, genotoxicity, reduction of cell division and expansion), thus limiting plant growth, development and survival [84]. In *Arabidopsis*, a NaCl or ABA treatment has been shown to induce the transcription of a range of abiotic stress-responsive genes (*ABI1, ABI2, KAT1, KAT2, DREB2A, RD29A* and *RD29B*). Using ChIP, this induction was found significantly correlated with an increase in the active mark H3K4me3 and a decrease in the repressive mark H3K9me2 [85]. Also, suggesting a link between different histone PTMs, mutations in the histone deacetylase HDA6 partially suppressed the H3K4me3 increase observed in response to stress, while the H3K9me2 decrease was not affected. Since histone acetylation and H3K4 trimethylation are often associated with gene activation [86], the repressive function of HDA6 may suggest that acetylation on histone H3 is required for K4 methylation to occur [87]. Unfortunately, because ChIP experiments were not normalised against the total H3 density, it is difficult to attribute the decrease in H3K9me2 to either the active removal of methyl groups by a histone demethylases or the active removal of nucleosome in response to a stress. Supporting the second possibility, the nucleosome density of two other drought stress-inducible genes (*RD20* and *RAP2.4*) was found to gradually decrease in response to drought stress [74].

In soya bean and in response to a high NaCl concentration, these histone methylation marks were also found altered at some salinity-induced transcription factors (i.e. MYB, b-ZIP and AP2/DREB family members) that were primarily identified by microarray analysis [88]. For some genes, their transcriptional induction was correlated with an increased level of histone acetylation and H3K4me3, accompanied or not with a reduced level of DNA methylation and H3K9me2 in various parts of the promoter or coding regions. For other genes, DNA methylation had no influence on histone methylation. This work perfectly reflects the heterogeneity of the effect of salinity on histone methylation and DNA methylation, and supports the role(s) of histone methylation changes in the expression of some transcription factors important for salinity tolerance.

As mentioned above, H3K4me3 was found to be involved in the transcriptional induction of stress-responsive gene upon salt stress exposure. In plants, the JmjC-domain-containing histone demethylases JMJ14, JMJ15 and JMJ18 have been reported to display an H3K4me2/3 demethylase activity as well as to regulate diverse aspects of chromatin function and development [89–95]. Recently, the overexpression of *JMJ15* was reported to preferentially down-regulate many stress-related genes preferentially marked by H3K4me2/3 and to enhance salt stress tolerance [96]. In contrast, the loss-of-function mutant was more sensitive to salt. Despite the fact that an increased JMJ15 level may regulate stress-responsive gene transcription programmes in *Arabidopsis*, the role of H3K4me3 resetting in these processes is still very elusive.

Besides histone lysine methylation, arginine methylation was also involved in establishing the transcriptional response to salt stress. The protein arginine methyltransferase 5 (PRMT5), also named Shk1 kinase-binding protein1 (SKB1), is a type II methyltransferase that catalyses symmetric H4R3 dimethylation, a repressive mark known to promote flowering through the repression of the floral repressor *FLOWERING LOCUS C* (*FLC*) in *Arabidopsis* [20, 97]. Interestingly, gain-of-function *SKB1* mutants showed an enhanced salt stress tolerance and sensitivity to ABA [98]. As a consequence of PRMT5 disassociating from chromatin, the H4R3sme2 level at stress-responsive genes was reduced during salt stress, resulting in their induced expression. Suggesting an additional function of PRMT5 on non-histone proteins, the methylation level of the U6 small nuclear ribonucleoprotein Sm-like4 (LSM4, a core protein of the spliceosome) was increased in response to salt stress and ABA. Since splicing defects were observed in the *prmt5/skb1* and *lsm4* mutants, with both of them being hypersensitive to salt stress, authors proposed that PRMT5 might mediate plant development and salt response by altering the methylation status of H4R3me2 and LSM4, linking transcriptional regulation to pre-mRNA splicing [98].

4.4. Histone methylation/demethylation in response to temperature

In plants, temperature stresses are classically classified into different types according to temperature exposure, which may be warm, high, chilling or freezing temperature. Due to global warming and because temperature stress greatly affects plant growth and development, immunity and circadian rhythm, and poses a serious threat to the global food supply, the genetic mechanisms of plant responses to heat have been well studied. Plants exposed to temperature stresses modulate the transcription of a large number of genes involved in distinct biochemical and physiological response pathways and networks of phytohormones or secondary metabolites, ultimately leading to increased tolerance to hazardous temperature stresses [99–102]. The role played by histone methylation during the plant response to a heat or a cold stress is discussed separately hereafter.

4.4.1. Heat stress

Heat stress during seed development decreases the seed size in many cereals, resulting in severe yield losses [103, 104]. In rice, a molecular mechanism involving the putative rice polycomb repressive complex 2 (PRC2) gene *Fertilization-Independent Endosperm1* (*OsFIE1*) was suggested as being a potential key component involved in regulating the thermal sensitivity of seed enlargement during endosperm development [105]. When developing seeds were exposed to a heat stress, both DNA methylation and H3K9 methylation were reduced on *OsFIE1* resulting in its derepression. Under heat stress, syncytial stage-specific MADS-box genes involved in seed size regulation were precociously repressed, due to the increased deposition of H3K27me3 silencing marks by the PRC2 complex [105]. In the unicellular green alga *Chlamydomonas reinhardtii*, histone modification was also affected by heat stress [106]. Indeed, after heat stress, the level of H3K4me1 was found decreased and the level of histone acetylation increased at promoter regions of active genes compared with inactive regions. As a hypothesis, authors proposed that upon heat stress, the heat shock transcription factor HSF1 might promote chromatin remodelling and RNAPII recruitment for transcription initiation/elongation [106]. Finally, while addressing molecular mechanisms of the response of cotton anthers to high temperature, two jmjC domain-containing genes, putatively involved in histone demethylation, were found significantly repressed during anther development under heat stress [107].

Because euchromatin is gene rich and usually transcriptionally active, investigation about the role of histone methylation in temperature stress acclimation was largely centred on euchromatin-associated coding regions. Focusing on the transcriptionally silent heterochromatin, mainly constituted of repetitive DNA sequences, some works demonstrated the transcriptional activation of normally silent transposable element embedded within heterochromatic regions under stress conditions [108]. Intriguingly, such activation under heat stress can occur without alteration of DNA methylation and with only minor changes in both H3K9me2 and H3K4me3 [109, 110]. In summary, these works suggest that temperature stress-mediated transcription of tandem-repeat elements might play a vital role in the adaptation of plants to temperature stimuli, offering an efficient mechanism by which heat or cold could promote the expression of some stress-responsive genes. Upon activation and when inserted into or very close to a gene, such transposable elements could interfere with the expression of this gene, giving rise to deleterious mutations, genetic instability or positive contribution to gene regulation and adaptation [111].

4.4.2. Cold stress

The increased tolerance of plants to cold is referred as 'cold acclimation'. Cold acclimation differs from vernalization, as the last one requires a long-term exposure to cold temperatures, while cold acclimation can be achieved in a couple of days under non-freezing low temperatures [112]. Locally, histone methylation changes in cold-responsive genes were addressed in *Arabidopsis*. Upon cold stress, the repressive mark H3K27me3 decreased, in both a histone occupancy-dependent (i.e. arising from the lowering nucleosome density) and -independent (i.e. as the result of the activity of a not yet identified histone demethylase) manner, on the cold-responsive genes, cold-regulated 15A (*COR15A*) and galactinol synthase 3 (*ATGOLS3*; [113]). Interestingly, the decrease in H3K27me3 upon stimulation occurred more gradually than their rapid transcriptional induction, so that their activation may not be inhibited by H3K27me3 itself but rather lead to the removal of H3K27me3. Also, while the transcription of *COR15A* and *ATGOLS3* was completely repressed to the initial level upon returning plants to normal growth conditions, the H3K27me3 decrease was maintained. Given that this decrease does not affect the transcriptional induction of *COR15A* and *ATGOLS3* upon re-exposure to cold temperatures, such chromatin change can so far only be view as a 'reminiscence' of a recent transcriptional activity and not as a stress memory implicated in a gene priming process.

In maize during cold stress, changes in histone modifications, including the heterochromatic marks H3K9me2 and DNA methylation, were assessed through a genome-wide approach [114]. The more detailed analysis of the two knob-associated tandem-repetitive sequences, the 180-bp repeat and the 350-bp repeat termed TR-1, demonstrated that their selectively and transiently cold-activated transcription was correlated with a decreased H3K9me2 and DNA methylation, together with an increased H3K9 acetylation. Such cold-induced transcriptional activation of tandem repeats is selective and transient, and the silencing state is recovered as the treatment continues.

5. Histone methylation as a memory mark of stress

In animals, the formation of memory immune cells after primary antigen recognition confers long-lasting resistance, resulting in an accelerated and a more effective immune response in case of second exposure. Despite the absence of such memory immune cells, plants often acquire a systemic immunity to further infections after a primary localized infection [115]. This requires the accumulation of the plant hormone SA in systemic tissues and is called systemic acquired resistance (reviewed in [116]. The SAR is also associated with gene priming in systemic tissues, in which defence genes will be expressed more rapidly and robustly in case of a second attack [117]. At the transcriptional level, gene expression is primarily influenced by the chromatin structure, which in turn is controlled partly by processes, often referred to as 'epigenetic' processes, which can be transmitted through mitosis and/or inherited through meiosis [118]. Therefore, chromatin remodelling through histone methylation offers a potential mechanism for short-/long-term stress memory within the lifespan of an individual, referred to as somatic memory, and/or across generations, referred to as transgenerational memory.

5.1. Somatic stress memory

In *Arabidopsis,* a priming event, either treatment with BTH or infection with *Pseudomonas syringae pv. maculicola,* systematically resulted in an increase in the level of H3K4me2/me3 at defence gene promoters (namely the WRKY transcription factors *WRKY6, WRKY29* and *WRKY52*) that are normally found on active genes, while the genes remain inactive [119]. More interestingly, this increase also occurs in leaves distal to localized foliar infection. Hence, even if the histone-modifying enzyme involved in this process remains unidentified, results from Jaskiewicz et al. [119] clearly suggest that histone methylation might create a 'memory' of the primary infection that is associated with an amplified reaction to a second stress stimulus. Further, Luna et al. [120] observed that promoters of SA-inducible *PR1, WRKY6* and *WRKY53* in the progeny of *Pst*-inoculated *Arabidopsis* plants were enriched with acetylated histone, while the promoter of the JA-inducible gene *PLANT DEFENSIN1.2* (*PDF1.2*) showed an increased level of H3K27me3. For decades, the signalling protein *NON EXPRESSOR OF PR1* (NPR1) has been implicated in mediating SAR induction [115] and also the crosstalk between SA- and JA/ET-dependent defence pathways, enabling plants to mount an appropriate defence reaction, depending on the nature of the attacker and the stage of infection [121, 122]. More recently, NPR1 has been proposed to play a critical role in the expression of the transgenerational SAR as progeny from *npr1* failed to develop transgenerational defence phenotypes and failed to present enrichment for H3K27me3 at the *PDF1.2* promoter [120]. Together, these findings suggested that one or more systemic signals are stored as an immune memory on defence-related gene promoters in the form of histone modifications, thus providing the plant with a life-long protection, which can be transmitted to subsequent generations.

Besides being involved in defence priming related to biotic stress, histone methylation was also proposed as a priming strategy against drought. To further explore the functional impact of histone methylation on biotic stress responses in *Arabidopsis,* Kim et al. [123] followed chromatin dynamics of several drought genes (*RD20, RD29A* and *AtGOLS2*) and a rehydration-inducible gene (*ProDH*) during drought and rehydration. As previously discussed, a strong correlation was observed between H3K4me3 enrichment (i.e. especially in gene bodies) and transcription for drought genes upon drought [74, 123]. Such a correlation was also detected for the rehydration gene upon rehydration [123]. Suggesting a memory role for H3K4me3, RNAPII rapidly disappeared after rehydration at drought genes, while H3K4me3 was gradually decreased. Concomitantly, by training plants with up to four successive drought treatments, Ding et al. [124] uncovered the existence of two distinct subsets of genes within the dehydration stress–response gene fraction. The 'non-trainable' genes (e.g. the ABA-independent *RD29A* and *COR15A*) have repetitively similar transcription rates during each stress treatment, while 'trainable' ones (e.g. the ABA-dependent *RD29B* and *RAB18*) increased the magnitude of their subsequent transcriptional response, relative to their initial stress response. Using ChIP, Ding et al. [124] observed that the H3K4me3 enrichment at 'trainable' genes, especially in gene bodies, was atypically retained from the preceding transcription after rehydration. Even more interestingly, the RNAPII phosphorylated at C-terminal domain (CTD) repeat serine 5 (Ser5P; associated with transcription initiation) was found stalled on these genes as a memory mark from a previously transcribed state. In contrast to 'trainable' genes, the stress-induced H3K4me3 and Ser5P enrichment at 'non-trainable' genes was decreased to its basal level during recovery [124]. Moreover, this transcriptional memory can persist in the absence of inducing signals at least for 5 days, but is lost after 7 days. Supporting a specific role for H3K4me3 in stress memory, other active chromatin marks such as acetylation of histones H3 and H4 were found rapidly increased at drought genes upon stress and decreased at comparable levels as before induction quickly after recovery [123, 124]. Consistent with an activating role of ATX1 at dehydration stress–response genes [68, 125, 126], dehydration-induced transcript levels were diminished in *atx1* plants [127]. However, ATX1 does not seem to have a critical impact on drought stress memory in *Arabidopsis*. Indeed, while being less increased than in wild-type plants, trainable genes still produced increased transcripts in trained, relative to untrained, *atx1* plants and retained high H3K4me3 levels during the watered recovery states. ATX1, ATX2, SDG25 and SET DOMAIN GROUP2/ATX-RELATED3 (SDG2/ATXR3) belong to the same class III of H3K4me3 methyltransferases and are thought to act, partially redundantly, as H3K4 methyltransferase [128–134]. More recently, SDG2 has been found to be essential for the full transcriptional activation of various hormone-responsive genes upon hormone treatment (i.e. including the ABA-dependant *RD29A*) via its H3K4 trimethyltransferase activity [135]. It is, therefore, likely that the other class III HKMT might also contribute to the drought stress memory. In summary, these results suggest that in addition to be a good marker of gene activation when found around promoter and 5′ regions of genes, H3K4me3 might also play a role in establishing a transcriptional short-term somatic memory of drought stress when found in gene bodies.

Using a large-scale approach, the distribution of H3K4me2, H3K4me3, H3K9me2 and H3K27me3 was analysed in *Arabidopsis* seedlings, which have been treated with mild salt stress in the seedling stage, resulting in an increased tolerance upon an additional salt stress application [136]. At low resolution in primed seedlings, H3K4me2 and H3K4me3 most commonly consisted of higher peaks of pre-existing enriched histone modification domains, named islands, whereas H3K9me2 produced the least differences. By contrast, majority of differences in H3K27me3 resulted from a higher number of islands with lower genome coverage. At high resolution, changes in H3K27me3 were already detectable a few hours after salt addition, suggesting that demethylation of H3K27me3 operates at a speed that is comparable to that of transcriptional regulation. Interestingly, this effect fades over time; however, it is still clearly visible after a 10-day-growth period in control conditions. In response to a second stress treatment, genes with high responsiveness, such as *HKT1* (i.e. encoding a root-specific Na transporter) and *PIP2E* (i.e. encoding a plasma membrane aquaporin), experienced a decrease of H3K27me3, whereas genes with lower responsiveness, such as *GH3.1* and *GH3.3* (i.e. encoding auxin and JA-amino acid-conjugating enzymes, respectively), experienced an increase of H3K27me3. Conversely, another group reported that in plants that have experienced several exposures to dehydration stress no significant change in the level of H3K27me3 could be detected on trainable and non-trainable genes, or during transcriptionally active/inactive gene states [137]. However, the high H3K27me3 level present at inactive dehydration stress memory genes did not interfere with the transition to an active transcription and with the accumulation of H3K4me3 [138]. Together, the function of H3K27me3 in genes that dynamically change transcription seems to depend on the type of environmental stimuli.

In contrast to H3K27me3, the higher level of H3K4me3 retained at the trainable gene *RD29B*, when its transcription is low, further supports the idea that H3K4me3 works as a 'memory' histone mark of a previously active state [138]. Generally, H3K4me3 and H3K27me3 play antagonistic roles in gene transcription and are therefore mutually repulsive at developmental genes [37, 139]. Seemingly, the presence of both marks, referred as 'bivalent domains', was first described in mammalian stem cells and was proposed to represent a pluripotent chromatin state that poises genes for activation upon appropriate developmental cues [140, 141]. Further work is required to determine whether H3K4me3 and H3K27me3 co-exist at certain genes whose expression is rapidly altered in response to environmental stimuli.

Because plant stress research has traditionally focused on single stresses, we separately described priming in response to abiotic or biotic environmental cues. However, in nature, plants are constantly exposed to mild environmental stresses during their lifetime. While testing how different environmental histories can affect the response of the plant to a subsequent biotic stress, Singh et al. [142] reported that *Arabidopsis* plants exposed to a recurrent abiotic stress (i.e. heat, cold or salt) were more resistant to *Pst* than plants grown in a more stable environment. This enhanced resistance was due to the priming of commonly used marker genes of pattern-triggered immunity (PTI; *WRKY53*, *FLG22-INDUCED RECEPTOR KINASE1* (*FRK1*) and *NDR1/HIN1-LIKE10* (*NHL10*)). Indeed, enrichment for epigenetic marks associated with transcriptional activation, such as H3K4me2 and H3K4me3, at PTI-responsive genes was observed after the exposure to recurrent stress, resulting in an enrichment of RNA polymerase II and a primed transcription in response to a subsequent bacterial infection. Collectively, these works on somatic stress memory promote the idea that, in plant, the environmental history can shape/modulate the response to stress, providing a mechanistic link between histone methylation and gene priming.

5.2. Transgenerational stress memory

The transgenerational stress memory refers to the transmittance of certain environmental responses from one generation to the next, thus providing the offspring of environmentally challenged plants with an adaptive advantage for better fitness (i.e. improve plant stress tolerance and impart developmental flexibility; [143]). Compared with DNA methylation and RNA interference (RNAi), very few studies suggest the involvement of histone methyltransferases and histone methylation changes in this process [144]. In *Arabidopsis*, changes in DNA methylation, histone modifications and gene expression were followed in the progeny of plants exposed to salt stress over one generation [145]. Although the DNA from the progeny of plants exposed to salt stress was globally hypomethylated, the majority of genes and promoters causing methylation changes were hypermethylated and lowly expressed. In addition, DNA hypermethylation was correlated with an increased level of the repressive mark H3K9me2. Among these hypermethylated genes, a large number was encoding different histone methyltransferases, such as the *Arabidopsis SU(VAR)3-9* homologues *SUVH2, SUVH5, SUVH8,* involved in H3K9 methylation, or the PRC2 subunit *CURLY LEAF* (*CLF*), involved in H3K27 methylation. Following this work, the progeny of heat-stressed plants was used to explore epigenetic variations under both normal and stressed conditions, in comparison to the progeny of control plants [146]. Similarly to salt stress, the progeny of plants exposed to heat stress had a global decrease of genomic DNA methylation and a reduced expression of several *SUVH* genes, which correlated with their enrichment in H3K9me2. Together, the hypermethylation of SUVH genes in the progeny of stressed plants may represent a protective mechanism against hypermethylation of the entire genome. Interestingly, in both works [145, 146], the transposon expression was elevated in the progeny of stressed plants. Because main targets of the SUVH pathway are transposable elements [24], the authors proposed that a decrease in the expression of *SUVH* genes might contribute to transposon activation, which at opportune times can create intragenomic potential upon transposition to facilitate adaptation in response to environmental changes [147]. In summary, these works suggested a role for histone methylation in the inheritance of stress memory; however, whether histone methylation changes are heritable through multiple generations and whether they sustain the acquisition of adaptive traits is still a matter of debate [148]. However, the evidence to date favours the view that stress-induced transgenerational changes in chromatin might increase the survival chances of the plant species, rather than each individual, by broadening the phenotypic plasticity and the genetic variation within the population [149, 150].

6. Discussion and perspectives

Recent advances, especially in *Arabidopsis*, have uncovered that chromatin remodelling through histone methylation changes are not only restricted to developmental needs but also an integral part of the very complex cascade of events that lead to abiotic/biotic stress tolerance, resistance and short-/long-term memory. Currently, a preliminary view is emerging, indicating that histone methylation changes, providing specific chromatin configurations, can be classified into several interrelated categories when involved in stress responses (Fig. 1): (*i*) histone methylation changes that are basally present on stress-related genes to establish a 'permissive' chromatin state that may either limit the spreading of repressive chromatin marks and/or potentiate a rapid transcriptional induction upon need; (*ii*) histone methylation changes that are transiently induced from an inactive or a permissive chromatin state by stress, to either facilitate the transcriptional initiation and/or reinforce transcription of stress-responding genes, and finally, histone methylation changes that are established in response to a stress; (*iii*) maintained for a certain time during the lifespan of an individual (i.e. somatic memory) or (*iv*) transmitted to one or more subsequent generations (i.e. transgenerational memory).

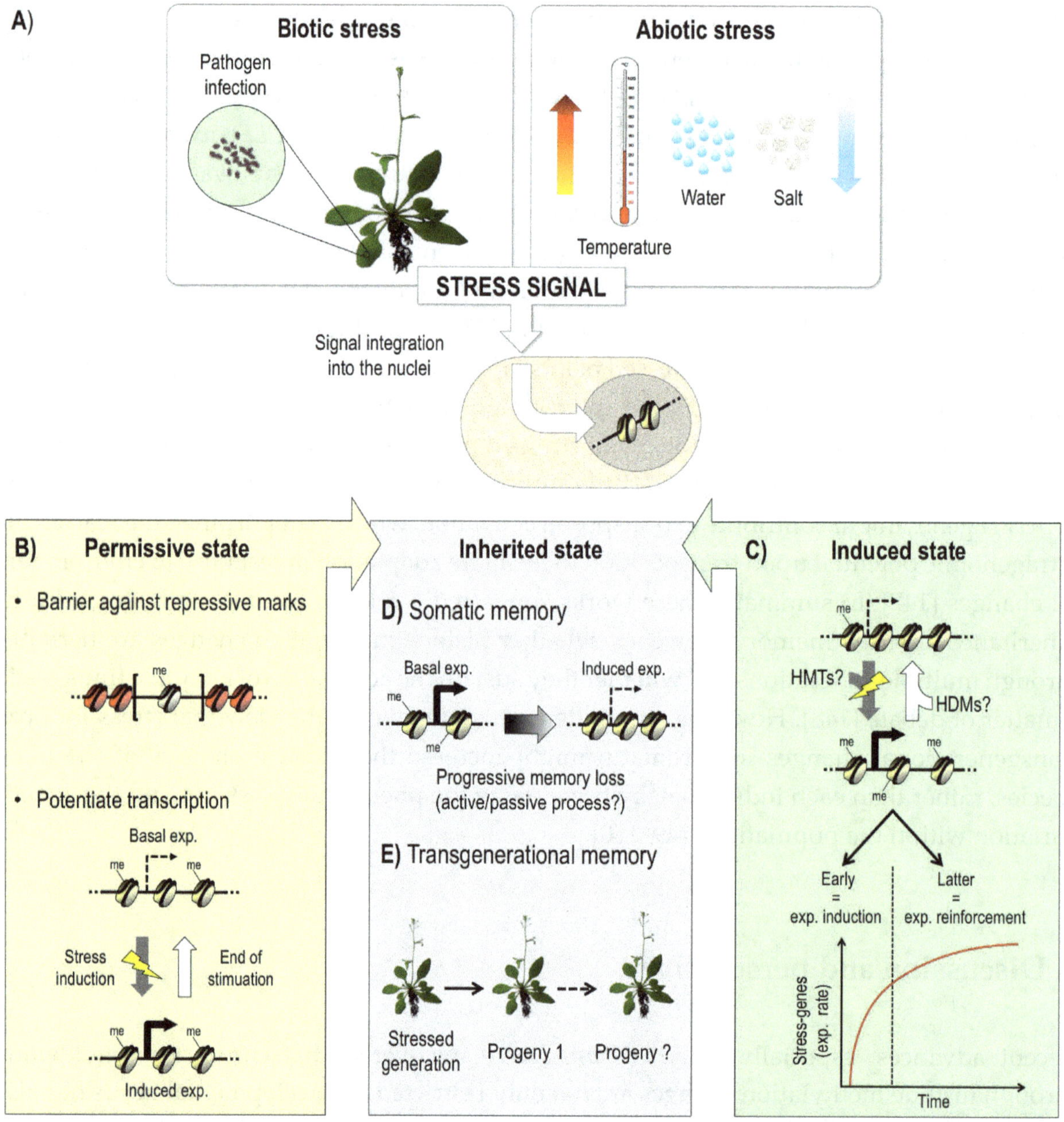

Figure 1. Hypothetical model depicting the role of histone methylation/demethylation in regulating plant stress responses. (A) Plants exposed either to biotic or abiotic stresses integrate the signal stress into the cell nuclei, where it affects the chromatin structure through histone methylation changes. According to our knowledge, these histone methylation changes can be classified into several interrelated categories. **(B)** The permissive state represents a more loosened chromatin state that will either offer a protection against repressive marks (represented with nucleosomes in red) and/or potentiates a rapid transcriptional induction upon stress induction. **(C)** The induced state represents histone methylation changes that are transiently induced by a stress signal. If methylation changes occur early during the stress-response process, they might participate to the transcriptional induction of stress-responding genes, while if they occur later, they might reinforce the transcription of stress-responding genes. Both permissive and the induced states can be maintained allowing a faster and/or stronger transcriptional induction of stress-responding genes upon a subsequent challenge. The memorized chromatin state can be maintained **(D)** for a certain time during the lifespan of an individual and referred as the somatic memory, or **(E)** transmitted to one or more subsequent generations and referred as the transgenerational memory.

Nonetheless, this emerging view is facing many gaps, inaccuracies and divergences, mainly related to numerous difficulties inherent to the study of such a dynamic and acute process. In this respect, plants in nature are usually challenged simultaneously by different kinds of stresses. Responses to these stress combinations are largely controlled by different signalling pathways that can interact in a non-additive manner, producing effects that could not have been predicted from the study of either stress individually [151, 152]. The occurrence of simultaneous biotic and abiotic stresses introduces an added degree of complexity that requires stresses to be imposed simultaneously and to treat each set of environmental conditions as an entirely new stress. For this reason and to clarify the mechanism behind the regulation of stress responses by histone methylation changes, there is a strong necessity to intensify our investigations. For instance, the correlation between histone methylation/demethylation and stress responses remains elusive and clarifications will require in-depth dynamic approaches based on comparative analyses of both epigenomes and transcriptomes during stress responses. In parallel, current knowledge about the corresponding histone-modifying enzymes is still largely missing. This lack of knowledge is pending on the identification of different stress-responsive histone modifiers and will require large-scale screens and genetic analyses for the sensitivity of different histone methyltransferases/demethylases mutants to various stresses, combined or not. Among other factors governing stress-induced chromatin changes, almost nothing is known about the specific reader/effector that will recognize particular histone methylation sites in order to determine their functional and structural outcome. An effort in this direction will most likely benefit the comprehensive understanding of the fundamental mechanisms connecting histone methylation changes with the modulation of transcription of stress-responsive genes, subsequently enabling plant to withstand stress. Higher-resolution chromatin studies are undoubtedly required to reveal the targeted stress-responsive genes and the specific sites of histone methylation/demethylation. Nevertheless, investigation of the direct effects of histone methylation/demethylation in plants is difficult. One reason is that plant genomes harbour high-copy number of histone genes (e.g. the *Arabidopsis* genome comprises 47 genes that encode 33 different core histone proteins; www.chromdb.org) and the incorporation/modification of such variants can result in the formation of chromatins with particular properties and functions [153–155]. Although ChIP assays have proven valuable in helping to identify histone methylation changes, many antibodies used to detect these changes have been so far unable to distinguish between different variants. New technologies (e.g. generation of mutants with point mutations targeting amino acid in the N-terminal tail of histone using the clustered regularly interspaced short palindromic repeats (CRISPR)/CRISPER-associated (Cas) system) [156] will need to be explored to unravel histone methylation changes of specific histone variants and their functions. Another challenge is that plants consist of many functionally specialized tissues and cell types, each with its own unique epigenome, transcriptome and proteome. Until now, histone methylation changes induced by stresses were exclusively addressed in entire plant or organs, meaning that the obtained profiles most likely reflect the consensus of multiple tissue- or cell-specific profiles that may differ. New methods allowing the mapping of chromatin features in specific tissue/cell types such as the one described by Wang and Deal [157] will be decisive for determining the cell-/tissue-specific chromatin alterations involved in a particular stress response. Finally, as plants have finite resources that must be balanced between growth and defence against stresses, often resulting in a growth or yield penalty, histone methylation changes in response to stress should be integrated in a more global developmental view, taking into account the involvement of several histone methyltransferases/demethylases in various processes such as root growth, flowering time, floral organogenesis, gametophyte or embryo formation [33]. Finally, understanding such regulatory network is an essential step to provide both novel paradigms and potential tools for further exploitation towards sustainable agriculture.

Acknowledgements

This work was supported by the Centre National de la Recherche Scientifique, the Agence Nationale de la Recherche (ANR-12-BSV2-0013-02), the European Commission (FP7-PEOPLE-2013-ITN, grant number 607880), the National Council of Science and Technology of Mexico (CONACYT) and the Program 'Estancias Posdoctorales y Sabaticas al Extranjero' (Fellowship number 237557. BOBADILLA-LANDEY Roberto; CVU:162391).

Author details

Roberto Bobadilla and Alexandre Berr*

*Address all correspondence to: Alexandre.Berr@ibmp-cnrs.unistra.fr

Institut de Biologie Moléculaire des Plantes, CNRS UPR2357, Université de Strasbourg, Strasbourg, France

References

[1] Osakabe Y, Yamaguchi-Shinozaki K, Shinozaki K and Tran LS. (2013) Sensing the environment: key roles of membrane-localized kinases in plant perception and response to abiotic stress. J Exp Bot. 64(2):445–458.

[2] Pieterse CMJ, Leon-Reyes A, Van der Ent S and Van Wees SCM. (2009) Networking by small-molecule hormones in plant immunity. Nat Chem Biol. 5(5):308–316.

[3] Nakashima K, Ito Y and Yamaguchi-Shinozaki K. (2009) Transcriptional regulatory networks in response to abiotic stresses in *Arabidopsis* and grasses. Plant Physiol. 149(1):88–95.

[4] van Verk MC, Gatz C and Linthorst HJM. (2009) Transcriptional regulation of plant defense responses. Adv Bot Res. 51:397–438.

[5] Riechmann JL and Ratcliffe OJ. (2000) A genomic perspective on plant transcription factors. Curr Opin Plant Biol. 3(5):423–434.

[6] Moore JW, Loake GJ and Spoel SH. (2011) Transcription dynamics in plant immunity. Plant Cell. 23(8):2809–2820.

[7] Luger K, Mäder AW, Richmond RK, Sargent DF and Richmond TJ. (1997) Crystal structure of the nucleosome core particle at 2.8 Å resolution. Nature. 389(6648):251–260.

[8] Andrews AJ and Luger K. (2011) Nucleosome structure(s) and stability: variations on a theme. Annu Rev Biophys. 40:99–117.

[9] Nelissen H, Boccardi TM, Himanen K and Van Lijsebettens M. (2007) Impact of core histone modifications on transcriptional regulation and plant growth. Crit Rev Plant Sci. 26(5-6):243–263.

[10] Shen WH and Xu L. (2009) Chromatin remodeling in stem cell maintenance in *Arabidopsis thaliana*. Mol Plant. 2(4):600–609.

[11] Berr A and Shen WH. (2010) Molecular mechanisms in epigenetic regulation of plant growth and development. In: Pua E. C. and Davey M. R. (Eds.) Plant Dev Biol – Biotechnol Perspect. 2:325–344.

[12] Arents G and Moudrianakis EN. (1995) The histone fold: a ubiquitous architectural motif utilized in DNA compaction and protein dimerization. Proc Natl Acad Sci USA. 92(24):11170–11174.

[13] Davey CA, Sargent DF, Luger K, Maeder AW and Richmond TJ. (2002) Solvent mediated interactions in the structure of the nucleosome core particle at 1.9 Å resolution. J Mol Biol. 319(5):1097–1113.

[14] Marks P, Rifkind RA, Richon VM, Breslow R, Miller T and Kelly WK. (2001) Histone deacetylases and cancer: causes and therapies. Nat Rev Cancer. 1(3):194–202.

[15] Mutskov V, Gerber D, Angelov D, Ausio J, Workman J and Dimitrov S. (1998) Persistent interactions of core histone tails with nucleosomal DNA following acetylation and transcription factor binding. Mol Cell Biol. 18(11):6293–6304.

[16] Strahl BD and Allis CD. (2000) The language of covalent histone modifications. Nature. 403(6765):41–45.

[17] Jaskiewicz M, Peterhansel C and Conrath U. (2011) Detection of histone modifications in plant leaves. J Vis Exp. (55).

[18] Zhang X. 2012. Chromatin modifications in plants. In: Wendel JF, editor. Plant genome diversity. Vienna: Springer. p. 237–56.

[19] Guccione E, Bassi C, Casadio F, Martinato F, Cesaroni M, Schuchlautz H, Lüscher Band Amati B. (2007) Methylation of histone H3R2 by PRMT6 and H3K4 by an MLL complex are mutually exclusive. Nature. 449(7164):933–937.

[20] Wang X, Zhang Y, Ma Q, Zhang Z, Xue Y, Bao S and Chong K. (2007) SKB1-mediated symmetric dimethylation of histone H4R3 controls flowering time in *Arabidopsis*. EMBO J. 26(7):1934–1941.

[21] Zhao Q, Rank G, Tan YT, Li H, Moritz RL, Simpson RJ, Cerruti L, Curtis DJ, Patel DJ, Allis CD, Cunningham JM and Jane SM. (2009) PRMT5-mediated methylation of histone H4R3 recruits DNMT3A, coupling histone and DNA methylation in gene silencing. Nat Struct Mol Biol. 16(3):304–311.

[22] Wang H, Huang ZQ, Xia L, Feng Q, Erdjument-Bromage H, Strahl BD, Briggs SD, Allis CD, Wong J, Tempst P and Zhang Y. (2001) Methylation of histone H4 at arginine 3 facilitating transcriptional activation by nuclear hormone receptor. Science. 293(5531):853–857.

[23] Strahl BD, Briggs SD, Brame CJ, Caldwell JA, Koh SS, Ma H, Cook RG, Shabanowitz J, Hunt DF, Stallcup MR and Allis CD. (2001) Methylation of histone H4 at arginine 3 occurs in vivo and is mediated by the nuclear receptor coactivator PRMT1. Curr Biol. 11(12):996–1000.

[24] Thorstensen T, Grini PE and Aalen RB. (2011) SET domain proteins in plant development. Biochim Biophys Acta. 1809(8):407–420.

[25] Di Lorenzo A and Bedford MT. (2011) Histone arginine methylation. FEBS Lett. 585(13):2024–2031.

[26] Pedersen MT and Helin K. (2010) Histone demethylases in development and disease. Trends in Cell Biol. 20(11):662–671.

[27] Springer NM, Napoli CA, Selinger DA, Pandey R, Cone KC, Chandler VL, Kaeppler HF and Kaeppler SM. (2003) Comparative analysis of SET domain proteins in maize and *Arabidopsis* reveals multiple duplications preceding the divergence of monocots and dicots. Plant Physiol. 132(2):907–925.

[28] Ng DW, Wang T, Chandrasekharan MB, Aramayo R, Kertbundit S and Hall TC. (2007) Plant SET domain-containing proteins: structure, function and regulation. Biochim Biophys Acta. 1769(5-6):316–329.

[29] Pontvianne F, Blevins T and Pikaard CS. (2010) *Arabidopsis* histone lysine methyltransferases. In: Delseny M. And Kader J. C. (eds) Advances in botanical research. London: Academic Press Ltd-Elsevier Science Ltd. 1–22.

[30] Aquea F, Vega A, Timmermann T, Poupin MJ and Arce-Johnson P. (2011) Genome-wide analysis of the SET DOMAIN GROUP family in grapevine. Plant Cell Rep. 30(6):1087–1097.

[31] Huang Y, Liu C, Shen WH and Ruan Y. (2011) Phylogenetic analysis and classification of the *Brassica rapa* SET-domain protein family. BMC Plant Biol. 11:175.

[32] Aiese-Cigliano R, Sanseverino W, Cremona G, Ercolano MR, Conicella C and Consiglio FM. (2013) Genome-wide analysis of histone modifiers in tomato: gaining an insight into their developmental roles. BMC Genomics. 14:57.

[33] Berr A, Shafiq S and Shen WH. (2011) Histone modifications in transcriptional activation during plant development. Biochim Biophys Acta. 1809(10):567–576.

[34] Hammond-Kosack KE and Jones JDG. (2000) Response to plant pathogens. In: Buchannan B. Gruissem W. Jones R. (eds). Biochemistry and molecular biology of plants. Rockville, MD: American Society of Plant Physiologists. 1102–1157.

[35] Jones JDG and Dangl JL. (2006) The plant immune system. Nature. 444(7117):323–329.

[36] Mosher RA, Durrant WE, Wang D, Song J and Dong X. (2006) A comprehensive structure-function analysis of *Arabidopsis* SNI1 defines essential regions and transcriptional repressor activity. Plant Cell. 18(7):1750–1765.

[37] Roudier F, Ahmed I, Bérard C, Sarazin A, Mary-Huard T, Cortijo S, Bouyer D, Caillieux E, Duvernois-Berthet E, Al-Shikhley L, Giraut L, Després B, Drevensek S, Barneche F, Dèrozier S, Brunaud V, Aubourg S, Schnittger A, Bowler C, Martin-Magniette M L, Robin S, Caboche M and Colot V. (2011). Integrative epigenomic mapping defines four main chromatin states in *Arabidopsis*. EMBO J. 30:1928–1938.

[38] Alvarez-Venegas R, Al Abdallat A, Guo M, Alfano JR and Avramova Z (2007). Epigenetic control of a transcription factor at the cross section of two antagonistic pathways. Epigenetics. 2(2):106–113.

[39] Gruner K, Griebel T, Návarová H, Attaran E and Zeier J. (2013) Reprogramming of plants during systemic acquired resistance. Front Plant Sci. 4:252.

[40] Alvarez-Venegas R, Sadder M, Hlavacka A, Baluska F, Xia YN, Lu GQ, et al. (2006) The *Arabidopsis* homolog of trithorax, ATX1, binds phosphatidylinositol 5-phosphate, and the two regulate a common set of target genes. Proc Natl Acad Sci USA. 103(15): 6049–6054.

[41] Li J, Brader G, Kariola T and Palva ET. (2006) WRKY70 modulates the selection of signaling pathways in plant defense. Plant J. 46(3):477–491.

[42] Xu L, Zhao Z, Dong A, Soubigou-Taconnat L, Renou JP, Steinmetz A and Shen WH. (2008) Di- and tri- but not monomethylation on histone H3 lysine 36 marks active transcription of genes involved in flowering time regulation and other processes in *Arabidopsis thaliana*. Mol Cell Biol. 28(4):1348–1360.

[43] Palma K, Thorgrimsen S, Malinovsky FG, Fiil BK, Nielsen HB, Brodersen P, et al. (2010) Autoimmunity in Arabidopsis *acd11* Is Mediated by Epigenetic Regulation of an Immune Receptor. PLoS Pathog 6(10): e1001137. doi:10.1371/journal.ppat.1001137.

[44] De-La-Peña C, Rangel-Cano A and Alvarez-Venegas R. (2012) Regulation of disease-responsive genes mediated by epigenetic factors: interaction of *Arabidopsis-Pseudomonas*. Mol Plant Pathol. 13(4):388–398.

[45] Tsuchiya T and Eulgem T. (2011) EMSY-like genes are required for full RPP7-mediated race-specific immunity and basal defense in *Arabidopsis*. Mol Plant Microbe Interact. 24(12):1573–1581.

[46] Tsuchiya T and Eulgem T. (2013) An alternative polyadenylation mechanism coopted to the *Arabidopsis* RPP7 gene through intronic retrotransposon domestication. Proc Natl Acad Sci U S A. 110(37):E3535-E3543.

[47] Musselman CA and Kutateladze TG. (2011) Handpicking epigenetic marks with PHD fingers. Nucleic Acids Res. 39(21):9061–9071.

[48] Tsuchiya T and Eulgem T. (2013) Mutations in EDM2 selectively affect silencing states of transposons and induce plant developmental plasticity. Sci Rep. 3:1701.

[49] Li T, Chen X, Zhong X, Zhao Y, Liu X, Zhou S, Cheng S and Zhou DX. (2013) Jumonji C domain protein JMJ705-mediated removal of histone H3 lysine 27 trimethylation is involved in defense-related gene activation in rice. Plant Cell. 25(11):4725–4736.

[50] Pieterse CM, Poelman EH, Van Wees SC and Dicke M. (2013) Induced plant responses to microbes and insects. Front Plant Sci. 4:475.

[51] Berr A, McCallum EJ, Alioua A, Heintz D, Heitz T and Shen WH. (2010). *Arabidopsis* histone methyltransferase SET DOMAIN GROUP8 mediates induction of the jasmonate/ethylene pathway genes in plant defense response to necrotrophic fungi. Plant Physiol. 154(3):1403–1414.

[52] Berr A, Ménard R, Heitz T and Shen WH. (2012) Chromatin modification and remodelling: a regulatory landscape for the control of *Arabidopsis* defence responses upon pathogen attack. Cell Microbiol. 14(6):829–839.

[53] Grossniklaus U and Paro R. (2014) Transcriptional silencing by polycomb-group proteins. Cold Spring Harb Perspect Biol. 6(11):a019331.

[54] Wang W, Vinocur B and Altman A. (2007) Plant responses to drought, salinity and extreme temperatures towards genetic engineering for stress tolerance. Planta. 218(1):1–14.

[55] Gao JP, Chao DY and Lin HX. (2007) Understanding abiotic stress tolerance mechanisms: recent studies on stress response in rice. J Integr Plant Biol 49(6):742–750.

[56] Mittler R. (2006) Abiotic stress, the field environment and stress combination." Trends Plant Sci. 11(1):15–19.

[57] Santner A, Calderon-Villalobos LI and Estelle M. (2009) Plant hormones are versatile chemical regulators of plant growth. Nat Chem Biol. 5(5):301–307

[58] Cutler SR, Rodriguez PL, Finkelstein RR and Abrams SR. (2010) Abscisic acid: emergence of a core signaling network. Annu Rev Plant Biol. 61:651–679.

[59] Fujita M, Fujita Y, Noutoshi Y, Takahashi F, Narusaka Y, Yamaguchi-Shinozaki K and Shinozaki K. (2006) Crosstalk between abiotic and biotic stress responses: a current view from the points of convergence in the stress signaling networks. Curr Opin Plant Biol. 9(4):436–442.

[60] Asselbergh B, De Vieesschauwer D and Hofte M. (2008) Global switches and fine-tuning – ABA modulates plant pathogen defense. Mol Plant Microbe Interact. 21(6): 709–719.

[61] Yasuda M, Ishikawa A, Jikumaru Y, Seki M, Umezawa T, Asami T, Maruyama-Nakashita A, Kudo T, Shinozaki K, Yoshida S and Nakashita H. (2008) Antagonistic interaction between systemic acquired resistance and the abscisic acid-mediated abiotic stress response in *Arabidopsis*. Plant Cell. 20(6):1678–1692.

[62] Atkinson NJ and Urwin PE. (2012) The interaction of plant biotic and abiotic stresses: from genes to the field. J Exp Bot. 63(10):3523–3543.

[63] Rajjou L, Duval M, Gallardo K, Catusse J, Bally J, Job C and Job D. (2012) Seed germination and vigor. Annu Rev Plant Biol. 63:507–533.

[64] Lopez-Molina L, Mongrand S and Chua NH. (2001) A postgermination developmental arrest checkpoint is mediated by abscisic acid and requires the ABI5 transcription factor in *Arabidopsis*. Proc Natl Acad Sci USA. 98(8):4782–4787.

[65] Perruc E, Kinoshita N, Lopez-Molina L. (2007). The role of chromatin-remodeling factor PKL in balancing osmotic stress responses during *Arabidopsis* seed germination. Plant J. 52(5):927–936.

[66] Raghavendra AS, Gonugunta VK, Christmann A and Grill E. (2010) ABA perception and signalling. Trends Plant Sci. 15(7):395–401.

[67] Koornneef M, Leon-Kloosterziel K, Schwartz SH and Zeevaart JAD. (1998) The genetic and molecular dissection of abscisic acid biosynthesis and signal transduction in *Arabidopsis*. Plant Physiol Biochem. 36:83–89.

[68] Ding Y, Avramova Z and Fromm M. (2011) The Arabidopsis trithorax-like factor ATX1 functions in dehydration stress responses via ABA-dependent and ABA-independent pathways. Plant J. 66(5):735–744.

[69] Ramachandra RA, Chaitanya KV and Vivekanandan M. (2004) Drought induced responses of photosynthesis and antioxidant metabolism in higher plants. J Plant Physiol. 161(11):1189–1202.

[70] Araus JL, Slafer GA, Royo C and Serret MD. (2008) Breeding for yield potential and stress adaptation in cereals. Crit Rev Plant Sci. 27:377–412.

[71] Hirayama T and Shinozaki K. (2010) Research on plant abiotic stress responses in the post-genome era: past, present and future. Plant J. 61(6):1041–1052.

[72] Fukao T and Xiong L. (2013) Genetic mechanisms conferring adaptation to submergence and drought in rice: simple or complex? Curr Opin Plant Biol. 16(2):196–204.

[73] Huang D, Wu W, Abrams SR and Cutler AJ. (2008) The relationship of drought-related gene expression in *Arabidopsis thaliana* to hormonal and environmental factors. J Exp Bot. 59(11):2991–3007.

[74] Kim JM, To TK, Ishida J, Morosawa T, Kawashima M, Matsui A, Toyoda T, Kimura H, Shinozaki K and Seki M. (2008). Alterations of lysine modifications on the histone H3 N-tail under drought stress conditions in *Arabidopsis thaliana*. Plant Cell Physiol. 49:1580–1588.

[75] Kim JM, Sasaki T, Ueda M, Sako K and Seki M. (2015) Chromatin changes in response to drought, salinity, heat, and cold stresses in plants. Front Plant Sci. 6:114.

[76] van Dijk K, Ding Y, Malkaram S, Riethoven JJ, Liu R, Yang J, Laczko P, Chen H, Xia Y, Ladunga I, Avramova Z and Fromm M. (2010) Dynamic changes in genome-wide histone H3 lysine 4 methylation patterns in response to dehydration stress in *Arabidopsis thaliana*. BMC Plant Biol. 10:238.

[77] Papaefthimiou D and Tsaftaris A. (2012) Significant induction by drought of HvPKDM7-1, a gene encoding a jumonji-like histone demethylase homologue in barley (H. Vulgare). Acta Physiol Plant. 34(3):1187–1198.

[78] Govind G, Vokkaliga Thammegowda H, Jayaker Kalaiarasi P, Iyer D, Muthappa S, Nese S and Makarla U. (2009) Identification and functional validation of a unique set of drought induced genes preferentially expressed in response to gradual water stress in peanut. Mol Genet Genomics. 281(6):591–605.

[79] Rando OJ and Chang HY. (2009) Genome-wide views of chromatin structure. Annu Rev Biochem. 78:245–271.

[80] Zong W, Zhong X, You J and Xiong L. (2013). Genome-wide profiling of histone H3K4-tri-methylation and gene expression in rice under drought stress. Plant Mol Biol 81(1-2):175–188.

[81] Hsu FC and Shih MC. (2013) Plant defense after flooding. Plant Signal Behav. 8(11):e26922.

[82] Tsuji H, Nakazono M, Saisho D, Tsutsumi N and Hirai A. (2000). Transcript levels of the nuclear-encoded respiratory genes in rice decrease by oxygen deprivation: evidence for involvement of calcium in expression of the alternative oxidase 1a gene. FEBS Lett. 471(2-3):201–204.

[83] Tsuji H, Saika H, Tsutsumi N, Hirai A and Nakazono M. (2006) Dynamic and reversible changes in histone H3-Lys4 methylation and H3 acetylation occurring at submergence-inducible genes in rice. Plant Cell Physiol. 47(7):995–1003.

[84] Carillo P, Annunziata MG, Pontecorvo G, Fuggi A and Woodrow P. (2011) Salinity stress and salt tolerance. In: Arun S, editor. Abiotic Stress in Plants - Mechanisms and Adaptations. Croatia: InTech. pp. 22–38.

[85] Chen LT, Luo M, Wang YY and Wu K. (2010). Involvement of *Arabidopsis* histone deacetylase HDA6 in ABA and salt stress response. J Exp Bot. 61(2):3345–3353.

[86] He G, Elling A and Deng X. (2011) The epigenome and plant development. Rev Plant Biol. 62(1):411.

[87] Nightingale KP, Gendreizig S, White DA, Bradbury C, Hollfelder F and Turner BM. (2007) Cross-talk between histone modifications in response to histone deacetylase inhibitors: MLL4 links histone H3 acetylation and histone H3K4 methylation. J Biol Chem. 282(7):4408–4416.

[88] Song Y1, Ji D, Li S, Wang P, Li Q and Xiang F. (2012) The dynamic changes of DNA methylation and histone modifications of salt responsive transcription factor genes in soybean. PLoS One. 7(7):e41274.

[89] Deleris A, Greenberg MV, Ausin I, Law RW, Moissiard G, Schubert D and Jacobsen SE. (2010) Involvement of a Jumonji-C domain-containing histone demethylase in DRM2-mediated maintenance of DNA methylation. EMBO Rep. 11(12):950–955.

[90] Lu F, Cui X, Zhang S, Liu C and Cao X. (2010) JMJ14 is an H3K4 demethylase regulating flowering time in *Arabidopsis*. Cell Res. 20(3):387–390.

[91] Searle IR, Pontes O, Melnyk CW, Smith LM and Baulcombe DC. (2010) JMJ14, a JmjC domain protein, is required for RNA silencing and cell-to-cell movement of an RNA silencing signal in *Arabidopsis*. Genes Dev. 24(10):986–991.

[92] Le Masson I, Jauvion V, Bouteiller N, Rivard M, Elmayan T and Vaucheret H. (2012) Mutations in the *Arabidopsis* H3K4me2/3 demethylase JMJ14 suppress posttranscriptional gene silencing by decreasing transgene transcription. Plant Cell. 24(9):3603–3612.

[93] Yang H, Mo H, Fan D, Cao Y, Cui S and Ma L. (2012) Overexpression of a histone H3K4 demethylase, JMJ15, accelerates flowering time in *Arabidopsis*. Plant Cell Rep. 31(7):1297–1308.

[94] Yang H, Han Z, Cao Y, Fan D, Li H, Mo H, Feng Y, Liu L, Wang Z, Yue Y, Cui S, Chen S, Chai J and Ma L. (2012) A companion cell-dominant and developmentally regulated H3K4 demethylase controls flowering time in *Arabidopsis* via the repression of FLC expression. PLoS Genet. 8(4):e1002664.

[95] Cui X, Jin P, Cui X, Gu L, Lu Z, Xue Y, Wei L, Qi J, Song X, Luo M, An G and Cao X. (2013) Control of transposon activity by a histone H3K4 demethylase in rice. Proc Natl Acad Sci USA. 110(5):1953–1958.

[96] Shen Y, Conde E, Silva N, Audonnet L, Servet C, Wei W and Zhou DX. (2014) Over-expression of histone H3K4 demethylase gene JMJ15 enhances salt tolerance in *Arabidopsis*. Front Plant Sci. 5:290.

[97] Schmitz RJ, Sung S and Amasino RM. (2008) Histone arginine methylation is required for vernalization-induced epigenetic silencing of FLC in winter-annual *Arabidopsis thaliana*. Proc Natl Acad Sci USA. 105(2):411–416.

[98] Zhang Z, Zhang S, Zhang Y, Wang X, Li D, Li Q, Yue M, Li Q, Zhang YE, Xu Y, Xue Y, Chong K and Bao S. (2011) *Arabidopsis* floral initiator SKB1 confers high salt tolerance by regulating transcription and pre-mRNA splicing through altering histone H4R3 and small nuclear ribonucleoprotein LSM4 methylation. Plant Cell. 23(1):396–411.

[99] Thomashow MF. (1999) PLANT COLD ACCLIMATION: freezing tolerance genes and regulatory mechanisms. Annu Rev Plant Physiol Plant Mol Biol. 50:571–599.

[100] Chinnusamy V, Zhu J and Zhu JK. (2007) Cold stress regulation of gene expression in plants. Trends Plant Sci. 12(10):444–451.

[101] Bokszczanin KL and Fragkostefanakis S. (2013). Perspectives on deciphering mechanisms underlying plant heat stress response and thermotolerance. Front Plant Sci. 4:315.

[102] Qu AL, Ding YF, Jiang Q and Zhu C. (2013). Molecular mechanisms of the plant heat stress response. Biochem Biophys Res Commun. 432(2):203–207.

[103] Peng S, Huang J, Sheehy JE, Laza RC, Visperas RM, Zhong X, Centeno GS, Khush GS and Cassman KG. (2004) Rice yields decline with higher night temperature from global warming. Proc Natl Acad Sci USA 101(27):9971–9975.

[104] Lobell DB, Schlenker W and Costa-Roberts J. (2011) Climate trends and global crop production since 1980. Science. 333:616–620.

[105] Folsom JJ, Begcy K, Hao X, Wang D and Walia H. (2014) Rice fertilization-independent endosperm1 regulates seed size under heat stress by controlling early endosperm development. Plant Physiol. 165(1):238–248.

[106] Strenkert D, Schmollinger S and Schroda M. (2013) Heat shock factor 1 counteracts epigenetic silencing of nuclear transgenes in *Chlamydomonas reinhardtii*. Nucleic Acids Res. 41(10):5273–5289.

[107] Min L, Li Y, Hu Q, Zhu L, Gao W, Wu Y, Ding Y, Liu S, Yang X and Zhang X. (2014). Sugar and auxin signalling pathways respond to high-temperature stress during

anther development as revealed by transcript profiling analysis in cotton. Plant Physiol. 164(3):1293–1308.

[108] Grandbastien MA. (1998) Activation of plant retrotransposons under stress conditions. Trends Science. 3(5):181–187. Plants

[109] Pecinka A, Dinh HQ, Baubec T, Rosa M, Lettner N and Mittelsten Scheid O. (2010) Epigenetic regulation of repetitive elements is attenuated by prolonged heat stress in *Arabidopsis*. Plant Cell. 22(9):3118–3129.

[110] Tittel-Elmer M, Bucher E, Broger L, Mathieu O, Paszkowski J and Vaillant I. (2010) Stress-induced activation of heterochromatic transcription. PLoS Genet. 6(10):e1001175.

[111] Lippman Z, Gendrel AV, Black M, Vaughn MW, Dedhia N, McCombie WR, Lavine K, Mittal V, May B, Kasschau KD, Carrington JC, Doerge RW, Colot V and Martienssen R. (2004) Role of transposable elements in heterochromatin and epigenetic control. Nature. 430(6998):471–476.

[112] Sung S and Amasino RM. (2005) Remembering winter: toward a molecular understanding of vernalization. Annu Rev Plant Biol. 56:491–508.

[113] Kwon CS, Lee D, Choi G and Chung WI. (2009) Histone occupancy-dependent and -independent removal of H3K27 trimethylation at cold-responsive genes in *Arabidopsis*. Plant J. 60(1):112–121.

[114] Hu Y, Zhang L, He S, Huang M, Tan J, Zhao L, Yan S, Li H, Zhou K, Liang Y and Li L. (2012) Cold stress selectively unsilences tandem repeats in heterochromatin associated with accumulation of H3K9ac. Plant Cell Environ. 35(12):2130–2142.

[115] Durrant WE and Dong X. (2004) Systemic acquired resistance. Annu Rev Phytopathol 42:185–209.

[116] Mishina TE and Zeier J. (2007) Pathogen-associated molecular pattern recognition rather than development of tissue necrosis contributes to bacterial induction of systemic acquired resistance in *Arabidopsis*. Plant J. 50(3):500–513.

[117] Conrath U. (2011) Molecular aspects of defense priming. Trends Plant Sci. 16(10):524–531.

[118] Heard E and Martienssen RA. (2014) Transgenerational epigenetic inheritance: myths and mechanisms. Cell. 157(1):95–109.

[119] Jaskiewicz M, Conrath U and Peterhansel C. (2011) Chromatin modification acts as a memory for systemic acquired resistance in the plant stress response. Embo Reports. 12(1):50–55.

[120] Luna E, Bruce TJ, Roberts MR, Flors V and Ton J. (2012) Next-generation systemic acquired resistance. Plant Physiol. 158(2):844–853.

[121] Spoel SH, Koornneef A, Claessens SM, Korzelius JP, Van Pelt JA, Mueller MJ, Buchala AJ, Métraux JP, Brown R, Kazan K, Van Loon LC, Dong X and Pieterse CM. (2003) NPR1 modulates cross-talk between salicylate- and jasmonate-dependent defense pathways through a novel function in the cytosol. Plant Cell. 15(3):760–770.

[122] Koornneef A and Pieterse CM. (2008) Cross talk in defense signaling. Plant Physiol. 146(3):839–844.

[123] Kim JM, To TK, Ishida J, Matsui A, Kimura H and Seki M. (2012) Transition of chromatin status during the process of recovery from drought stress in *Arabidopsis thaliana*. Plant Cell Physiol. 53:847–856.

[124] Ding Y, Fromm M and Avramova Z. (2012) Multiple exposures to drought 'train' transcriptional responses in *Arabidopsis*. Nat Commun. 3:740.

[125] Ding Y, Lapko H, Ndamukong I, Xia Y, Al-Abdallat A, Lalithambika S, Sadder M, Saleh A, Fromm M, Riethoven JJ, Lu G and Avramova Z. (2009) The *Arabidopsis* chromatin modifier ATX1, the myotubularin-like AtMTM and the response to drought. Plant Signal Behav. 4(11):1049–1058.

[126] Ding Y, Avramova Z and Fromm M. (2011) Two distinct roles of ARABIDOPSIS HOMOLOG OF TRITHORAX1 (ATX1) at promoters and within transcribed regions of ATX1-regulated genes. Plant Cell. 23(1):350–363.

[127] Ding Y, Ndamukong I, Xu Z, Lapko H, Fromm M and Avramova Z. (2012) ATX1-generated H3K4me3 is required for efficient elongation of transcription, not initiation, at ATX1-regulated genes. PLoS Genet. 8(12):e1003111.

[128] Saleh A, Alvarez-Venegas R, Yilmaz M, Le O, Hou G, Sadder M, Al-Abdallat A, Xia Y, Lu G, Ladunga I and Avramova Z. (2008) The highly similar *Arabidopsis* homologues of trithorax ATX1 and ATX2 encode proteins with divergent biochemical functions. Plant Cell. 20(3):568–579.

[129] Avramova Z. (2009) Evolution and pleiotropy of TRITHORAX function in *Arabidopsis*. Int J Dev Biol. 53(2-3):371–381.

[130] Berr A, Xu L, Gao J, Cognat V, Steinmetz A, Dong A and Shen WH. (2009) SET DOMAIN GROUP25 encodes a histone methyltransferase and is involved in FLOWERING LOCUS C activation and repression of flowering. Plant Physiol. 151(3):1476–1485.

[131] Tamada Y, Yun JY, Woo SC and Amasino RM. (2009) ARABIDOPSIS TRITHORAX-RELATED7 is required for methylation of lysine 4 of histone H3 and for transcriptional activation of FLOWERING LOCUS C. Plant Cell. 21(10):3257–3269.

[132] Berr A, McCallum EJ, Ménard R, Meyer D, Fuchs J, Dong A and Shen WH. (2010) *Arabidopsis* SET DOMAIN GROUP2 is required for H3K4 trimethylation and is crucial for both sporophyte and gametophyte development. Plant Cell. 22(10):3232–3248.

[133] Guo L, Yu Y, Law JA and Zhang X. (2010) SET DOMAIN GROUP2 is the major histone H3 lysine [corrected] 4 trimethyltransferase in *Arabidopsis*. Proc Natl Acad Sci USA. 107(43):18557–18562.

[134] Yun JY, Tamada Y, Kang YE and Amasino RM. (2012) *Arabidopsis* trithorax-related3/SET domain GROUP2 is required for the winter-annual habit of *Arabidopsis thaliana*. Plant Cell Physiol. 53(5):834–846.

[135] Kim S, Lee J, Yang JY, Jung C and Chua NH. (2013) *Arabidopsis* histone methyltransferase SET DOMAIN GROUP2 is required for regulation of various hormone responsive genes. J Plant Biol. 56(1):39–48.

[136] Sani E, Herzyk P, Perrella G, Colot V and Amtmann A. (2013). Hyperosmotic priming of *Arabidopsis* seedlings establishes a long-term somatic memory accompanied by specific changes of the epigenome. Genome Biol. 14(6):R59.

[137] Liu N, Ding Y, Fromm M and Avramova Z. (2014) Different gene-specific mechanisms determine the 'revised-response' memory transcription patterns of a subset of *A. thaliana* dehydration stress responding genes. Nucleic Acids Res. 42(9):5556–5566.

[138] Liu N, Fromm M and Avramova Z. (2014) H3K27me3 and H3K4me3 chromatin environment at super-induced dehydration stress memory genes of *Arabidopsis thaliana*. Mol Plant. 7(3):502–513.

[139] Bouyer D, Roudier F, Heese M, Andersen ED, Gey D, Nowack MK, Goodrich J, Renou JP, Grini PE, Colot V and Schnittger A. (2011) Polycomb repressive complex 2 controls the embryo-to-seedling phase transition. PLoS Genet. 7(3):e1002014.

[140] Azuara V, Perry P, Sauer S, Spivakov M, Jorgensen HF, John RM, Gouti M, Casanova M, Warnes G, Merkenschlager M and Fisher AG. (2006). Chromatin signatures of pluripotent cell lines. Nat Cell Biol. 8(5):532–538.

[141] Bernstein BE, Mikkelsen TS, Xie X, Kamal M, Huebert DJ, Cuff J, Fry B, Meissner A, Wernig M, Plath K, Jaenisch R, Wagschal A, Feil R, Schreiber SL and Lander ES. (2006) A bivalent chromatin structure marks key developmental genes in embryonic stem cells. Cell. 125(2):315–326.

[142] Singh P, Yekondi S, Chen PW, Tsai CH, Yu CW, Wu K, Zimmerli L. (2014) Environmental history modulates *Arabidopsis* pattern-triggered immunity in a HISTONE ACETYLTRANSFERASE1-Dependent manner. Plant Cell. 26(6):2676–2688.

[143] Bruce TJA, Matthes MC, Napier JA and Pickett JA. (2007) Stressful 'memories' of plants: evidence and possible mechanisms. Plant Sci. 173(6):603–608.

[144] Liu J, Feng L, Li J and He Z. (2015) Genetic and epigenetic control of plant heat responses. Front Plant Sci. 6:267.

[145] Bilichak A, Ilnystkyy Y, Hollunder J and Kovalchuk I. (2012) The progeny of *Arabidopsis thaliana* plants exposed to salt exhibit changes in DNA methylation, histone modifications and gene expression. PLoS One. 7(1):e30515.

[146] Migicovsky Z, Yao Y and Kovalchuk I. (2014) Transgenerational phenotypic and epigenetic changes in response to heat stress in *Arabidopsis thaliana*. Plant Signal Behav. 9(2):e27971.

[147] Oliver KR, McComb JA and Greene WK. (2013) Transposable elements: powerful contributors to angiosperm evolution and diversity. Genome Biol Evol. 5(10):1886–1901.

[148] Iwasaki M and Paszkowski J. (2014) Epigenetic memory in plants. EMBO J. 33(18): 1987–1998.

[149] Becker C and Weigel D. (2012) Epigenetic variation: origin and transgenerational inheritance. Curr Opin Plant Biol. 15(5):562–567.

[150] Weigel D and Colot V. (2012) Epialleles in plant evolution. Genome Biol. 13(10):249.

[151] Rizhsky L, Liang HJ, Shuman J, Shulaev V, Davletova S and Mittler R. (2004) When defense pathways collide. The response of *Arabidopsis* to a combination of drought and heat stress. Plant Physiol. 134(4):1683–1696.

[152] Asselbergh B, Achuo AE, Hofte M and Van Gijsegem F. (2008) Abscisic acid deficiency leads to rapid activation of tomato defence responses upon infection with *Erwinia chrysanthemi*. Mol Plant Pathol. 9(1):11–24.

[153] She W and Baroux C. (2014) Chromatin dynamics during plant sexual reproduction. Front Plant Sci. 5:354.

[154] Talbert PB and Henikoff S. (2014) Environmental responses mediated by histone variants. Trends Cell Biol. 24(11):642–650.

[155] Probst AV and Mittelsten Scheid O. (2015) Stress-induced structural changes in plant chromatin. Curr Opin Plant Biol. 27:8–16.

[156] Sander JD and Joung JK. (2014) CRISPR-Cas systems for editing, regulating and targeting genomes. Nat Biotechnol. 32(4):347–355.

[157] Wang D and Deal RB. (2015) Epigenome profiling of specific plant cell types using a streamlined INTACT protocol and ChIP-seq. Methods Mol Biol. 1284:3–25.

3

Glycation of Plant Proteins under Environmental Stress — Methodological Approaches, Potential Mechanisms and Biological Role

Tatiana Bilova, Uta Greifenhagen, Gagan Paudel, Elena Lukasheva, Dominic Brauch, Natalia Osmolovskaya, Elena Tarakhovskaya, Gerd Ulrich Balcke, Alain Tissier, Thomas Vogt, Carsten Milkowski, Claudia Birkemeyer, Ludger Wessjohann and Andrej Frolov

Abstract

Environmental stress is one of the major factors reducing crop productivity. Due to the oncoming climate changes, the effects of drought and high light on plants play an increasing role in modern agriculture. These changes are accompanied with a progressing contamination of soils with heavy metals. Independent of their nature, environmental alterations result in development of oxidative stress, i.e. increase of reactive oxygen species (ROS) contents, and metabolic adjustment, i.e. accumulation of soluble primary metabolites (amino acids and sugars). However, a simultaneous increase of ROS and sugar concentrations ultimately results in protein glycation, i.e. non-enzymatic interaction of reducing sugars or their degradation products (α-dicarbonyls) with proteins. The eventually resulting advanced glycation end-products (AGEs) are known to be toxic and pro-inflammatory in mammals. Recently, their presence was unambiguously demonstrated *in vivo* in stressed *Arabidopsis thaliana* plants. Currently, information on protein targets, modification sites therein, mediators and mechanisms of plant glycation are being intensively studied. In this chapter, we comprehensively review the methodological approaches for plant glycation research and discuss potential mechanisms of AGE formation under stress conditions. On the basis of these patterns and additional *in vitro* experiments, the pathways and mechanisms of plant glycation can be proposed.

Keywords: Advanced glycation end-products (AGEs), Ageing, Environmental stress, Glycation, Proteomics

1. Introduction

1.1. Environmental stress, ROS and protein glycation

1.1.1. Environmental stress and ROS formation

Environmental stress is one of the major factors reducing the productivity of crop plants all over the world [1]. Drought, high light, salinity and increased heavy metal soil contents, as well as extreme temperature, represent its important manifestations [2]. On the physiological level, extreme environmental conditions ultimately results in decrease in the CO_2 assimilation rate and in growth inhibition [3]. Simultaneous accumulation of reduced equivalents results in an overload of the chloroplast and mitochondrial electron transport chains and enhanced production of the reactive oxygen species (ROS), i.e. singlet oxygen ($^1O_2$), superoxide radical anion ($O_2^{\cdot-}$), peroxide ion (O_2^{2-}), hydrogen peroxide (H_2O_2), various hydroperoxides and hydroxide radical ($OH^{\cdot}$) [4]. When ROS production overwhelms their detoxication, oxidative stress develops [5].

Thus, transfer of electrons to molecular oxygen (O_2) from ubisemiquinone in mitochondria and thylakoid membrane-bound primary electron acceptor of photosystem I (PSI) in chloroplasts yields $O_2^{\cdot-}$ (and, when further reduction occurs, O_2^{2-}), further converted to H_2O_2 by superoxide dismutase (SOD) activity (predominantly Mn- and CuZn-SOD in mitochondria and chloroplasts, respectively) [6,7]. The radical oxygen species can abstract protons from (bis-)allylic methylenes of polyunsaturated fatty acids (PUFAs) [8]. The subsequent capture of O_2 molecule by the resulting carbon-centered radical yields a peroxyl radical, that is able to initiate a chain reaction of lipid peroxidation [9]. The PUFAs can be directly attacked by the protonated form of $O_2^{\cdot-}$ ($HO_2^{\cdot}$) [10], thus the content of lipid hydroperoxides is one of the most reliable markers of oxidative stress.

The hydroperoxides can be easily involved in the Fenton reaction, i.e. transition metal ion-mediated reduction, yielding $OH^{\cdot}$, i.e. one of the most short-living and toxic ROS, directly and irreversibly modifying lipids, proteins and nucleic acids [11]. The metal ions oxidized during Fenton reaction are reduced *in vivo* by cellular antioxidants or $O_2^{\cdot-}$ (Haber-Weiss reaction), that considerably increases the production of $OH^{\cdot}$ [12]. As environmental stress is accompanied with a strong upregulation of mono- and oligosaccharides in all plant tissues [13], metal-catalyzed oxidation of sugars (so-called monosaccharide autoxidation) [14] also might be enhanced under the conditions of oxidative stress. The resulting products – hydroxycarbonyls and α-oxocarbonyls – are the potent protein modification agents and can induce essential changes in their structure and function [15]. These reactions, termed protein glycation, i.e. modification of proteins by carbonyl compounds (carbohydrates and α-oxocarbonyls), is believed to be an important factor in stress-related protein damage [16].

1.1.2. Protein glycation

In the first step of this process (usually termed "early glycation"), reducing sugars, aldoses and ketoses reversibly interact with amino groups resulting in the very labile N/O-acetal intermediates: aldoamines and ketoamines, respectively (Figure 1). These compounds easily condense yielding aldimines and ketoimines (Schiff bases), which undergo Amadori [17] and Heyns [18] rearrangements, variants of the acyloin shift. Amadori rearrangement involves proton transfer from C1 to C2 via the enol/enamine intermediate yielding *N*-substituted 1-amino-deoxy-ketoses, the Amadori products (Figure 1A). Similarly, in the case of ketoamines, in course of Heyns rearrangement [18], a proton migrates from C2 to C1 forming 2-amino-deoxyaldosyl adducts, often referred to as Heyns products (Figure 1B) [19]. Both Amadori and Heyns products are termed as "early glycation products", the first relatively stable intermediates of glycation [20].

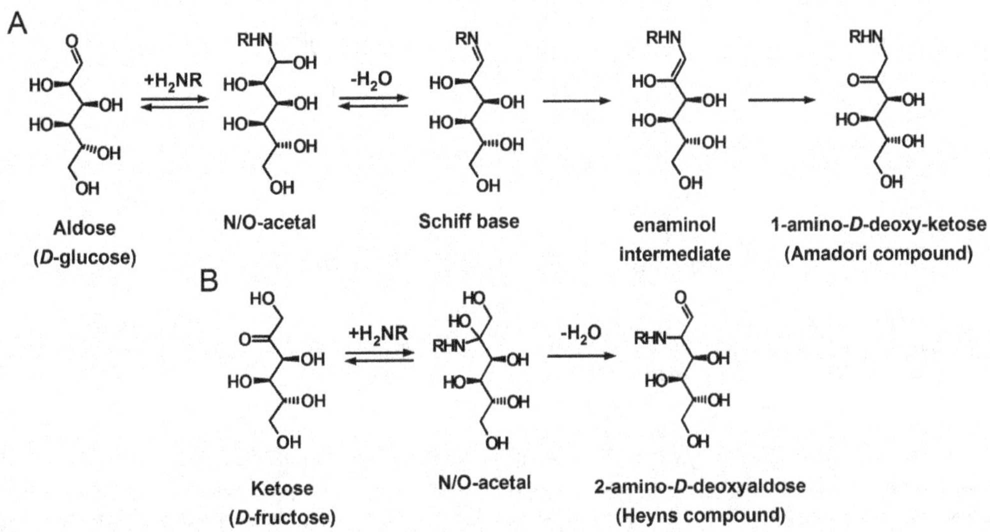

Figure 1. Formation of N/O-acetal and Schiff intermediates with subsequent Amadori rearrangement (A), and ketamine intermediates with subsequent Heyns rearrangement (B, not all intermediates shown).

These early glycation products, as well as free sugars, readily autoxidize (by the mechanisms similar to those described for free sugars) with formation of highly reactive α-dicarbonyl compounds (presumably glyoxal, methylglyoxal and various osones) – potent reactive intermediates of advanced glycation [21]. Depending on the structure of the carbohydrate moiety involved in this degradation, i.e. free sugars, or protein-bound early glycation products, two principle advanced glycation pathways, namely "oxidative glycosylation" and "glycoxidation", respectively, are distinguished [14,22,23]. The interaction of α-dicarbonyls with lysyl amino and arginyl guanidino side chain groups results in formation of so-called advanced glycation end-products (AGEs) – protein Maillard reaction compounds accumulating during thermal processing of food (Figure 2) [24], but also endogenously, e.g. under the conditions of persisting hyperglycemia.

1.1.3. Advanced Glycation End-products (AGEs)

AGEs represent a highly heterogenic group of compounds, varying greatly in their stability. Thus, the term "advanced glycation end-products" is, to high extent, conventional: some AGEs

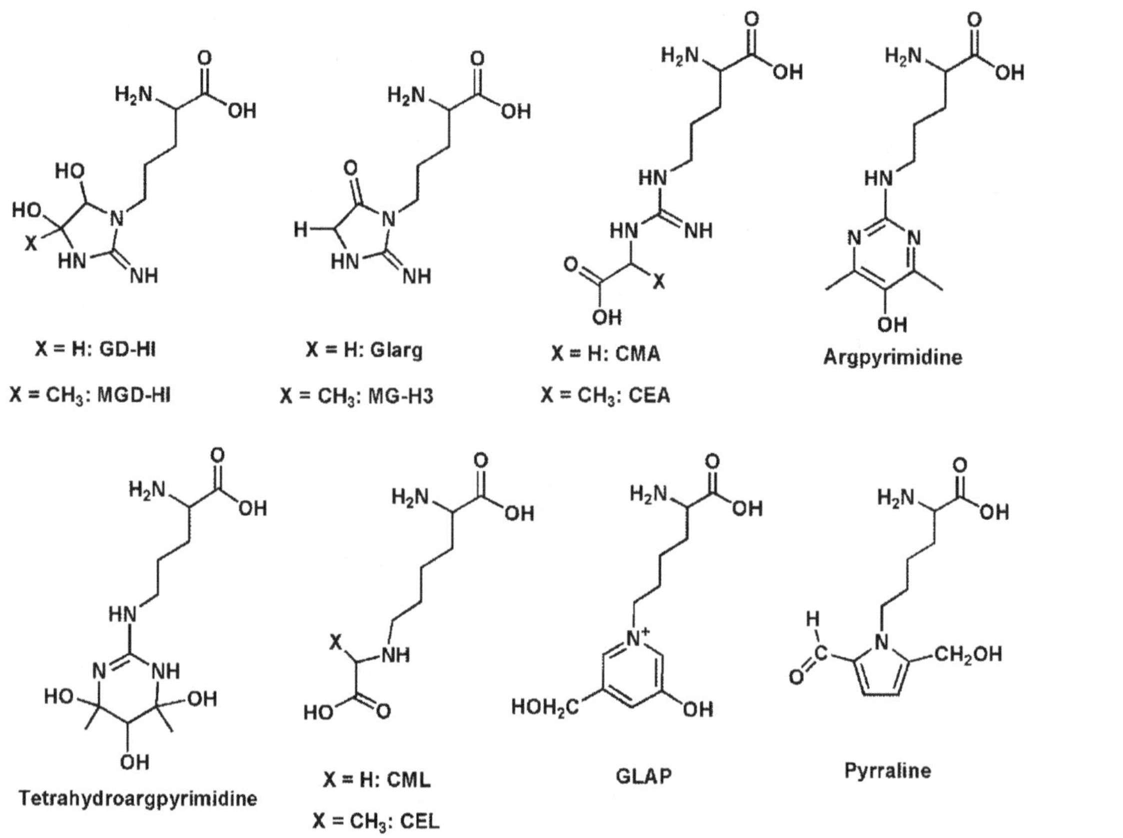

Figure 2. AGEs detected *in vivo*: GD-HI, glyoxal-derived dihydroxyimidazolidine; MGD-HI, methylglyoxal-derived dihydroxyimidazolidine; MG-H, methylglyoxal-derived hydroimidazolone; Glarg, glyoxal-derived hydroimidazolone; CML, N^{ε}-carboxymethyllysine; CEL, N^{ε}-carboxyethyllysine; GLAP, glyceraldehyde-derived pyridinium compound.

are still reactive and can be easily involved in further reactions [25]. In the past few decades, several AGEs were comprehensively characterized (Figure 2). Among lysine-derived modifications, N^{ε}-(carboxymethyl)lysine (CML) [26], N^{ε}-(carboxyethyl)lysine (CEL) [27], ε-(2-formyl-5-hydroxymethyl-pyrrolyl)-*L*-norleucine (pyrraline) [28] and glyceraldehyde-derived pyridinium compound (GLAP) [29] are the best-characterized. Not less attention was paid to the modifications of arginine. Thus, Schwarzenbolz and coworkers reported 1-(4-amino-4-carboxybutyl)2-imino-5-oxo-imidazolidine (Glarg) as a product of the reaction of arginine with glyoxal [30] and yielding N^{δ}-carboxyethylarginine (CMA) upon hydrolysis at 37°C [31]. Methylglyoxal was shown to form isomeric methylglyoxal-derived hydroimidazolones (MG-Hs) with N^{δ}-(5-methyl-4-oxo-5-hydroimidazolinone-2-yl)-*L*-ornithine (MG-H1) as the major isomer [32]. Hydrolysis of 2-amino-5-(2-amino-4-hydro-4-methyl-5-imidazolon-1-yl)pentanoic acid (MG-H3) yields carboxyethyl-*L*-arginine (CEA) [33]. Sequential modification of arginine with two methylglyoxal molecules results in N^{δ}-(5-hydroxy-4,6-dimethylpyrimidine-2-yl)-*L*-ornithine (argpyrimidine, Argpyr) [34] and N^{δ}-(4-carboxy-4,6-dimethyl-5,6-dihydroxy-1,4,5,6-tetrahydropyrimidine-2-yl)-*L*-ornithine (tetra-hydroargpyrimidine, TH-Argpyr) [35].

Upon their absorption in human intestine, AGEs interact with endothelial and macrophage pattern recognition receptors for AGEs (e.g. RAGEs) and trigger NF-κB-mediated expression of pro-inflammatory species (e.g. adhesion molecules, including vascular cell adhesion molecule-1 and intercellular adhesion molecule-1) [36] and foster the development of inflammatory diseases – e.g. atherosclerosis and type 2 diabetes mellitus [37]. AGEs of different chemical structure and origin, most often CML, pentosidine and hydroimidazolones, are known to be the ligands of RAGEs and to trigger inflammatory response [38].

Surprisingly, other reports showed mammalian serum and urinary concentrations of AGEs to be independent from dietary intake of thermally processed foods [39]. Moreover, the levels of CML and fluorescent AGEs in the plasma of vegetarian individuals were higher in comparison to those in the omnivorous individuals [40], even though the vegetarian diet had lower contents of lysine- and arginine-containing proteins. Remarkably, this effect was stronger in plasma of long-term vegetarians [41]. These facts indicate a high relevance of protein glycation (both early and advanced) in plants. Obviously, this explains the presence of multiple efficient anti-glycative enzymatic systems, like glyoxalase I and II [42], ribulosamine/erythrulosamine 3-kinase [43], acylamino acid-releasing enzyme [44].

Recently, Bechtold and co-workers reported an increase in the total contents of individual AGE classes upon the application of experimental environmental stress [16]. Thus, it is obvious that environmental changes are accompanied with enhanced generation of AGEs in plant tissues. In other words, due to the continuously altering growth conditions, AGEs might accumulate in plants during their life span, causing stress-related changes of the plant proteome and its physiological state. Important to note, that due to the dramatically different metabolic background (i.e. other patterns of carbohydrates, as well as high contents of potential antioxidants and carbonyl traps), pathways of glycation in plants may differ from those described in mammalians. However, no information about the proteins and biochemical pathways affected by such glycation reactions (i.e. its structural and functional patterns) was available until very recently. The most recent studies from our labs on the protein glycation patterns of model plants in the absence and presence of environmental stresses, as well as the impact of protein glycation in plant ageing, are added to this chapter.

2. Methodological approaches for the study of protein glycation in plants

2.1. Proteomics in plant glycation research

Recently, using liquid chromatography-tandem mass spectrometry (LC-MS/MS)-based approach, Bechtold and co-workers estimated the total levels of individual AGE classes in *Arabidopsis thaliana* leaf tissues [16]. Their methodology relied on the tandem mass spectrometric (MS/MS) analysis of individual amino acids in tissue hydrolyzates by multiple reaction monitoring (MRM) [45]. Though this method provides high sensitivity and specificity, this strategy could not provide information about the identity and modification patterns of individual glycated proteins. However, this question can be addressed by the methods of LC-MS-based proteomics – the technique based on the analysis of protein enzymatic hydrolyzates

[46]. Indeed, in its classical implementation, unbiased LC-MS approach combines high separation efficiency of the high-performance or ultra-high-performance liquid chromatography (HPLC or UPLC, respectively) and outstanding resolution and mass accuracy of high-resolution mass spectrometry – Orbitrap- and quadrupole-time-of-flight (QqTOF)-MS [47,48]. It dramatically increases the analytical resolution (i.e. the number of analytes, annotated in one experiment) of the whole method. Moreover, due to the superior separation, low-abundant post-translationally modified peptides are less amenable to the ion suppression effects, accompanying electrospray ionization (ESI) [49]. Obviously, this analytical strategy seems to be the optimal for the study of such heterogenic and low-abundant set of modifications as AGEs.

2.2. Protein isolation and proteolysis

Depending on the target protein fraction (soluble or total), aqueous (aq.), extraction can be performed in the absence and presence of phenol, respectively [50]. In the first case, however, even the extracts of green parts contain high amounts of soluble metabolites that might inhibit the activity of proteases used for digestion. Such incompleteness of proteolysis can be observed as well, when proteins are extracted from plant parts rich in anti-nutritive (i.e. protease inhibiting or denaturing) phenolics like insoluble condensed tannins in seeds [51]. Therefore, the extracts can be purified by gel filtration chromatography and/or ultrafiltration using Centricon or Vivaspin centrifugation devices [52] prior to the determination of protein concentration. Alternatively, the proteins can be isolated by phenol extraction. In this case, phenolics contaminants can be removed by addition of 1–5% of soluble or insoluble polyvinylpyrrolidone [50]. While purified aqueous extracts can be easily digested by proteases in the presence of only deoxycholate as a denaturizing agent [53], the dried proteins isolated with phenol (containing also the fraction of hydrophobic membrane proteins) can be reconstituted only in the presence of both chaotropic compounds (urea, thiourea) and strong detergents. Conventional detergents, such as sodium dodecyl sulfate (SDS) or 3-[(3-cholamidopropyl)dimethylammonio]-1-propanesulfonate (CHAPS), co-elute with proteolytic peptides and disturb ESI. This can be avoided by application of the detergents which do not impact protease activity and can be easily destroyed upon the digest. For example, *Rapi*Gest™ SF Surfactant (Waters Corporation, Milford, US) or Progenta Protea Biosciences, Inc., Morgantown, US) can be applied [54]. The detergents can be dissolved in the lysis buffer (7 mol/L urea and 2 mol/L in 50 mmol/L Tris-HCl, pH 7.5) and destroyed directly after proteolysis at pH 2 for 20–45 min (Figure 3).

The completeness of the digestion can be controlled by the SDS-polyacrylamide gel electrophoresis (PAGE). Performing SDS-PAGE additionally prior to proteolysis provides the opportunity to validate the Bradford assay results by relative quantification of total lane densities and, if necessary, normalize the results of the LC-MS-based label-free quantification. It is important to stress that the use of chaotropic agents, acid-labile detergents and strong acids for their cleavage ultimately require RP-based solid phase extraction (SPE) after completion of the digest and verification of its completeness by SDS-PAGE.

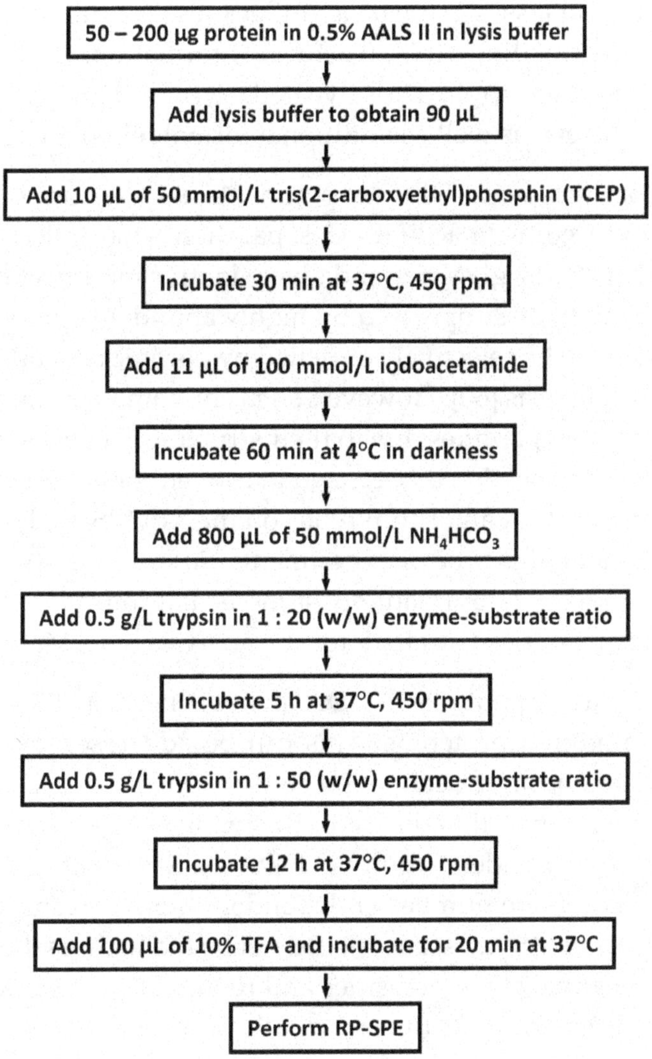

Figure 3. Protocol for tryptic digestion of the total plant protein obtained by phenol extraction. The procedure employed tryptic digestion in the presence of an acid-labile detergent ALLS II, required to ensure efficient solubilisation of membrane proteins.

2.3. LC-MS analysis of glycated peptide mixtures

The proteolytic digests are, typically, complex mixtures. Therefore, for successful detection and identification of their components (i.e. peptides), adequate chromatographic and mass separation techniques need to be applied. The selection of a strategy for LC-MS analysis (in terms of throughput, sensitivity, separation efficiency and reproducibility) depends on the aims of particular research. In the absolute majority of cases, LC-MS analyses rely on so-called data-dependent acquisition experiments (DDA) [55]. These complex experiments comprise survey MS scan (typically performed in Orbitrap- or TOF mass analyzer), and multiple dependent ones – MS/MS, relying on linear ion trap (LIT) or QqTOF analysis [56,57]. Modern instruments provide a possibility for introduction of the second dependent scan. Thus, hybrid LTQ-Orbitrap instruments support multi-step activation (MSA) experiments, comprising an additional MS/MS scan with a low-energy collision-induced dissociation (CID) [58]. The modification-specific neutral losses, appearing under these conditions, trigger a high-energy dependent MS/MS scan providing rich structural information. In the past decade, these experiments were successfully applied to glycated tryptic digests using Amadori-specific losses of two and three waters, as well as additional formaldehyde molecule [59].

In the most straightforward way, DDA experiments are performed without additional analytical procedures preceding a RP-HPLC separation – so-called shotgun proteomics. However, as well as other PTMs, glycative and glycoxidative modifications are low abundant. Hence, their ionization might be suppressed by highly abundant species. To avoid this, early glycated tryptic peptides can be selectively enriched by boronic acid affinity chromatography (BAC) before LC-MS/MS analysis [60]. However, such enrichment is not possible for the whole fraction of advanced glycated peptides due to their structural heterogeneity. Therefore, these low-abundant species need to be directly detected in conventional data-dependent acquisition (DDA) experiments. As the DDA algorithm relies on the MS/MS analysis of the most intense signals in each time segment, this type of experiments suffers from so-called undersampling, i.e. missed fragmentation of low-abundant quasi-molecular ions [61]. Because of this reason, shotgun proteomics is not a desired strategy for the analysis of PTMs.

Thus, the number of co-eluting peptides in DDA analyses needs to be reduced to increase the coverage of the AGE-modified proteome. This can be addressed by three approaches: *(i)* introduction of enrichment or pre-fractionation step, *(ii)* gas phase fractionation (GPF) on the MS level and *(iii)* use of retention time-based exclusion lists. In all cases, for reliable identification of AGE-containing peptides, multiple DDA experiments are required. A special attention needs to be paid on the number of dependent MS/MS scans and the total duration of cycle. Thus, the number of MS/MS scans should not be too high, as the quasi-molecular ions of AGE peptides have relatively low intensities and require, therefore, longer fill times in LIT and accumulation time TOF. The cycle times typically need to be shorter, than in conventional shotgun DDA experiments, as the peak widths of low abundant peptides are smaller.

The generalized analytical strategy might comprise both qualitative and quantitative approaches, i.e. identification of glycated peptides in DDA experiments with their subsequent label-free quantification in additional full-scan MS experiments (Figure 4). For identification of glycation sites, early glycated peptides can be selectively enriched [60], while the analysis of AGE-containing species might rely on two-dimensional liquid chromatography (LC × LC) [62]. BAC is a well-established analytical tool to enrich Amadori and Heyns products from mammalian tissues (predominantly plasma) [56]. However, for application to the study of plant glycation, this method requires some optimization. Thus, the protein extract must be effectively washed prior to digestion (e.g. by ultrafiltration) to remove the co-extracted carbohydrate-related metabolites (mono-, oligosaccharides, sugar esters and glycosides of (poly)phenolics). Due to their *cis*-diol groups, these metabolites saturate the binding sites of the affinity column and reduce the peptide enrichment efficiency. Moreover, phenylpropanoids, as well as their sugar esters and glycosides, are well-retained on reversed phase [63] and might co-elute with peptides in course of RP-HPLC separation, leading to essential ion suppression during the ESI process and shift of peptide signals in the lower part on the instrument dynamic range. Thus, the analysis of early glycation proteins in total extracts that can not be efficiently ultrafiltrated is a challenging task.

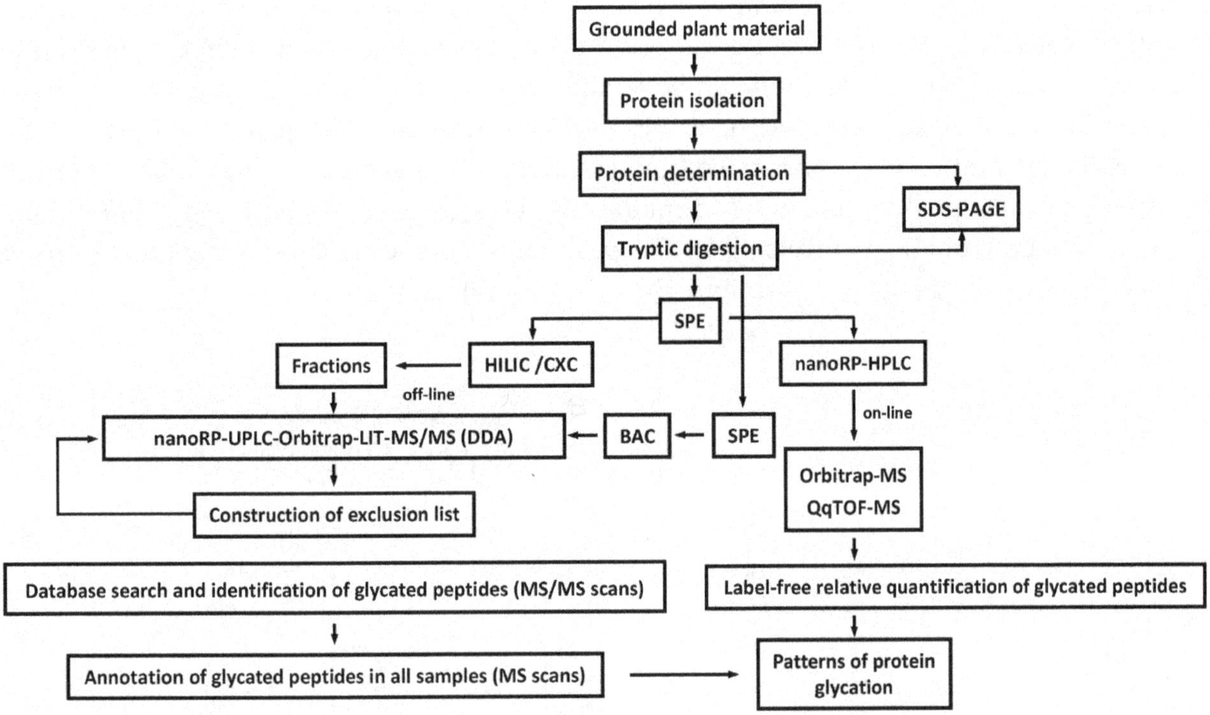

Figure 4. Experimental workflow for the analysis of glycated proteome. The analytical strategy comprises separation and MS/MS-based identification of modified peptides, as well as label-free quantification of corresponding glycated sites.

Pre-fractionation is usually introduced as an (relative to the RP) orthogonal separation procedure – most often cation exchange or hydrophilic interaction chromatography (HILIC) [62,64]. This approach essentially decreases sample complexity, facilitating, thereby, the fragmentation of low-abundant species. Similarly, fractionation can be applied on the MS level. Thus, simultaneously formed quasi-molecular ions (i.e. originating from the co-eluting species) can be fractioned by their *m/z* values in the quadrupole filter preceding LIT (GPF approach). In this case, however, repeated injections are necessary to cover the whole mass range that increases the overall analysis duration. For the most of applications, three GPF segments (*m/z* 400–600, 600–800 and 800–1500) are sufficient to significantly improve protein identification rates [57,64]. Additionally, based on the results of one DDA experiment, all unmodified peptides can be excluded from fragmentation in following ones on software level by generation of a corresponding "exclusion list" in the instrumental method [57]. However, the combination of these approaches is required for in-depth proteome analysis. The peptides can be identified by database search using SEQUEST or Mascot search engines.

Early glycated tryptic peptides can be annotated in high-resolution Orbitrap-MS experiments by the *m/z* value and charge of the corresponding quasi-molecular ion, and retention time in corresponding extracted ion chromatograms (XICs) as shown in Figure 5A [57]. Unambiguous identification of these peptides relies on characteristic series of b- and y-ions with consideration of the losses of three water molecules (pyrylium ions) and additional loss of formaldehyde (fulylium ions) [59] (Figure 5A). Identification and quantification of AGE-modified peptides are based on the same principle [65]. The only difference in comparison to the early glycation species is the absence of the neutral losses accompanying parent and fragment ions. Additionally, the glycation state of peptides can be confirmed by characteristic fragments in the low *m/z* range [61,66]. This information can be further used for label-free quantification of individual AGE-modified peptides and, hence, specific glycation sites. The quantification typically relies on the integration of the annotated signals and qualitative comparisons of obtained peak areas. Thereby, the peak areas can be normalized to the signals of unmodified peptides in the same sample or in quality controls (QC). The peak integration can be performed by means of the vendor software packages – Xcalibur Quan Browser or LCquan.

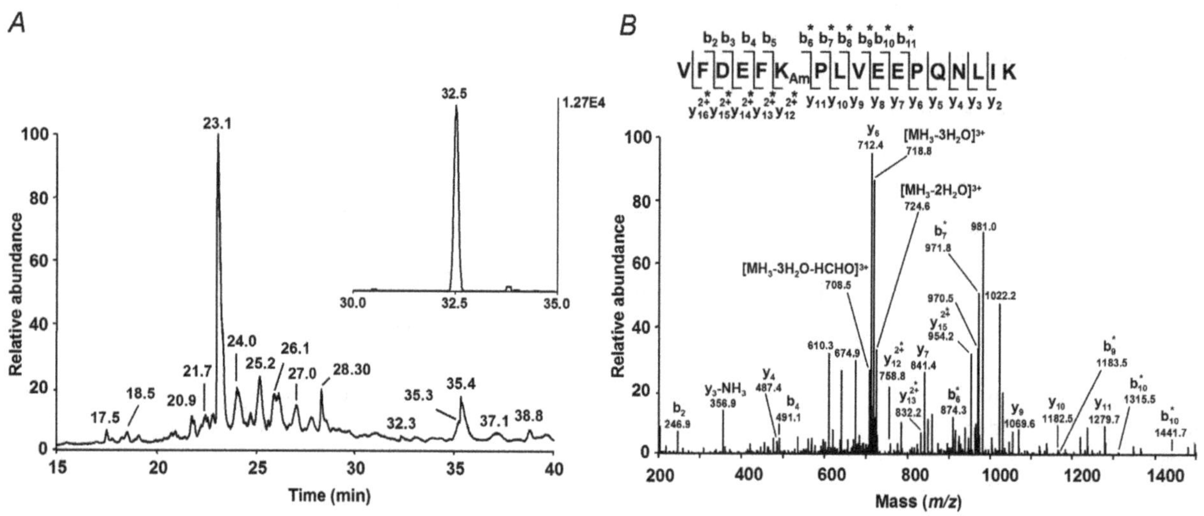

Figure 5. Annotation of Amadori peptides by nanoUPLC-ESI-LTQ-Orbitrap-MS/MS DDA experiments. Reversed phase ESI-Orbitrap-MS total ion chromatogram and the segment of the extracted ion chromatogram (insert) for *m/z* 736.39 ± 0.02 in the t_R range of 30–35 min (A) and an ESI-LTQ-MS/MS spectrum at *m/z* 736.4 corresponding to the $[M+3H]^{3+}$ ion of the peptide VFDEFK$_{Am}$PLVEEPQNLIK (B) acquired at 32.5 min of the same DDA experiment. Pyrylium b and y fragment ions are marked with asterisk [61].

The physiological role of glycation can be assessed by the system biology software tools. Thus, for the grouping of AGE-modified proteins by their functions, the mapping software MapMan (Max-Planck Institute of Molecular Plant Physiology, Potsdam-Golm, Germany, http://mapman.gabipd.org) can be used. The functional annotation of proteins might give insight in biological effects of AGEs in plants and provide the material for future biological studies. Thus, the proteomic data can be complemented by the result of transcriptional analysis and determination of enzymatic activities. Afterwards, the functional role of glycation in respect of particular proteins can be confirmed by the experiments with corresponding mutants.

3. *In vivo* glycation of plant proteins

3.1. Plant protein glycation patterns

The possibility of plant protein glycation is considered since the beginning of the past decade, when Yamauchi and co-workers proposed the formation of AGEs as one of the possible mechanisms underlying inactivation of ribulose bisphosphate carboxylase/oxidase (RuBisCO) by high light [67]. Thereby, they proposed ascorbic acid as a possible precursor of AGEs. Indeed, this highly abundant compound in plant tissues can easily autoxidize and is recognized, therefore, as a potent glycation agent [68]. However, besides ascorbic acid, photosynthetically active leaf tissues contain high amounts of highly reactive pentoses, tetroses and trioses, as well as their phosphorylated forms, that might be even more reactive [69]. Probably, these sugars could be an important factor of light-dependent glycation. Recently, Bechtold and co-workers provided *in vivo* confirmation of this assumption: the authors found that tissue fructosyl lysine contents (determined by LC-MS/MS in exhaustive enzymatic protein hydrolyzates) are approximately four-fold higher in the day time in comparison to the dark period. It was not, however, the case for AGEs – just minor changes in their contents were observed during the day.

Implementation of the proteomic approach resulted in identification of proteins involved in glycation and exact modification sites therein. This strategy allowed identification of several hundreds of polypeptides containing early glycation moieties. Interestingly, the number of modification sites was higher in *Arabidopsis thaliana* in comparison to its close relative *Brassica napus*. Thus, glycation patterns might vary between species, although in both cases they are dominated by triose- and tetrose-derived products, accompanied with less abundant groups of pentose-modified sites, while hexose-derived modifications (typically the most representative in mammals) were less abundant.

Surprisingly, in plant proteins, the numbers of AGE-modified residues are essentially higher in comparison to the early glycated sites: approximately three- and seven-fold differences were observed for *A. thaliana* and *B. napus* proteome, respectively. It is the principle difference from glycation in mammals: though thousands of early glycated proteins were identified in human plasma and red blood cell membranes [64,70], only several dozens were proved to be AGE-modified [61,66]. Interestingly, the AGE modification sites in plant proteins are not accompanied with their early glycated counterparts, and are mostly originating from glyoxal and methylglyoxal. This situation differs drastically from the observations done with mammals. Indeed, several confirmed AGE sites in blood proteins (at least those representing the major plasma polypeptide human serum albumin, HSA) resembled the early glycated residues, indicating glycoxidation as an important pathway of AGE formation *in vivo*. The absence of such glycation sites in plant proteome clearly indicates the early glycation products as unlikely precursors of AGEs in plants. In this context, oxidative glycosylation rather than glycoxidation might be the predominant AGE formation pathway in plants. Remarkably, the number of early glycated lysyl residues was not only absolutely but also relatively (in comparison to the number of AGE-modified sites) lower in *B. napus*, than in *A. thaliana*. Most probably, it indicates higher activities of deglycation enzymes in the former plant.

It was shown in mammalians that the proteins controlling gene expression (e.g. transcription factors or the molecules involved in protein metabolism) can be the targets of glycation [71]. The same was demonstrated for plants. This might indicate the involvement of AGE formation in the regulation of gene expression on the levels of transcription and protein biosynthesis. This can be explained by the role of protein degradation in AGE metabolism and high representation of arginyl residues in the transcription factors that makes these molecules highly amenable to interaction with α-dicarbonyls [71].

3.2. Protein glycation and environmental stress

3.2.1. Experimental models for the study of plant glycation

The main environmental stresses the plant can encounter in its habitat are high irradiation, contamination with heavy metals or metalloids and drought. It is necessary to take into account that drought is a general manifestation accompanying water deficit and is characteristic for the response to some other environmental factors, like high salt contents in soil or extremely low or high temperatures [72]. Obviously, for the study of any stress-specific response, selection of an appropriate model is of the principle importance. In this context, the researcher needs to be able to define all stress parameters by the selected experimental setup. This can be relatively easily achieved for a high light stress by using a phytotron equipped with the lamps providing required light intensity and complete climate control. In this case, a soil model can be applied (Figure 6) [73]. However, this approach does not provide the conditions equal for all plants, when a heavy metal stress is applied. That is why, growth of plants in aqueous (aq.) culture with a subsequent addition of a heavy metal salt to a growth medium seems to be a more adequate solution [74] (Figure 6).

Figure 6. Experimental models of environmental stress.

As far as the drought stress is concerned, the water deficit can be modeled both in soil and in aqueous systems. According to Boyer [75], this water deficit can be expressed as a decrease of water potential (Ψ_w), i.e., as a difference of water potentials in a solution and pure water divided by partial molar volume of water. However, the soil-based approaches do not allow experimental definition of Ψ_w, as in this case it depends from the water absorption by plant. In aqueous culture, osmotically active substances, like commonly used sorbitol or polyethylene glycol (PEG), can affect the function of root epidermis [72]. Because of this, agar-based PEG infusion model, introduced in the past decade by van der Weele and co-workers [76], and additionally optimized to mature plants (Figure 6) seems to be more suited for this purpose. The confirmation of the stress development might rely on the determination of native leaf fluorescence [77], hydrogen peroxide contents [16] and some further parameters.

3.2.2. Stress-related glycation of plant proteins

Light is one of the most essential factors determining plant growth, development and survival [78]. However, a continuous exposure of plants to the high doses of sun irradiation might exceed the capacity of the mesophyll photosynthetic apparatus and trigger development of the oxidative stress [7]. In the context of the protein Maillard chemistry, discussed above, it is logical to assume that enhanced ROS and monosaccharide production would lead to the increase of AGE formation in plant green tissues. As was proposed earlier, such modifications of the RuBisCO polypeptide chains might impact inactivation of the enzyme with high light [67]. Similar mechanisms might accompany the development of drought and metal stress. Moreover, the tissue metal-scavenging mechanisms include activation of the enzyme phyto-chelatinsynthase, requiring GSH as a substrate [79]. Hence, glyoxalase system, critical for detoxication of methylglyoxal (MGO) and utilizing GSH as a substrate as well [80], can not perform efficient scavenging of this dicarbonyl under stress conditions, which might stimulate enhanced glycation by MGO in the proteins of metal-treated plants.

The effect of stress on the formation of AGEs differs on the qualitative and quantitative levels. Thus, in most cases, relatively low number of stress-specific glycation sites can be detected in the proteins of stressed plants, and such sites are representing mostly the molecules involved in transcription and protein degradation (i.e. those known to be upregulated under stress conditions) [81]. This could be explained by the activation of some unknown enzymatic systems eliminating either AGE precursors, or AGE-modified proteins, or their early glycated precursors, i.e. Amadori and Heyns compounds. Indeed, activation of the glyoxalase system (comprising enzymes glyoxalase I and II) is well documented during environmental stress [82]. Additionally, although in mammals advanced glycation decreases the rates of proteolysis, in plants it can be not the case, that was confirmed by *in vitro* experiments [67]. Thus, acylamino acid-releasing enzyme might impact in plant protein degradation pathways [44]. However, the pathways, involved in enhanced degradation of glycated proteins in plants still need to be studied. Finally, Amadori and Heyns products can be degraded via phosphorylation pathway [43]. Besides, the possibility of the further reactions of AGEs to form new unknown products also needs to be considered [83].

Compared to the qualitative alterations, the quantitative stress-related changes in glycated proteome are much more pronounced. Thus, several AGE classes were at least two-fold increased even after short application of light stress [16]. This tendency could be followed on the level of individual glycation sites. However, changes in representation of a certain AGE moiety need to be verified on the level of the abundance of corresponding protein, i.e. the abundance of specific modification sites need to be considered together with the data on gene expression on the RNA and protein levels. Moreover, the changes in protein degradation rates need to be taken into account.

As was demonstrated in the pioneer study of Bechtold and co-workers, stress-related upregulation of early glycation products is much stronger in comparison to AGEs [16], which was confirmed on the level of individual glycation sites. Thus, stress conditions lead to the considerable increase of the contents of Amadori and Heyns adducts, while the response on the level of advanced glycation is much lower. Moreover, stress-specific AGEs are dominated by α-dicarbonyl-derived products, and only few could originate from Amadori or Heyns products. The negligible role of early glycation products in AGE formation additionally supports the existence of powerful Amadori/Heyns product-degrading enzymatic mechanisms in plant tissues. Moreover, it might indicate the presence of a potent stress-inducible anti-glycation system scavenging or/and reducing α-dicarbonyls. This assumption can be supported by the absence of changes in the carbonylated proteome, as well as glyoxal and methylglyoxal contents throughout the stress development. Thus, scavenging of these advanced glycation intermediates by the amino functioned metabolites might be the most probable scenario.

3.3. *In vitro* modeling of plant glycation reactions

Due to their high photosynthetic activity, green parts of plants are characterized with high contents of carbohydrates. Thus, for *B. napus* leaf tissue, these contents were more than 1 μmol/g f.w. for the major sugars (glucose, fructose, sucrose, Asc and DHA) and were in the range of 10–60 nmol/g f.w. for such compounds as glucose-6-phosphate, ribose and arabinose that are known to highly-reactive [84,85]. Hence, metabolic background of plant glycation differs essentially from that reported for mammals. Thus, animal tissues are rich in glucose (that is known to be one of the weakest glycation agents), while the contents of more reactive sugars are negligible [86]. This strong prevalence of one potential glycation agents dramatically affects the mammalian glycation patterns, which are strongly dominated by the protein fructosamines, i.e. glucose-derived Amadori compounds [64]. In contrast, photosynthetically active tissues are rich in trioses, tetroses and pentoses as well, that might affect the early glycation patterns. Moreover, these highly reactive carbohydrates might be much more susceptible to autoxidation, i.e. represent potent precursors of α-dicarbonyls and, hence, AGEs. Obviously, the ability of certain sugars to act as a glycation agent depends not only on its tissue content but also on its ability to react with proteins. However, the reactivities of individual plant carbohydrates towards protein lysyl and arginyl residues, as well as their glycation potential, are still unknown.

To address this question, *in vitro* glycation models based on synthetic peptides were established recently [24,83]. Ideally, such model peptides are N-terminally protected and contain only one residue (lysine or arginine, typically in mid position) susceptible to glycation. Essential advantage of such systems is their relative simplicity: the products can be separated and analyzed by tandem mass spectrometry (MS^n). Evaluation of peptide tandem mass spectra can be complemented by adequate methods of carbohydrate and α-dicarbonyl analysis [87,88]. These peptide models can be potentially applied to the assessment of reactivity of individual glycation agents (i.e. sugars and dicarbonyls) and determination of their glycation potential. Such analyses might rely on integration of selected XICs (characteristic for individual glycation products) at specific retention times. This would give a possibility for screening of plant sugars for their ability to form certain Amadori/Heyns compounds and AGEs. This would essentially impact in understanding of plant glycation pathways.

4. Conclusions

To conclude, protein glycation is a common post-translational modification in plants. Despite this, essential differences in comparison to mammalian glycation patterns were observed. Thus, glycation patterns are strongly dominated by AGEs, while the number of Amadori-modified lysyl residues is at least one order of magnitude lower compared to human plasma proteome. Moreover, individual AGE-modified sites are not represented by their Amadori/Heyns counterparts. It indicates autoxidation of free sugars rather than glycoxidation (i.e. AGE formation from early glycated products – one of the main glycation mechanisms in mammalians) as the major pathway of advanced glycation in plants. Environmental stress considerably affects glycation patterns, mostly on the quantitative level. However, due to the high heterogeneity of potential plant glycation agents, a high variability of glycation pathways and mechanisms can be expected. To clarify these pathways, simple *in vitro* models based on synthetic peptides can be used.

Acknowledgements

We thank the "Deutsche Forschungsgemeinschaft" (DFG, FR-3117/2-1) for financial support to AF, GP, DB and TB, the Ernst-Schering Foundation to UG, the Deutscher Akademischer Austauschdienst (DAAD Program "Dmitry Mendeleev") to ET, TB and EL, St. Petersburg University (grant No. 1.38.233.2014) to NO and EL are gratefully acknowledged, as well as the state of Saxony-Anhalt and SKWP (Agrochemical Institute Piesteritz) to LAW. This work was conducted under the auspices of the Science Campus Plant Based Bioeconomy (www.sciencecampus-halle.de).

Author details

Tatiana Bilova[1,2,3], Uta Greifenhagen[2], Gagan Paudel[1,2], Elena Lukasheva[3], Dominic Brauch[2], Natalia Osmolovskaya[3], Elena Tarakhovskaya[3], Gerd Ulrich Balcke[4], Alain Tissier[4], Thomas Vogt[4], Carsten Milkowski[5], Claudia Birkemeyer[2], Ludger Wessjohann[1] and Andrej Frolov[1,2*]

*Address all correspondence to: afrolov@ipb-halle.de

1 Leibniz Institute of Plant Biochemistry, Department of Bioorganic Chemistry, Halle/Saale, Germany

2 Universität Leipzig, Faculty of Chemistry and Mineralogy, Leipzig, Germany

3 St. Petersburg State University, Department of Plant Physiology and Biochemistry, Saint-Petersburg, Russia

4 Leibniz Institute of Plant Biochemistry, Department of Cell and Metabolic Biology, Halle/Saale, Germany

5 Martin Luther University Halle-Wittenberg, Interdisciplinary Center for Crop Plant Research (IZN), Halle/Saale, Germany

References

[1] Rodziewicz P, Swarcewicz B, Chmelewska K, Wojakowska A, Stobiecki M. Influence of abiotic stress on plant proteome and metabolome. Acta Physiol Plant 2014;36:1–19.

[2] Cruz de Carvalho MH. Drought stress and reactive oxygen species: production, scavenging and signaling. Plant Signal Behav 2008;3:156–65.

[3] Chaves MM, Flexas J, Pinheiro C. Photosynthesis under drought and salt stress: regulation mechanisms from whole plant to cell. Ann. Bot 2009;103:551–60.

[4] Kar RK. Plant responses to water stress: role of reactive oxygen species. Plant Signal Behav 2011;6:1741–5.

[5] Gill SS, Tuteja N. Reactive oxygen species and antioxidant machinery in abiotic stress tolerance in crop plants. Plant Physiol Biochem 2010;48:909–30.

[6] Puntarulo S, Sanchez RA, Boveris A. Hydrogen peroxide metabolism in soybean embryonic axes at the onset of germination. Plant Physiol 1988;86:626–30.

[7] Scarpeci TE, Zanor MI, Carrillo N, Mueller-Roeber B, Valle EM. Generation of superoxide anion in chloroplasts of *Arabidopsis thaliana* during active photosynthesis: a focus on rapidly induced genes. Plant Mol. Biol 2008;66:361–78.

[8] Repetto M, Semp J, Boveris A. Lipid peroxidation: chemical mechanism, biological implications and analytical determination. In: Lipid Peroxidation: Chemical Mechanism, Biological Implications and Analytical Determination. Editor Catala A. InTech 2012.

[9] Halliwell B, Gutteridge JM. Oxygen toxicity, oxygen radicals, transition metals and disease. Biochem. J 1984;219:1–14.

[10] Bielski BH, Arudi RL, Sutherland MW. A study of the reactivity of $HO_2/O_2^{\cdot-}$ with unsaturated fatty acids. J Biol Chem 1983;258:4759–61.

[11] Berlett BS, Stadtman ER. Protein oxidation in aging, disease, and oxidative stress. J Biol Chem 1997;272:20313–6.

[12] Kehrer JP. The Haber-Weiss reaction and mechanisms of toxicity. Toxicology 2000;149:43–50.

[13] Arbona V, Manzi M, Ollas C, Gomez-Cadenas A. Metabolomics as a tool to investigate abiotic stress tolerance in plants. Int J Mol Sci 2013;14:4885–911.

[14] Wolff SP, Dean RT. Glucose autoxidation and protein modification. The potential role of 'autoxidative glycosylation' in diabetes. Biochem J 1987;245:243–50.

[15] Fedorova M, Bollineni RC, Hoffmann R. Protein carbonylation as a major hallmark of oxidative damage: update of analytical strategies. Mass Spectrom Rev 2014;33:79–97.

[16] Bechtold U, Rabbani N, Mullineaux PM, Thornalley PJ. Quantitative measurement of specific biomarkers for protein oxidation, nitration and glycation in Arabidopsis leaves. Plant J 2009;59:661–71.

[17] Hodge JE. The Amadori rearrangement. Adv Carbohydr Chem 1955;10:169–205.

[18] Heyns K, Noack H. Die Umsetzung von D-Fructose mit L-Lysine und L-Arginin und deren Beziehung zu nichtenzymatischen Bräunungsreaktionen. Chem Ber 1962;95:720–27.

[19] Pilkova L, Pokorny J, Davidek J. Browning reactions of Heyns rearrangement products. Die Nahrung 1990;34:759–64.

[20] Kislinger T, Humeny A, Pischetsrieder M. Analysis of protein glycation products by matrix-assisted laser desorption ionization time-of-flight mass spectrometry. Curr Med Chem 2004;11:2185–93.

[21] Wolff SP, Dean RT. Aldehydes and dicarbonyls in non-enzymic glycosylation of proteins. Biochem J 1988;249:618–9.

[22] Wells-Knecht KJ, Zyzak DV, Litchfield JE, Thorpe SR, Baynes JW. Mechanism of autoxidative glycosylation: identification of glyoxal and arabinose as intermediates in the autoxidative modification of proteins by glucose. Biochemistry 1995;34:3702–9.

[23] Wells-Knecht MC, Thorpe SR, Baynes JW. Pathways of formation of glycoxidation products during glycation of collagen. Biochemistry 1995;34:15134–41.

[24] Greifenhagen U, Frolov A, Hoffmann R. Oxidative degradation of *N*-fructosylamine-substituted peptides in heated aqueous systems. Amino Acids 2015;47:1065–76.

[25] Klopfer A, Spanneberg R, Glomb MA. Formation of arginine modifications in a model system of N^{α}-tert-butoxycarbonyl (Boc)-arginine with methylglyoxal. J Agric Food Chem 2011;59:394–401.

[26] Ahmed MU, Thorpe SR, Baynes JW. Identification of N^{ε}-(carboxymethyl)lysine as a degradation product of fructoselysine in glycated protein. J Biol Chem 1986;261:4889–94.

[27] Ahmed MU, Brinkmann FE, Degenhardt TP, Thorpe SR, Baynes JW. N^{ε}-(carboxyethyl)lysine, a product of the chemical modification of proteins by methylglyoxal, increases with age in human lens proteins. Biochem J 1997;324(Pt 2):565–70.

[28] Foerster A, Henle T. Glycation in food and metabolic transit of dietary AGEs (advanced glycation end-products): studies on the urinary excretion of pyrraline. Biochem Soc Trans 2003;31:1383–5.

[29] Usui T, Shimohira K, Watanabe H, Hayase F. Detection and determination of glyceraldehyde-derived pyridinium-type advanced glycation end product in streptozotocin-induced diabetic rats. Biosci Biotechnol Biochem 2007;71:442–8.

[30] Schwarzenbolz U, Henle T, Haeßner R, Klostermeyer H. On the reaction of glyoxal with proteins. Z Lebensm Unters Forsch A 1997;205:121–4.

[31] Glomb MA, Lang G. Isolation and characterization of glyoxal-arginine modifications. J Agric Food Chem 2001;49:1493–501.

[32] Henle T, Walter AW, Haeßner R, Klostermeyer H. Detection and identification of a protein-bound imidazolone resulting from the reaction of arginine residues and methylglyoxal. Z Lebensm Unters Forsch 1994;199:55–8.

[33] Gruber P, Hofmann T. Chemoselective synthesis of peptides containing major advanced glycation end-products of lysine and arginine. J Pept Res 2005;66:111–24.ε

[34] Shipanova IN, Glomb MA, Nagaraj RH. Protein modification by methylglyoxal: chemical nature and synthetic mechanism of a major fluorescent adduct. Arch Biochem Biophys 1997;344:29–36.

[35] Oya T, Hattori N, Mizuno Y, Miyata S, Maeda S, Osawa T, Uchida K. Methylglyoxal modification of protein. Chemical and immunochemical characterization of methylglyoxal-arginine adducts. J Biol Chem 1999;274:18492–502.

[36] Goldin A, Beckman JA, Schmidt AM, Creager MA. Advanced glycation end products: sparking the development of diabetic vascular injury. Circulation 2006;114:597–605.

[37] Nettleton JA, Steffen LM, Schulze MB, Jenny NS, Barr RG, Bertoni AG, Jacobs DR, Jr. Associations between markers of subclinical atherosclerosis and dietary patterns derived by principal components analysis and reduced rank regression in the Multi-Ethnic Study of Atherosclerosis (MESA). Am J Clin Nutr 2007;85:1615–25.

[38] Uribarri J, Cai W, Sandu O, Peppa M, Goldberg T, Vlassara H. Diet-derived advanced glycation end products are major contributors to the body's AGE pool and induce inflammation in healthy subjects. Ann NY Acad Sci 2005;1043:461–6.

[39] Maza MP, Bravo A, Leiva L, Gattas V, Petermann M, Garrido F, Bunout D, Hirsch S, Barrera G, Fernandez M. Fluorescent serum and urinary advanced glycoxidation end-products in non- diabetic subjects. Biol Res 2007;40:203–12.

[40] Krajcovicova-Kudlackova M, Sebekova K, Schinzel R, Klvanova J. Advanced glycation end products and nutrition. Physiol Res 2002;51:313–6.

[41] Sebekova K, Krajcoviova-Kudlackova M, Schinzel R, Faist V, Klvanova J, Heidland A. Plasma levels of advanced glycation end products in healthy, long-term vegetarians and subjects on a western mixed diet. Eur J Nutr 2001;40:275–81.

[42] Singla-Pareek SL, Yadav SK, Pareek A, Reddy MK, Sopory SK. Enhancing salt tolerance in a crop plant by overexpression of glyoxalase II. Transgenic Res 2008;17:171–80.

[43] Fortpied J, Gemayel R, Stroobant V, Van SE: Plant ribulosamine/erythrulosamine 3-kinase, a putative protein-repair enzyme. Biochem J 2005;388:795–802.

[44] Yamauchi Y, Ejiri Y, Toyoda Y, Tanaka K. Identification and biochemical characterization of plant acylamino acid-releasing enzyme. J Biochem 2003;134:251–7.

[45] Thornalley PJ, Battah S, Ahmed N, Karachalias N, Agalou S, Babaei-Jadidi R, Dawnay A. Quantitative screening of advanced glycation endproducts in cellular and extracellular proteins by tandem mass spectrometry. Biochem J 2003;375:581–92.

[46] Mann M, Jensen ON. Proteomic analysis of post-translational modifications. Nat Biotechnol 2003;21:255–61.

[47] Chernushevich IV, Loboda AV, Thomson BA. An introduction to quadrupole-time-of-flight mass spectrometry. J Mass Spectrom 2001;36:849–65.

[48] Zubarev RA, Makarov A. Orbitrap mass spectrometry. Anal Chem 2013;85:5288–96.

[49] Taylor PJ. Matrix effects: the Achilles heel of quantitative high-performance liquid chromatography-electrospray-tandem mass spectrometry. Clin Biochem 2005;38:328–34.

[50] Isaacson T, Damasceno CM, Saravanan RS, He Y, Catala C, Saladie M, Rose JK. Sample extraction techniques for enhanced proteomic analysis of plant tissues. Nat Protoc 2006;1:769–74.

[51] Ikeda K, Sakaguch T, Kusano T, Yasumoto K. Endogenous factors affecting protein digestibility in buckwheat. Cereal Chem 1991;68:424–7.

[52] Esquivel MG, Ferreira RB, Teixeira AR. Protein degradation in C3 and C4 plants with particular reference to ribulose bisphosphate carboxylase and glycolate oxidase. J Exp Bot 1998;49:807–16.

[53] Bollineni RC, Fedorova M, Bluher M, Hoffmann R. Carbonylated plasma proteins as potential biomarkers of obesity induced type 2 diabetes mellitus. J Proteome Res 2014;13:5081–93.

[54] Li M, Powell MJ, Razunguzwa TT, O'Doherty GA. A general approach to anionic acid-labile surfactants with tunable properties. J Org Chem 2010;75:6149–53.

[55] Liu H, Sadygov RG, Yates JR, III. A model for random sampling and estimation of relative protein abundance in shotgun proteomics. Anal Chem 2004;76:4193–201.

[56] Frolov A, Hoffmann R. Identification and relative quantification of specific glycation sites in human serum albumin. Anal Bioanal Chem 2010;397:2349–56.

[57] Frolov A, Bluher M, Hoffmann R. Glycation sites of human plasma proteins are affected to different extents by hyperglycemic conditions in type 2 diabetes mellitus. Anal Bioanal Chem 2014.

[58] Zhang Q, Petyuk VA, Schepmoes AA, Orton DJ, Monroe ME, Yang F, Smith RD, Metz TO. Analysis of non-enzymatically glycated peptides: neutral-loss-triggered MS(3) versus multi-stage activation tandem mass spectrometry. Rapid Commun. Mass Spectrom 2008;22:3027–34.

[59] Frolov A, Hoffmann P, Hoffmann R. Fragmentation behavior of glycated peptides derived from *D*-glucose, *D*-fructose and *D*-ribose in tandem mass spectrometry. J Mass Spectrom 2006;41:1459–69.

[60] Frolov A, Hoffmann R. Analysis of amadori peptides enriched by boronic acid affinity chromatography. Ann NY Acad Sci 2008;1126:253–6.

[61] Greifenhagen U, Nguyen VD, Moschner J, Giannis A, Frolov A, Hoffmann R. Sensitive and site-specific identification of carboxymethylated and carboxyethylated peptides in tryptic digests of proteins and human plasma. J Proteome Res 2015; 14(2): 768-77.

[62] Bollineni RC, Hoffmann R, Fedorova M: Identification of protein carbonylation sites by two-dimensional liquid chromatography in combination with M. J Proteomics 2011;74:2338–50.

[63] Frolov A, Henning A, Bottcher C, Tissier A, Strack D. An UPLC-MS/MS method for the simultaneous identification and quantitation of cell wall phenolics in *Brassica napus* seeds. J Agric Food Chem 2013;61:1219–27.

[64] Zhang Q, Monroe ME, Schepmoes AA, Clauss TR, Gritsenko MA, Meng D, Petyuk VA, Smith RD, Metz TO. Comprehensive identification of glycated peptides and

their glycation motifs in plasma and erythrocytes of control and diabetic subjects. J Proteome Res 2011;10:3076–88.

[65] Ehrlich H, Hanke T, Simon P, Born R, Fischer C, Frolov A, Langrock T, Hoffmann R, Schwarzenbolz U, Henle T, Bazhenov VV, Worch H. Carboxymethylation of the fibrillar collagen with respect to formation of hydroxyapatite. J Biomed Mater Res B Appl Biomater 2010;92:542–51.

[66] Schmidt R, Böhme D, Singer D, Frolov A. Specific tandem mass spectrometric detection of AGE-modified arginine residues in peptides. J Mass Spectrom 2015;50:613–24.

[67] Yamauchi Y, Ejiri Y, Tanaka K. Glycation by ascorbic acid causes loss of activity of ribulose-1,5-bisphosphate carboxylase/oxygenase and its increased susceptibility to proteases. Plant Cell Physiol 2002;43:1334–41.

[68] Linetsky M, Shipova E, Cheng R, Ortwerth BJ: Glycation by ascorbic acid oxidation products leads to the aggregation of lens proteins. Biochim Biophys Acta 2008;1782:22–34.

[69] Karachalias N, Babaei-Jadidi R, Ahmed N, Thornalley PJ. Accumulation of fructosyl-lysine and advanced glycation end products in the kidney, retina and peripheral nerve of streptozotocin-induced diabetic rats. Biochem Soc Trans 2003;31:1423–5.

[70] Zhang Q, Tang N, Schepmoes AA, Phillips LS, Smith RD, Metz TO. Proteomic profiling of nonenzymatically glycated proteins in human plasma and erythrocyte membranes. J Proteome Res 2008;7:2025–32.

[71] Rabbani N, Thornalley PJ. The dicarbonyl proteome: proteins susceptible to dicarbonyl glycation at functional sites in health, aging, and disease. Ann NY Acad Sci 2008;1126:124–7.

[72] Verslues PE, Agarwal M, Katiyar-Agarwal S, Zhu J, Zhu JK. Methods and concepts in quantifying resistance to drought, salt and freezing, abiotic stresses that affect plant water status. Plant J 2006;45:523–39.

[73] Vanderauwera S, Zimmermann P, Rombauts S, Vandenabeele S, Langebartels C, Gruissem W, Inze D, Van BF. Genome-wide analysis of hydrogen peroxide-regulated gene expression in Arabidopsis reveals a high light-induced transcriptional cluster involved in anthocyanin biosynthesis. Plant Physiol 2005;139:806–21.

[74] Gomes MP, Marques TCLISM, Carneiro MMLC, Soares AM. Anatomical characteristics and nutrient uptake and distribution associated with the Cd-phytoremediation capacity of Eucaliptus camaldulenses Dehnh. J Soil Sci Plant Nutr 2012;12:481–96.

[75] Boyer JS. Plant productivity and environment. Science 1982;218:443–8.

[76] van der Weele CM, Spollen WG, Sharp RE, Baskin TI. Growth of Arabidopsis thaliana seedlings under water deficit studied by control of water potential in nutrient-agar media. J Exp Bot 2000;51:1555–62.

[77] Kitao M, Utsugi H, Kuramoto S, Tabuchi R, Fujimoto K, Lihpai S. Light-dependent photosynthetic characteristics indicated by chlorophyll fluorescence in five mangrove species native to Pohnpei Island, Micronesia. Physiol Plant 2003;117:376–82.

[78] Fankhauser C, Chory J. Light control of plant development. Annu Rev Cell Dev Biol 1997;13:203–29.

[79] Sharma SS, Dietz KJ. The significance of amino acids and amino acid-derived molecules in plant responses and adaptation to heavy metal stress. J Exp Bot 2006;57:711–26.

[80] Yadav SK, Singla-Pareek SL, Sopory SK. An overview on the role of methylglyoxal and glyoxalases in plants. Drug Metabol Drug Interact 2008;23:51–68.

[81] Gong F, Hu X, Wang W. Proteomic analysis of crop plants under abiotic stress conditions: where to focus our research? Front Plant Sci 2015;6:418.

[82] Kaur C, Ghosh A, Pareek A, Sopory SK, Singla-Pareek SL. Glyoxalases and stress tolerance in plants. Biochem Soc Trans 2014;42:485–90.

[83] Frolov A, Schmidt R, Spiller S, Greifenhagen U, Hoffmann R. Formation of arginine-derived advanced glycation end products in peptide-glucose mixtures during boiling. J Agric Food Chem 2014;62:3626–35.

[84] Bucala R, Tracey KJ, Cerami A. Advanced glycosylation products quench nitric oxide and mediate defective endothelium-dependent vasodilatation in experimental diabetes. J Clin Invest 1991;87:432–8.

[85] Chevalier F, Chobert JM, Dalgalarrondo M, Choiset Y, Haertle T. Maillard glycation of beta-lactoglobulin induces conformation changes. Nahrung 2002;46:58–63.

[86] Brown LD, Cavalli C, Harwood JE, Casadei A, Teng CC, Traggiai C, Serra G, Bevilacqua G, Battaglia FC. Plasma concentrations of carbohydrates and sugar alcohols in term newborns after milk feeding. Pediatr Res 2008;64:189–93.

[87] Henning C, Liehr K, Girndt M, Ulrich C, Glomb MA. Extending the spectrum of alpha-dicarbonyl compounds in vivo. J Biol Chem 2014;289:28676–88.

[88] Milkovska-Stamenova S, Schmidt R, Frolov A, Birkemeyer C. GC-MS method for the quantitation of carbohydrate intermediates in glycation systems. J Agric Food Chem 2015;63:5911–9.

4

Small RNAs in Plant Response to Abiotic Stress

By Shirin Mirlohi and Yuke He

Abstract

Small noncoding (nc) RNAs (sRNAs) are the important posttranscriptional regulatory factors in gene regulatory networks. They are involved in many important processes of plant development and stress responses. Increasingly research data reveal that microRNAs (miRNAs), heterochromatic small interfering RNAs (hc-siRNAs), trans-acting small interfering RNAs (ta-siRNAs), natural antisense small interfering RNAs (nat-siRNAs), repeat-associated small interfering RNAs (ra-siRNAs), and the piwi-interacting RNAs (piRNAs) are involved in heat stress, salt stress, cold stress, and drought stress, which are found in metazoans. Some small RNAs are required for plant thermotolerance and salt tolerance. These findings facilitate our investigation of the genetic basis of plant adaptability to various environmental stresses and the genetic manipulation of plant tolerance to many abiotic stresses. This chapter highlights the recent advances in understanding the crucial roles of sRNAs in plant responses to heat, drought, salinity, and cold and proposes the potential technologies and strategies used to identify abiotic-stress-regulated sRNAs in addition to the recent advances and methods for validation and analysis of their target genes.

Keywords: Abiotic stress, microRNAs, Plants, Small RNAs, Thermotolerance

1. Introduction

Environmental stresses, such as heat, drought, salinity, nutrient deficiency, and low temperature, are the major natural limiting factors for plant growth and crop productivity and thus are the major causes of crop losses worldwide. In recent years, much progress has been made in unraveling the complex and sophisticated molecular mechanisms by which plants have evolved during periods of environmental stresses, and a great deal of attention has been paid to identifying these stress-responsive proteins and their relevant gene networks. Plant-stress responses depend on the precise expression of the genes and their accurate regulation, which is attained by multiple mechanisms at different levels such as transcriptional, posttranscriptional, and posttranslational regulations. Although studies have been mostly focused on the transcriptional level of regulatory mechanisms so far, recent results lead us to the point that posttranscriptional events also play a very important role in gene expression regulation in major scenarios of a plant life, from developmental processes to stress responses. Small noncoding RNAs (sRNAs) are the important posttranscriptional regulatory factors in gene regulatory networks. They are involved in many important processes of plant development and stress responses.

sRNAs are roughly divided into different categories based on the genomic origins of their precursors: microRNAs (miRNAs), trans-acting small interfering RNAs (ta-siRNAs), and natural antisense small interfering RNAs (nat-siRNAs). These sRNAs are loaded into RNA-induced silencing complexes (RISC) and regulate the expression of their relative target genes negatively by affecting the mRNA levels, chromatin remodeling, and DNA methylation. Understanding of sRNA-guided stress regulatory networks should provide us with new tools and vision for the genetic improvement of plant stress tolerance and eventually developing more stress-resistant plants in future.

This chapter highlights the recent advances in understanding the crucial roles of sRNAs in plant responses to heat, drought, salinity, nutrient deficiency and low-temperature stresses, and proposes potential technologies and strategies used to identify abiotic stress-regulated sRNAs in addition to the recent advances and methods for validation and analysis of their target genes.

2. Small noncoding RNAs

The discovery of RNA interference (RNAi) in the late 1990s has been a tornado for the past decade in terms of surprising geneticists for it changed the earlier understanding of the RNA field and the complexity of posttranscriptional control and epigenetic regulation caused by small RNAs. In 1995, Guo and Kemphues used antisense RNA sequence to block the par-1 mRNA in *Caenorhabditis elegans* when they figured out par-1 mRNA is repressed by par-1 mRNA itself [1].

Posttranscriptional gene silencing (PTGS) by RNAs was first reported in plants in 1996. The lin-4 gene, which is known to be essential for the timing of larval development in *C. elegans,* is controlled by a short RNA, which is not translated to any proteins but has a partially complementary sequence to the 3′ region of lin-4 transcript that inhibited the translation of lin-4 to protein [2]. Hence, this discovery even changed the concept of PTGS mentioned in high school biology textbooks as it questioned the central dogma proposed by Francis Crick in 1956, which stated that RNAs carry the biological information encoded in DNA molecules and they subsequently provide the code for translation into proteins [3-4]. This discovery called so much attention that was introduced as the breakthrough of the year when it was published in the journal *Science* in 2002 by Couzin and changed the basic concepts about the gene expression and RNA functionality [5]. This was in agreement with the results published by three independent labs that discovered miRNAs in the model plant *Arabidopsis thaliana* [6-8]. They reported many miRNAs in plants, most of which have a very conserved sequence among different species. Small RNAs could be much more than what was thought. They can influence almost all the functions in a cell by targeting the transcription factors and key genes.

One of the main reasons that small RNAs called so much attention was that soon after their discovery, the target genes for these small RNAs were reported to be crucial in leaf or flower development, which was consistent with the previous reports [9-13]. They have been found to have an impact on almost all the biological processes in eukaryotic cells as they have a wide range of target genes, which are corresponded to some of the previously identified regulatory genes and transcription factors that proved to play key roles such as in controlling cellular metabolism, growth and differentiation, phase transition timing and leaf patterning, and defense mechanisms against biotic and abiotic stresses in case of plants. These 18–25-nucleotide (nt) RNAs are categorized into many different classes based on their size, their biogenesis pathway, and their mode of action.

sRNAs are short nucleic acid sequences that give rise to the assembly of protein–RNA complexes, which later are able to repress the expression of their identified target genes by sequence-specific base pairing. This silencing of the target sequence can occur through several ways by (1) reducing their rates of transcription, (2) reducing the stability of their mRNAs in the cell, or (3) reducing the translation of their mRNAs into protein.

Although much of the work on ncRNAs field has been focused on small RNAs of under 40 nucleotides long, there are larger ncRNAs called mRNA-like ncRNAs (or mlncRNAs)that have received much less attention and have been reported to play some roles in some of the plant functions such as phosphate starvation response and nodulation. The article by Rymarquis et al. explains about them [14].

The generation of sRNAs involves a set of evolutionary conserved proteins, such as Dicer (DCR) or Dicer-like (DCL), Argonaute (AGO), and RNA-dependent RNA polymerase (RDR), which all together form the RNA silencing machinery in plants. The DCLs have been the most studied enzyme so far, which, in *Arabidopsis thaliana*, are classified into four groups: DCL1 acts during miRNA metabolism, DCL2 is responsible for the viral resistance, DCL3 triggers the transcriptional silencing, and DCL4 cooperates in posttranscriptional silencing and ta-siRNA metabolism (dissecting *Arabidopsis thaliana* dicer function in small RNA processing, gene silencing, DNA methylation patterning, and nature genetics). Plant genomes encode only one of the three known classes of AGO proteins, namely AGO1, which is involved in both miRNA and sRNA biogenesis.

The sRNAs are categorized into different classes based on their size, their biogenesis pathway, and their mode of action to at least six groups, including microRNAs (miRNAs), heterochromatic small interfering RNAs (hc-siRNAs), trans-acting small interfering RNAs (ta-siRNAs), natural antisense small interfering RNAs (nat-siRNAs), repeat-associated small interfering RNAs (ra-siRNAs), and the piwi-interacting RNAs (piRNAs), which are found in metazoans.

2.1. miRNAs

Typically, miRNAs are derived from single-stranded RNA precursors that are transcribed by RNA polymerase II from MiRNA genes called primary microRNA transcript (pri-miRNA), which are capable of forming a self-complementary fold-back structure named hairpin or stem loop in which the mature miRNA could reside on the 3′ or 5′ end (Table 1). This imperfect double-stranded structure is further recognized and processed by DCL1 in association with other protein factors [15]. This gives birth to the miRNA/miRNA* duplex which based on the thermodynamic features will have a different fate but usually the pre-miRNA strand is loaded onto an AGO1-containing, RNA-induced silencing complex (RISC) and the miRNA* strand lives for a short time in the cell. Mature microRNAs (miRNAs), which are the so-called hairpin-derived RNAs, are 20–24 nt long and single stranded while miRNA genes are 70–300 nt long. Mature miRNAs help the target recognition and cleavage in cooperation with AGO1 and miRISC. The first cleavage by DCL1 generates a stem-loop intermediate, called the precursor miRNA (pre-miRNA), and the second cleavage by DCL1 releases the miRNA duplex, one strand of which is known as mature miRNA and the other strand is known as miRNA* (miRNA star).

Class	Full Name	Originating Loci	Function	Biogenesis
miRNA	microRNA	*MIRNA* genes	Repress target gene expression through mRNA cleavage and translational repression	The fold-back structures of long ssRNA transcripts are cleaved by Dicers
siRNA	short-interfering RNA	Repeats, transposons, and retroelements (endogenous). Transgenes and viral RNAs (exogenous)	Silence repeats and transposons through RNA-dependent DNA methylation and chromatin modification	RDR-generated dsRNAs are cleaved by Dicers
ta-siRNA	trans-acting siRNA	TAS loci	Repress target gene expression through mRNA cleavage	TAS transcripts are cleaved by miRNAs, transcribed by RDR into dsRNA, and then processed by Dicers
nat-siRNA	natural antisense transcript-derived siRNA	Loci producing pairs of sense-antisense transcripts	Stressed-induced nat-siRNA to repress target gene expression through mRNA cleavage	The dsRNA derived from overlapping transcripts is cleaved by Dicers
piRNA	piwi-interacting RNA	Repeats, transposons, and retroelements	Germ-line-specific piRNA to suppress repeats and transposons in flies and mammals	ssRNA derived from transposons is cleaved by PIWI protein

Table 1. Small RNAs involved in plant response to abiotic stresses

Mammals use only one class of RNAse III enzyme, Dicer, to generate both miRNAs and siRNAs. In plants, there are a variety of specialized DCL endonucleases, which are classified into 10 categories. DCL1 is involved in miRNA biogenesis pathway while other DCLs participate in various aspects of sRNA-mediated generation or gene silencing pathway (Figure 1) [16].

All the information about the reported miRNAs and their sequences and annotations are stored in a database called miRBase (www.mirbase.org), which is updated on a regular basis with the new published data in the literature [17]. So far, there are 205 precursors and 384 mature miRNAs reported in the model plant *Arabidopsis thaliana*, which is about 1–2% of its genome. Theoretically, the perfect base pairing between miRNAs and complementary target mRNAs helps the process of finding target genes for each miRNA. By computational methods, there are a plenty of databases that are able to predict the potential target genes for each newly found miRNA.

2.2. siRNAs

siRNAs were first identified in 1999 in plants [18], and later there were many reports about diverse sets of endogenous siRNAs in plants as well as in animals (Table 1) [6] [8][19][20-22]. Most of the plant siRNAs are around 24 nt in length that are excised from the long double-stranded RNA duplexes or transcripts generated from inverted repeat regions [23-24]. The sources of these double-stranded sequences that eventually trigger biogenesis of siRNAs could be endogenous or exogenous. Endogenous plant siRNAs can be classified into several categories, including miRNA-induced trans-acting siRNAs (tasiRNAs), natural antisense siRNAs (nat-siRNAs), cis-acting siRNAs (casiRNAs), heterochromatic siRNAs, and many other unclassified small RNAs [25]. In plants, ta-siRNAs are generated from the genomic loci named TAS genes, which are transcribed by RNA-pol II. The generation of ta-siRNAs is triggered by an miRNA, which cleaves a nonprotein-coding transcript of a tasiRNA gene [26-29]. In plants, there are eight TAS loci reported so far, which belong to four families (TAS1–4). TAS1 and TAS2 families are cleaved by miR173 with the association of AGO1. TAS3 family transcripts are cleaved by the guidance of miR390 and AGO7 and usually target the auxin response factor (ARF) transcripts. TAS4 transcript is cleaved by miR828 guided together with AGO1 and they usually target myeloblastosis (MYB) transcription factors [30]. These cleaved RNAs are then processed by the suppressor of gene silencing 3 (SGS3) and copied into double-stranded RNAs by RNA-dependent RNA polymerase 6 (RDR6). DCL4 cleaves them in multiple rounds so that it finally gives rise to the 21-nt ta-siRNAs. ta-siRNAs are loaded onto AGO1 complex to degrade the target mRNAs [31].

The other class of siRNAs called nat-siRNAs are separated into two groups: cis-nat-siRNAs that are generated from two RNAs, which were transcribed from the same loci but opposite strands, and trans-nat-siRNAs, which were transcripts from different loci [32]. RDR6 and DCL2 are involved in generating 24-nt nat-siRNAs, and RDR6 and DCL1 are involved in generating 21-nt nat-siRNAs. trans-nat-siRNAs are transcripts from different loci but processed by the same proteins (RDR6 and DCL2). ta-siRNAs cleave the target mRNAs by being partially or fully complementary with them.

Heterochromatic siRNAs mostly originate from transposable elements or repeats and their mode of action is slightly different from miRNAs and ta-siRNAs, as they modulate the histone modification at their homologous regions in the genome and inhibit the gene expression at the transcriptional level.

2.2.1. nat-siRNAs (natural antisense siRNAs)

Natural antisense transcripts (NATs) are small RNA molecules, which are endogenous and show partial or entire complementarity to other transcripts (Table 1). cis-NATs are categorized in the nat-siRNAs group and are transcribed from the same genomic loci but in the opposite strand of DNA as their sense transcripts. This class of NATs is very common in eukaryotes (17–30% of the genes encode complementary cis-NATs in animals and plants) [33-37]. In animals, NATs are involved in alternative splicing, DNA methylation, RNA editing, and genomic imprinting [38-41]. In plants, several cis-NATs are involved in gene regulatory mechanisms [42-43]. There are already some reports about the identified cis-NATs in *Arabidopsis* and rice on the genome-wide scale [44-46].

2.2.2. ta-siRNA (trans-acting short interfering RNAs)

ta-siRNAs are 21 nt in length and are reported to be found only in plants so far (Table 1). ta-siRNAs originate from a noncoding RNA precursor, which is initially targeted to be cleaved by an miRNA molecule. RNA-dependent RNA polymerase converts the cleaved products into double-stranded RNA molecules, which are later cleaved again into 21-nt ta-siRNAs. Hence, the formation of these RNAs is determined by the presence of both miRNA (Dicer-Like1, Argonaute1, HYPONASTIC LEAVES1, and HUA ENHANCER 1) and siRNA (RNA-dependent RNA polymerase 6 and DCL4) biosynthesis pathways components. ta-siRNAs can guide cleavage of target mRNAs and regulate gene expression at the posttranscriptional level like plant miRNAs.

2.3. Small RNAs in abiotic stress

Abiotic stress is known to be one of the attention-calling factors globally, which causes a considerable yield loss each year. Hence, much effort has been made in understanding the complex stress-response mechanisms, especially in the identification of stress-responsive protein-coding genes. But, in recent years, after the discovery of small noncoding RNAs, they have been found to be involved in plant stress responses and indeed very functional players in these pathways. These small RNAs regulate the gene expression in different levels and hence are entangled within all the vital pathways in the plant development, metabolism, and stress response.

As sessile organisms, plants have evolved their specific adaptation and acclimation mechanisms in order to survive during the hard spell. To do the morphological and physiological adaptations to abiotic stresses, the plant needs to manage the complicated rearrangement of gene expression networks, which are controlled at transcriptional and post-transcriptional levels. The concern about future food shortages makes it imperative to better understand the genetic control of stress tolerance networks and pathways and to use this knowledge to increase the total tolerance of important crop species. As the first step, we have to understand the complex responses of the plants to stress, from changes in molecular level to physiological level.

With the help of high-throughput gene expression analysis, there have been many reports about the modulated genes and different small RNAs under abiotic stresses. The difference in the expression level of these genes could indicate that it might be responsive to stress condition and, as a result, it can help the plant to survive the hard condition. Many studies have been published in this regard in various plant species, some of which having an economic importance, like rice, wheat, legumes, barley, sugarcane, potato, and tomato as well as many other species. Some of the genes are induced after facing the stress conditions in these studies while some are downregulated, which is connected with the roles these genes play during the stress condition. Also, the respective miRNAs or other sRNAs that target these mRNAs show a different expression pattern during the stress condition. sRNAs, which are accumulated by stress, might downregulate their target genes and act as a negative regulator of stress tolerance; for instance, the genes involved in cell expansion and division should be downregulated as the plant needs to save energy in order to pass through the hard environmental condition. On the other hand, reduction of sRNA level might lead to upregulation of their target genes, mRNAs, which positively regulates the stress tolerance.

The molecular basis of plant tolerance to abiotic stresses and stress regulation of small RNAs has been studied using different methods to observe the altered expression of these molecules and their related target genes; for instance, the sequence analysis of small RNA libraries before and after the stress condition, microarray data analysis, mutagenesis, and RNAi. Their reports have identified numerous genes and sRNAs that are induced by applying different stress conditions, which is the material for the next step: making transgenic plants and check if these overexpressed transgenics could exhibit an improvement in stress tolerance. But the fact is that even though some of the genes and small RNAs show altered expression under stress, they do not play any role to make the plant more tolerant to the stress. And the reason is largely because of the complex genetic interactions underlying the plant tolerance toward stress, which are still to be understood.

From transcriptomic studies, we know that the stress conditions such as heat, drought, cold, and salt evoke the expression of an overlapping set of genes, suggesting that their signaling transduction pathways share common control points. Most of the genes, which are detected to be responsive to abiotic stresses, are usually the genes that regulate plant development and reproduction (as the plant faces the urge to save more energy for producing viable seeds rather than a high biomass), also senescence-related genes (as to recycle the nutrients from the old leaves to younger leaves and reproductive parts and wasting less water and energy for them), as well as the genes that are involved in the abscisic acid (ABA) pathway, which play a crucial role in plant growth and development pathway and redox pathway.

RNA interference technology is one of the potential reverse genetics tools for understanding the functional significance of these genes and their respective regulatory sRNAs. The information about stress-induced genes and sRNAs including the sequence and annotation could be found in the genome databases like National Centre for Biotechnology Information (NCBI) or stress complementary DNA (cDNA) databases. In addition to these stress-induced genes, the regulatory elements for these genes are also altered during stress condition including small regulatory RNAs. A number of these sRNAs, which are induced in different plant species and under different stress conditions, can be found in some recent review papers [47]. A vast amount of data has been published about the expression profiling of different sRNAs in various stress conditions. Although these expression-profiling experiments can provide us with some clues about the involvement of these sRNAs in gene regulation under those specific circumstances, to find the relevance of each of these sRNAs in imparting stress tolerance in plants can only be studied by functional genomic approaches like gene overexpression or downregulation. RNA interference technology using constructs transcribing self-complementary hairpin RNA is one of the reverse genetics approaches to downregulate genes in plants.

Another powerful technique to learn about gene functions in a developmental or physiological context in plants is by mutagenesis and to isolate the corresponding mutants with altered phenotypes. Various mutagenic agents, including chemical and biological, have been widely used in this regard, each of them with its own advantages and inconveniences.

For *Arabidopsis thaliana*, the genome sequence is publicly available; hence, relying on reverse genetics to understand the relevant roles of genes is currently a common practice. There are specific screening methods used in order to measure the effect of each stress on the overall plant physiology after the treatment as well as the methods used for inducing the specific stress condition in the plant, which have been studied and reviewed many times.

2.4. Drought stress

Drought stress is known as the most significant stress especially with regard to the climate change and global warming. It restricts plant growth and development severely, while tolerant plants are able to survive by several mechanisms such as consuming small amounts of water or keeping their stomata closed at a high rate under drought conditions.

Approximately two-thirds of the potential yield of major crops are lost every year due to the adverse growing environments [48]. A worldwide increase in arid areas, including the Mediterranean basin, has been predicted by the International Plant Protection Convention (IPPC) in 2001 and 2007. Water deficit also leads to salinity stress in many cases, which makes the growth situation even harder for the plants. Therefore, it is regarded as the most important abiotic stress and it is necessary to develop strategies toward sustainable use of water and improve plant-drought resistance [49]. Many genes have already been studied and reported to be involved in the drought-resistance response network in the plants. But in recent years, it has become clear that sRNAs play pivotal roles in stress responses as well in regulating the expression of resistance genes.

MiRNA-expression profiling under drought stress has now been performed in *Arabidopsis*, rice, and *Populus trichocarpa*, and many other plants under drought-stress conditions and some of the miRNAs were shown to be responsive toward this stress in different plants, some of which can be reviewed in the available literature [50-51].

Another group has worked on miRNA expression patterns of drought-resistant *Triticum turgidum* ssp. *Dicoccoides* in response to drought stress, using an miRNA microarray platform

[52]. MiR474, which targets proline dehydrogenase (PDH), was upregulated during drought stress in Zea maize [53]. Zhao et al. worked on miR169g and miR393 under drought condition in rice, while miR393 was conversely induced by drought [54]. Many other recent reports used several different methods to study the expression pattern changes of miRNAs in different plants under the stress condition [47].

2.5. Salt stress

Crops worldwide are threatened by excessive soil salinity due to the accumulation of salt delivered along with irrigation water and by coastal flooding and the high evapotranspiration rates caused by climate change. About 6% of the total arable land in the world is affected by excess salt [55] and it has been predicted to increase to about 30% of the world's arable land by 2025 and 50% by the year 2050 [56]. Several genes and pathways in plants are affected by salt stress [57]. Hence, the promising approach to address the problem of soil salinity is to increase the understanding of response of plants to salinity-related stress. These genes are mostly involved in signal transduction, activation of ion channels, and growth-factor-regulated modification of plant architecture, and, in particular, root morphology.

Besides the genes, numerous differentially regulated miRNAs have also been identified in salt-stressed plants. For instance, miR156, miR158, miR159, miR165, miR167, miR168, miR169, miR171, miR319, miR393, miR394, miR396, and miR397 were all reported to be overexpressed in response to salt stress in *Arabidopsis*, while the accumulation of miR398 was downregulated [50].

miR169 was also reported to be induced by high salinity stress [46]. The authors found a cis-acting ABA-responsive element (ABRE) in the upstream region of miR169n, which suggested that miR169n might be regulated by ABA. Another group used microarray experiments as a method to explore the miRNA profile of maize in different lines (salt-tolerant and salt-sensitive); finally, it was reported that the expression levels of miR156, miR164, miR167, and miR396 family members were downregulated considerably, while it was increased in miR162, miR168, miR395, and miR474 families after salt-shock in root tissue [58].

2.6. Cold stress

Some plants increase their tolerance to cold in order to deal with the low temperatures. This phenomenon is known as cold acclimation. In recent years, many cold-regulated genes have been identified in plants under cold stress. The C-repeat binding factor (CBF) cold-responsive pathway was considered as the most known cold tolerance pathway in plants [59]. There are three CBF/DREB1 family members, including CBF1, CBF2, and CBF3 (DREB1b, DREB1c, and DREB1a, respectively), encoding the DNA-binding proteins of Apetala2/ethylene responsive factor (AP2/ERF) family [60]. Also, the expression of many miRNAs in cold stress has been examined in different plants including *Arabidopsis thaliana*. Several miRNAs belonging to different families were reported to be upregulated under cold-stress condition in *Arabidopsis thaliana* (miR165/166, miR393, miR396, and miR408), while some other miRNAs (miR156/157, miR159/319, miR164, miR394, and miR398) were shown to be either transient or mildly regulated under cold-stress treatment condition [19][50]. In another report, the expression levels of miR168 and miR477 family members were increased after the cold-stress treatment, while miR156, miR475, and miR476 members were downregulated in *Populus* plants [55][19].

2.7. Heat stress

The average temperature of our planet is rising year by year because of the climate change. As a result, changes in the patterns of rainfall, droughts, and submergence stress are induced to the natural environments. Heat stress even alters the distribution and productivity of important crops negatively throughout the earth. A temperature rise of −5°C above the plant's optimum temperature is considered as a heat stress. It disrupts normal functions of cellular processes, may lead to delay in plant growth and development, and it might even result in death of the plant, but, usually, high temperatures result in water deficiency, which eventually leads to increase in salt concentration. Recent studies indicate that the projected global warming in the upcoming years will negatively affect the yield of important crops; hence, the necessity of focusing on gene networks and their regulatory components becomes obvious.

A major component with regard to responding to heat stress is the induction of heat shock proteins (HSPs), which get activated by heat shock transcription factors (HSFs). There are five classes of HSPs based on their molecular weights: HSP100, HSP90, HSP70, HSP60, and small heat shock proteins (sHSPs, 15–30 kDa). On the other hand, HSFs recognize heat stress elements on the promoter of heat stress-responsive genes (HSE: 5′-GAAnnTTC-3′). Plant HSFs are categorized into three classes based on their oligomerization domains (A, B, and C) [61].

However, the more upstream regulators of HSFs remain to be identified. Guan et al. have reported that miR398 is rapidly induced by being subjected to heat stress while its target genes (CSD1, CDS2, and CCS) are downregulated. They further reported that the expression levels of HSF and HSP genes in *csd1-*, *csd2-*, and *ccs*-mutant plants are increased under heat stress, and *csd1*, *csd2*, and *ccs* plants are more tolerant to heat stress than wild-type plants. They identified two HSFs, which act upstream of miR398, suggesting that this pathway is an essential regulatory loop for plant thermotolerance [62].

Based on deep sequencing experiments, Wang et al. suggest that there is a new class of small RNAs that originate from the chloroplast genome, which are responsive to heat stress [63]. They performed RNA sequencing (RNA-seq) and found 1031 cis-NATs in *Brassica rapa* based on the homology with *Arabidopsis* and 303 conserved cis-NATs, which correspond to the ones in *Arabidopsis* [64]. TAS1 (trans-acting siRNA precursor 1) targets, derived from small interfering RNAs named heat-induced TAS1 target1 (*HTT1*) and *HTT2*, are involved in thermotolerance [65]. *HTT1* and *HTT2* genes were highly upregulated in *Arabidopsis thaliana* seedlings in response to heat shock based on their microarray analysis. TAS1a has a trans-acting small interfering RNA, which targets the *HTT* genes. Overexpression of TAS1a accelerated the expression of TAS1-siRNAs and decreased the expression levels of *HTT* genes that eventually led to weaker thermotolerance. Conversely, stronger expression of *HTT1* and *HTT2* genes upregulated various *Hsf* genes, helping the plants to achieve a stronger thermotolerance. In HsfA1a-overexpression transgenic plants, which present a higher tolerance to heat stress, the *HTT* genes were upregulated. In the meantime, HsfA1a was shown to bind to the *HTT1* and

HTT2 promoter regions and activate them directly. Finally, they proposed that HTT1 interacts with Hsp70-14 and Hsp40, nuclear factor Y, and subunit C2 complex.

Wheat miRNAs showed differential expression in response to heat stress; by using Solexa high-throughput sequencing, Xin et al. cloned the small RNAs from wheat leaves treated by heat-stress gene [66]. Stief et al. also reported that miR156 is responsible for heat stress memory in *Arabidopsis* [67].

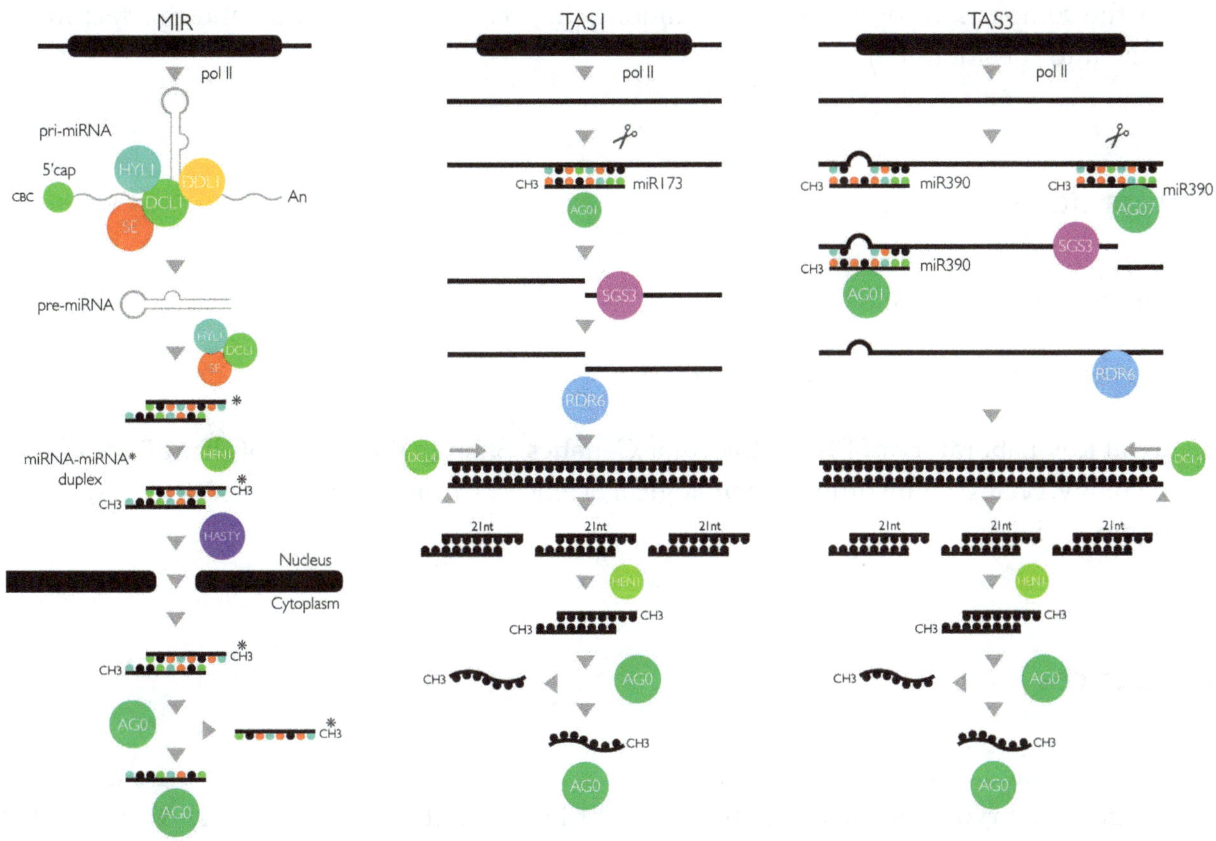

Figure 1. Biogenesis of miRNAs and ta-siRNAs

3. Perspective

Physiological responses to stress are controlled by expression of a large number of genes, many of which are regulated by microRNAs. At the molecular level, identification of stress-responsive genes is an initial step toward understanding plant stress response as pyramiding of different genes in the same plant is an option for achieving better stress tolerance. Although finding genes and sRNAs, which show induction by stress, is an important step toward stress tolerance improvement, most of the studies in which they use transgenics only show the importance of the introduced transgene and not the overall metabolic effects that the trans-host gets exposed to. On the other hand, the new stress-tolerant transgenic lines should have no or few undesired phenotypic changes plus a minimal yield penalty. In stress-tolerant transgenics, which are introduced so far, a constitutive promoter has been used for expressing the transgene in most of the cases. These transgenes must be utilized to overcome the problem of yield penalty and growth retardation in these experiments. Admittedly, most reports published on stress-tolerant transgenic plants are based on the limited characterization of the stress condition as well as the tolerant phenotypes. Adequate assays for phenotyping of the stress-tolerance trait must be undertaken under natural stress conditions. Overall, there is a lack of uniformity in the stress induction regimes applied by various research groups, which makes the comparisons of the responses among different reports difficult and this fact must be taken into consideration.

Author details

By Shirin Mirlohi and Yuke He*

*Address all correspondence to: ykhe@sibs.ac.cn

National Key Laboratory of Plant Molecular Genetics, Shanghai Institute of Plant Physiology and Ecology, Shanghai Institutes for Biological Sciences, Chinese Academy of Sciences, Shanghai, China

References

[1] S. Guo and K. J. Kemphues, "par-1, a gene required for establishing polarity in C. elegans embryos, encodes a putative Ser/Thr kinase that is asymmetrically distributed," *Cell*, vol. 81, no. 4, pp. 611–620, 1995.

[2] R. C. Lee, R. L. Feinbaum, and V. Ambros, "The C. elegans heterochronic gene lin-4 encodes small RNAs with antisense complementarity to lin-14," *Cell*, vol. 75, no. 5, pp. 843–854, 1993.

[3] F. Crick, "On protein synthesis," *Symp. Soc. Exp. Biol.*, vol. 12, pp. 138–163, 1958.

[4] F. Crick, "Central dogma of molecular biology," *Nature*, vol. 227, no. 5258, pp. 561–563, 1970.

[5] J. Couzin, "Breakthrough of the year. Small RNAs make big splash," *Science (New York, N.Y.)*, vol. 298, no. 5602, pp. 2296–2297, 2002.

[6] C. Llave, Z. Xie, K. D. Kasschau, and J. C. Carrington, "Cleavage of Scarecrow-like mRNA targets directed by a class of Arabidopsis miRNA," *Science*, vol. 297, no. 5589, pp. 2053–2056, 2002.

[7] W. Park, J. Li, R. Song, J. Messing, and X. Chen, "CARPEL FACTORY, a Dicer homolog, and HEN1, a novel protein, act in microRNA metabolism in Arabidopsis thaliana," *Curr. Biol.*, vol. 12, no. 17, pp. 1484–1495, 2002.

[8] B. J. Reinhart, E. G. Weinstein, M. W. Rhoades, B. Bartel, and D. P. Bartel, "MicroRNAs in plants," *Genes Dev.*, vol. 16, no. 13, pp. 1616–1626, 2002.

[9] C. Lu and N. Fedoroff, "A mutation in the Arabidopsis HYL1 gene encoding a dsRNA binding protein affects responses to abscisic acid, auxin, and cytokinin," *Plant Cell*, vol. 12, no. 12, pp. 2351–2366, 2000.

[10] X. Chen, J. Liu, Y. Cheng, and D. Jia, "HEN1 functions pleiotropically in Arabidopsis development and acts in C function in the flower," *Development*, vol. 129, no. 5, pp. 1085–1094, 2002.

[11] T. A. Golden, S. E. Schauer, J. D. Lang, S. Pien, A. R. Mushegian, U. Grossniklaus, D. W. Meinke, and A. Ray, "SHORT INTEGUMENTS1/SUSPENSOR1/CARPEL FACTORY, a Dicer homolog, is a maternal effect gene required for embryo development in Arabidopsis," *Plant Physiol.*, vol. 130, no. 2, pp. 808–822, 2002.

[12] S. E. Schauer, S. E. Jacobsen, D. W. Meinke, and A. Ray, "DICER-LIKE1: Blind men and elephants in Arabidopsis development," *Trends Plant Sci.*, vol. 7, no. 11. pp. 487–491, 2002.

[13] M.-H. Han, S. Goud, L. Song, and N. Fedoroff, "The Arabidopsis double-stranded RNA-binding protein HYL1 plays a role in microRNA-mediated gene regulation," *Proc. Natl. Acad. Sci. U. S. A.*, vol. 101, no. 4, pp. 1093–1098, 2004.

[14] L. A. Rymarquis, J. P. Kastenmayer, A. G. Hüttenhofer, and P. J. Green, "Diamonds in the rough: mRNA-like non-coding RNAs," *Trends Plant Sci.*, vol. 13, no. 7, pp. 329–334, 2008.

[15] A. S. Flynt and E. C. Lai, "Biological principles of microRNA-mediated regulation: Shared themes amid diversity," *Nat. Rev. Genet.*, vol. 9, no. 11, pp. 831–842, 2008.

[16] I. R. Henderson, X. Zhang, C. Lu, L. Johnson, B. C. Meyers, P. J. Green, and S. E. Jacobsen, "Dissecting Arabidopsis thaliana DICER function in small RNA processing, gene silencing and DNA methylation patterning," *Nat. Genet.*, vol. 38, no. 6, pp. 721–725, 2006.

[17] S. Griffiths-Jones, R. J. Grocock, S. van Dongen, A. Bateman, and A. J. Enright, "miRBase: MicroRNA sequences, targets and gene nomenclature," *Nucleic Acids Res.*, vol. 34, no. Database issue, pp. D140–D144, 2006.

[18] A. J. Hamilton and D. C. Baulcombe, "A species of small antisense RNA in posttranscriptional gene silencing in plants," *Science*, vol. 286, no. 5441, pp. 950–952, 1999.

[19] R. Sunkar and J.-K. Zhu, "Novel and stress-regulated microRNAs and other small RNAs from Arabidopsis," *Plant Cell*, vol. 16, no. 8, pp. 2001–2019, 2004.

[20] V. Ambros, "MicroRNA pathways in flies and worms: Growth, death, fat, stress, and timing," *Cell*, vol. 113, no. 6, pp. 673–676, 2003.

[21] A. Aravin, M. Lagos-Quintana, and A. Yalcin, "The Small RNA Profile during *Drosophila melanogaster* Development," *Dev. Cell*, vol. 5, no. 2, pp. 337–350, 2003.

[22] G. Tang, B. J. Reinhart, D. P. Bartel, and P. D. Zamore, "A biochemical framework for RNA silencing in plants," *Genes Dev.*, vol. 17, no. 1, pp. 49–63, 2003.

[23] X. Chen, "Small RNAs and their roles in plant development," *Annu. Rev. Cell Dev. Biol.*, vol. 25, pp. 21–44, 2009.

[24] O. Voinnet, "Origin, biogenesis, and activity of plant microRNAs," *Cell*, vol. 136, no. 4, pp. 669–87, 2009.

[25] S. Choudhuri, "Lesser known relatives of miRNA," *Biochem. Biophys. Res. Commun.*, vol. 388, no. 2. pp. 177–180, 2009.

[26] F. Vazquez, H. Vaucheret, R. Rajagopalan, C. Lepers, V. Gasciolli, A. C. Mallory, J. L. Hilbert, D. P. Bartel, and P. Crété, "Endogenous trans-acting siRNAs regulate the accumulation of arabidopsis mRNAs," *Mol. Cell*, vol. 16, no. 1, pp. 69–79, 2004.

[27] E. Allen, Z. Xie, A. M. Gustafson, and J. C. Carrington, "microRNA-directed phasing during trans-acting siRNA biogenesis in plants," *Cell*, vol. 121, no. 2, pp. 207–221, 2005.

[28] T. A. Montgomery, M. D. Howell, J. T. Cuperus, D. Li, J. E. Hansen, A. L. Alexander, E. J. Chapman, N. Fahlgren, E. Allen, and J. C. Carrington, "Specificity of ARGONAUTE7-miR390 interaction and dual functionality in TAS3 trans-acting siRNA formation," *Cell*, vol. 133, no. 1, pp. 128–141, 2008.

[29] T. A. Montgomery, S. J. Yoo, N. Fahlgren, S. D. Gilbert, M. D. Howell, C. M. Sullivan, A. Alexander, G. Nguyen, E. Allen, J. H. Ahn, and J. C. Carrington, "AGO1-miR173 complex initiates phased siRNA formation in plants," *Proc. Natl. Acad. Sci. U. S. A.*, vol. 105, no. 51, pp. 20055–20062, 2008.

[30] Q. J. Luo, A. Mittal, F. Jia, and C. D. Rock, "An autoregulatory feedback loop involving PAP1 and TAS4 in response to sugars in Arabidopsis," *Plant Mol. Biol.*, vol. 80, no. 1, pp. 117–129, 2012.

[31] E. Allen and M. D. Howell, "miRNAs in the biogenesis of trans-acting siRNAs in higher plants," *Semin. Cell Dev. Biol.*, vol. 21, no. 8, pp. 798–804, 2010.

[32] H. Jin, V. Vacic, T. Girke, S. Lonardi, and J.-K. Zhu, "Small RNAs and the regulation of cis-natural antisense transcripts in Arabidopsis," *BMC Mol. Biol.*, vol. 9, p. 6, 2008.

[33] J. Chen, M. Sun, W. J. Kent, X. Huang, H. Xie, W. Wang, G. Zhou, R. Z. Shi, and J. D. Rowley, "Over 20% of human transcripts might form sense-antisense pairs," *Nucleic Acids Res.*, vol. 32, no. 16, pp. 4812–4820, 2004.

[34] Y. Okazaki, M. Furuno, T. Kasukawa, J. Adachi, H. Bono, S. Kondo, I. Nikaido, N. Osato, R. Saito, H. Suzuki, I. Yamanaka, H. Kiyosawa, K. Yagi, Y. Tomaru, Y. Hasegawa, A. Nogami, C. Schönbach, T. Gojobori, R. Baldarelli, D. P. Hill, C. Bult, D. A. Hume, J. Quackenbush, L. M. Schriml, A. Kanapin, H. Matsuda, S. Batalov, K. W. Beisel, J. A. Blake, D. Bradt, V. Brusic, C. Chothia, L. E. Corbani, S. Cousins, E. Dalla, T. A. Dragani, C. F. Fletcher, A. Forrest, K. S. Frazer, T. Gaasterland, M. Gariboldi, C. Gissi, A. Godzik, J. Gough, S. Grimmond, S. Gustincich, N. Hirokawa, I. J. Jackson, E. D. Jarvis, A. Kanai, H. Kawaji, Y. Kawasawa, R. M. Kedzierski, B. L. King, A. Konagaya, I. V. Kurochkin, Y. Lee, B. Lenhard, P. A. Lyons, D. R. Maglott, L. Maltais, L. Marchionni, L. McKenzie, H. Miki, T. Nagashima, K. Numata, T. Okido, W. J. Pavan, G. Pertea, G. Pesole, N. Petrovsky, R. Pillai, J. U. Pontius, D. Qi, S. Ramachandran, T. Ravasi, J. C. Reed, D. J. Reed, J. Reid, B. Z. Ring, M. Ringwald, A. Sandelin, C. Schneider, C. A. M. Semple, M. Setou, K. Shimada, R. Sultana, Y. Takenaka, M. S. Taylor, R. D. Teasdale, M. Tomita, R. Verardo, L. Wagner, C. Wahlestedt, Y. Wang, Y. Watanabe, C. Wells, L. G. Wilming, A. Wynshaw-Boris, M. Yanagisawa, I. Yang, L. Yang, Z. Yuan, M. Zavolan, Y. Zhu, A. Zimmer, P. Carninci, N. Hayatsu, T. Hirozane-Kishikawa, H. Konno, M. Nakamura, N. Sakazume, K. Sato, T. Shiraki, K. Waki, J. Kawai, K. Aizawa, T. Arakawa, S. Fukuda, A. Hara, W. Hashizume, K. Imotani, Y. Ishii, M. Itoh, I. Kagawa, A. Miyazaki, K. Sakai, D. Sasaki, K. Shibata, A. Shinagawa, A. Yasunishi, M. Yoshino, R. Waterston, E. S. Lander, J. Rogers, E. Birney, and Y. Hayashizaki, "Analysis of the mouse transcriptome based on functional annotation of 60,770 full-length cDNAs," *Nature*, vol. 420, no. 6915, pp. 563–573, 2002.

[35] S. Misra, M. A. Crosby, C. J. Mungall, B. B. Matthews, K. S. Campbell, P. Hradecky, Y. Huang, J. S. Kaminker, G. H. Millburn, S. E. Prochnik, C. D. Smith, J. L. Tupy, E. J. Whitfied, L. Bayraktaroglu, B. P. Berman, B. R. Bettencourt, S. E. Celniker, A. D. N. J. de Grey, R. A. Drysdale, N. L. Harris, J. Richter, S. Russo, A. J. Schroeder, S. Q. Shu, M. Stapleton, C. Yamada, M. Ashburner, W. M. Gelbart, G. M. Rubin, and S. E. Lewis, "Annotation of the Drosophila melanogaster euchromatic genome: A systematic review," *Genome Biol.*, vol. 3, no. 12, p. RESEARCH0083, 2002.

[36] X.-J. Wang, T. Gaasterland, and N.-H. Chua, "Genome-wide prediction and identification of cis-natural antisense transcripts in Arabidopsis thaliana.," *Genome Biol.*, vol. 6, no. 4, p. R30, 2005.

[37] N. Osato, H. Yamada, K. Satoh, H. Ooka, M. Yamamoto, K. Suzuki, J. Kawai, P. Carninci, Y. Ohtomo, K. Murakami, K. Matsubara, S. Kikuchi, and Y. Hayashizaki, "Antisense transcripts with rice full-length cDNAs," *Genome Biol.*, vol. 5, no. 1, p. R5, 2003.

[38] S. H. Munroe and M. A. Lazar, "Inhibition of c-erbA mRNA splicing by a naturally occurring antisense RNA," *J. Biol. Chem.*, vol. 266, no. 33, pp. 22083–22086, 1991.

[39] C. Tufarelli, J. A. S. Stanley, D. Garrick, J. A. Sharpe, H. Ayyub, W. G. Wood, and D. R. Higgs, "Transcription of antisense RNA leading to gene silencing and methylation as a novel cause of human genetic disease," *Nat. Genet.*, vol. 34, no. 2, pp. 157–165, 2003.

[40] T. Moore, M. Constancia, M. Zubair, B. Bailleul, R. Feil, H. Sasaki, and W. Reik, "Multiple imprinted sense and antisense transcripts, differential methylation and tandem repeats in a putative imprinting control region upstream of mouse Igf2," *Proc. Natl. Acad. Sci. U. S. A.*, vol. 94, no. 23, pp. 12509–12514, 1997.

[41] N. T. Peters, J. A. Rohrbach, B. A. Zalewski, C. M. Byrkett, and J. C. Vaughn, "RNA editing and regulation of Drosophila 4f-rnp expression by sas-10 antisense read-through mRNA transcripts," *RNA*, vol. 9, no. 6, pp. 698–710, 2003.

[42] O. Borsani, J. Zhu, P. E. Verslues, R. Sunkar, and J.-K. Zhu, "Endogenous siRNAs derived from a pair of natural cis-antisense transcripts regulate salt tolerance in Arabidopsis," *Cell*, vol. 123, no. 7, pp. 1279–1291, 2005.

[43] S. Katiyar-Agarwal, R. Morgan, D. Dahlbeck, O. Borsani, A. Villegas, J.-K. Zhu, B. J. Staskawicz, and H. Jin, "A pathogen-inducible endogenous siRNA in plant immunity," *Proc. Natl. Acad. Sci. U. S. A.*, vol. 103, no. 47, pp. 18002–18007, 2006.

[44] X. Zhou, R. Sunkar, H. Jin, J.-K. Zhu, and W. Zhang, "Genome-wide identification and analysis of small RNAs originated from natural antisense transcripts in Oryza sativa," *Genome Res.*, vol. 19, no. 1, pp. 70–78, 2009.

[45] T. Lu, C. Zhu, G. Lu, Y. Guo, Y. Zhou, Z. Zhang, Y. Zhao, W. Li, Y. Lu, W. Tang, Q. Feng, and B. Han, "Strand-specific RNA-seq reveals widespread occurrence of novel cis-natural antisense transcripts in rice," *BMC Genomics*, vol. 13, no. 1, p. 721, 2012.

[46] X.-J. Wang, T. Gaasterland, and N.-H. Chua, "Genome-wide prediction and identification of cis-natural antisense transcripts in Arabidopsis thaliana," *Genome Biol.*, vol. 6, no. 4, p. R30, 2005.

[47] B. Khraiwesh, J.-K. Zhu, and J. Zhu, "Role of miRNAs and siRNAs in biotic and abiotic stress responses of plants," *Biochim. Biophys. Acta*, vol. 1819, no. 2, pp. 137–148, 2012.

[48] S. Bajaj, J. Targolli, L. F. Liu, T. H. D. Ho, and R. Wu, "Transgenic approaches to increase dehydration-stress tolerance in plants," *Mol. Breeding*, vol. 5, no. 6, pp. 493–503, 1999.

[49] M. M. Chaves, J. P. Maroco, and J. S. Pereira, "Understanding plant responses to drought - From genes to the whole plant," *Funct. Plant Biol.*, vol. 30, no. 3, pp. 239–264, 2003.

[50] H.-H. Liu, X. Tian, Y.-J. Li, C.-A. Wu, and C.-C. Zheng, "Microarray-based analysis of stress-regulated microRNAs in Arabidopsis thaliana," *RNA*, vol. 14, no. 5, pp. 836–843, 2008.

[51] L. Zhou, Y. Liu, Z. Liu, D. Kong, M. Duan, and L. Luo, "Genome-wide identification and analysis of drought-responsive microRNAs in Oryza sativa," *J. Exp. Bot.*, vol. 61, no. 15, pp. 4157–4168, 2010.

[52] M. Kantar, S. J. Lucas, and H. Budak, "miRNA expression patterns of Triticum dicoccoides in response to shock drought stress," *Planta*, vol. 233, no. 3, pp. 471–484, 2011.

[53] L. Wei, D. Zhang, F. Xiang, and Z. Zhang, "Differentially expressed miRNAs potentially involved in the regulation of defense mechanism to drought stress in maize seedlings," *Int. J. Plant Sci.*, vol. 170, no. 8, pp. 979–989, 2009.

[54] B. Zhao, R. Liang, L. Ge, W. Li, H. Xiao, H. Lin, K. Ruan, and Y. Jin, "Identification of drought-induced microRNAs in rice," *Biochem. Biophys. Res. Commun.*, vol. 354, no. 2, pp. 585–590, 2007.

[55] S. Lu, Y. H. Sun, and V. L. Chiang, "Stress-responsive microRNAs in Populus," *Plant J.*, vol. 55, no. 1, pp. 131–151, 2008.

[56] W. Wang, B. Vinocur, and A. Altman, "Plant responses to drought, salinity and extreme temperatures: Towards genetic engineering for stress tolerance," *Planta*, vol. 218, no. 1, pp. 1–14, 2003.

[57] J.-K. Zhu, "Salt and drought stress signal transduction in plants," *Annu. Rev. Plant Biol.*, vol. 53, pp. 247–273, 2002.

[58] D. Ding, L. Zhang, H. Wang, Z. Liu, Z. Zhang, and Y. Zheng, "Differential expression of miRNAs in response to salt stress in maize roots," *Ann. Bot.*, vol. 103, no. 1, pp. 29–38, 2009.

[59] H. A. Van Buskirk and M. F. Thomashow, "Arabidopsis transcription factors regulating cold acclimation," *Physiol. Plant.*, vol. 126, no. 1, pp. 72–80, 2006.

[60] S. J. Gilmour, D. G. Zarka, E. J. Stockinger, M. P. Salazar, J. M. Houghton, and M. F. Thomashow, "Low temperature regulation of the Arabidopsis CBF family of AP2 transcriptional activators as an early step in cold-induced COR gene expression," *Plant J.*, vol. 16, no. 4, pp. 433–442, 1998.

[61] P. von Koskull-Döring, K.-D. Scharf, and L. Nover, "The diversity of plant heat stress transcription factors," *Trends Plant Sci.*, vol. 12, no. 10, pp. 452–457, 2007.

[62] Q. Guan, X. Lu, H. Zeng, Y. Zhang, and J. Zhu, "Heat stress induction of miR398 triggers a regulatory loop that is critical for thermotolerance in Arabidopsis," *Plant J.*, vol. 74, no. 5, pp. 840–851, 2013.

[63] L. Wang, X. Yu, H. Wang, Y.-Z. Lu, M. de Ruiter, M. Prins, and Y.-K. He, "A novel class of heat-responsive small RNAs derived from the chloroplast genome of Chinese cabbage (Brassica rapa)," *BMC Genomics*, vol. 12, no. 1, p. 289, 2011.

[64] X. Yu, J. Yang, X. Li, X. Liu, C. Sun, F. Wu, and Y. He, "Global analysis of cis-natural antisense transcripts and their heat-responsive nat-siRNAs in Brassica rapa," *BMC Plant Biol.*, vol. 13, p. 208, 2013.

[65] S. Li, J. Liu, Z. Liu, X. Li, F. Wu, and Y. He, "Heat-induced TAS1 TARGET1 mediates thermotolerance via heat stress transcription factor A1a-directed pathways in Arabidopsis," *Plant Cell*, vol. 26, no. 4, pp. 1764–1780, 2014.

[66] M. Xin, Y. Wang, Y. Yao, C. Xie, H. Peng, Z. Ni, Q. Sun, "Diverse set of microRNAs are responsive to powdery mildew infection and heat stress in wheat (*Triticum aestivum* L.), " *BMC Plant Biol.*, vol. 10, p. 123, 2010.

[67] A. Stief, S. Altmann, K. Hoffmann, B. D. Pant, W.-R. Scheible, and I. Bäurle, "Arabidopsis miR156 regulates tolerance to recurring environmental stress through SPL transcription factors," Plant Cell, vol. 26, no. 4, pp. 1792–1807, 2014.

5

Role of Phytochelatins in Redox Caused Stress in Plants and Animals

Miguel Angel Merlos Rodrigo, Naser A. Anjum, Zbynek Heger, Ondrej Zitka, Adam Vojtech, Eduarda Pereira and Rene Kizek

Abstract

Varied environmental compartments (such as soil and water) potentially contaminated with different metals/metalloids can impact the health of both plants and animals/humans. Trace amounts of Cu, Mn, Mo, Ni and Zn are beneficial for higher plants, whereas, Cr, Cu, Co, Mn, Mo, Se, V and Zn are known as the micronutrient metal/metalloids for animals/humans. However, elevated levels of the metals/metalloids can cause severe toxic consequences in both plants and animals/humans. Common in plants and animals/humans, phytochelatins (PCs), the principal non-protein, S-rich, thiolate peptides, protect (through different mechanisms) cellular functions and metal/metalloid homeostasis by performing their chelation and/or detoxification. With the major aim of broadening the current knowledge on the subject, this chapter (a) overviews PCs' role and modulation separately in metal/metalloid-exposed plants and animals/humans; (b) discusses major methods for determination of PCs and bioassays for enzymes involved in PC synthesis; (c) evaluates the connection of PCs with bionanoparticles; and finally (d) highlights so far unexplored aspects in the present context.

Keywords: Phytochelatin, metal, glutathione, stress

1. Introduction

Anthropogenic activities have caused the release of a wide range of hazardous metals/metalloids (hereafter termed as 'metal/s') into the environment. In particular, increasing emissions of metals such as Cd, Hg and As into the environment pose an acute problem for all organisms. Metals, unlike organic contaminants, are not degradable and remain persistent in soils [1–3]. Once taken up, these metals can bring severe toxic consequences in cells due to their chemical similarity to replace the metals necessary for cellular functions. Nevertheless, metals at toxic levels have the capability to interact with several vital cellular biomolecules such as nuclear proteins and DNA, leading to excessive augmentation of reactive oxygen species (ROS) [4–6]. In addition, these metals generate ROS which in turn can cause neuro-toxicity, hepatotoxicity and nephrotoxicity in humans and animals [7, 8]. Notably, higher plants, algae, certain yeasts and animals are equipped with a repertoire of mechanisms to counteract metal toxicity. The key elements of these are chelation of metals by forming phytochelatins (PCs) and related cysteine-rich polypeptides [9–11]. PCs are produced from glutamine, cysteine and glycine and the process is catalysed by PC synthases known as γ-glutamylcysteine (γ-Glu-Cys) dipeptidyl transpeptidases [12, 11]. PCs have been identified in a wide variety of plant species, microorganisms and invertebrates. They are structurally related to glutathione (GSH) and were presumed to be the products of a biosynthetic pathway. Numerous physiological, biochemical and genetic studies have confirmed that GSH is the substrate for PC biosynthesis [13, 14]. The general structure of PCs is (γ-Glu-Cys)n-Gly, with increasing repetitions of the dipeptide Glu-Cys, where *n* can range from 2 to 11 but is typically no more than 5 [15]. Except glycine, other amino acid residues can be found on the C-terminal end of (γ-Glu-Cys)*n* peptides. In Figure 1, we show the general structure of PC and the major steps involved in its synthesis from GSH through PC synthase in response to high concentrations of toxic metals. Originally thought to be plant-specific, PC and PC synthases have now been reported in a few fungal taxa, such as the yeast *Schizosaccharomyces* sp. and the mycorrhizal ascomycete *Tuber melanosporum* [16, 17] and invertebrates belonging to the nematodes, annelids or plathyhelminths [18, 19, 4, 1, 20, 17, 21–24].

In the light of recent literature, the PCs' role and modulation are overviewed separately in metal-exposed plants and animals/humans and major methods for the determination of PCs and the bioassays for enzymes involved in PC synthesis are discussed hereunder. Additionally, connection of PCs with bionanoparticles is evaluated, and finally, major aspects so far unexplored in the present context are briefly highlighted.

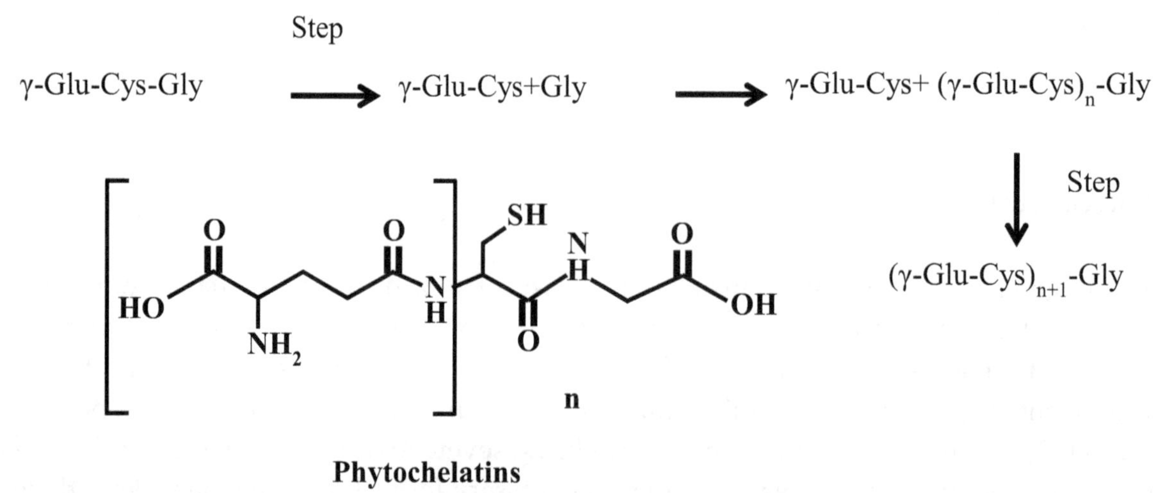

Figure 1. General structure of phytochelatins (PCs) and the major steps involved in its synthesis from glutathione (GSH) through a PC synthase in response to high concentrations of toxic metals.

2. Phytochelatins in metal/metalloid-exposed plants

Contamination by metals can be considered as one of the most critical threats to soil and water resources as well as to human health [25, 26]. In fact, the contamination of soils with toxic metals has often resulted from human activities, especially those related to accelerated rate of industrialization, intensive agriculture and extensive mining. Metal belongs to group of non-biodegradable, persistent inorganic chemical having cytotoxic, genotoxic and mutagenic effects on humans or animals and plants through influencing and tainting food chains, soil, irrigation or potable water and aquifers [27, 28, 6]. Chelation and sequestration of metals by particular ligands are the major mechanisms employed by plants to deal with metal stress. The two best-characterized metal-binding ligands in plant cells are the PC and metallothioneins (MTs) [29–33, 6, 34].

Figure 2 shows the scheme of metal-detoxification by PCs in a plant cell. PC, which has a higher affinity for Cd, is formed by the polymerization of 2–11 γ-EC moieties via PC synthase. Several studies confirm that in plants, both GSH and PC synthesis are increased after exposure to Cd and other metals [12, 35–41]. In Figure 3, we show both general functions of the PC and a model of complex between Cd^{+2} ion and one molecule of PC.

Gonzalez-Mendoza et al. showed that PC synthase gene (in coordination with the expression of metallothionein gene) is present in *Avicennia germinans* leaves, and that their expression increases in response to metal exposure, which supports the hypothesis that PC synthase and metallothionein are part of the metal-tolerance mechanisms in this species. In addition, these authors found that *A. germinans* has the ability to express both genes (*AvMT2* and *AvPCS*) as a coordinated response mechanism to avoid the toxic effects caused by non-essential metals. However, for essential metals such as Cu^{2+}, the results showed that *AvPCS* was the most active gene involved in the regulation of this metal in the leaves [42]. Recent study showed that *Lunularia cruciata* compartmentalizes Cd^{+2} in the vacuoles of the photosynthetic parenchyma by means of a PC-mediated detoxification strategy, and possesses a PC synthase that is activated by Cd and homeostatic concentrations of Fe(II) and Zn. *Arabidopsis thaliana* PC synthase displays a higher and broader response to several metals (such as Cd, Fe(II), Zn, Cu, Hg, Pb, As(III)) than *L. cruciata* PC synthase [35].

Naturally hyperaccumulating plants do not overproduce PCs as a part of their mechanism against toxic metals. This appears to be an inducible rather than a constitutive mechanism, observed especially in metal non-tolerant plants [43]. Some reports have argued against the roles of PC in some metal-tolerant plants based on the effects of buthionine-S-sulphoximine and PCs/metal concentrations [44]. Several studies on plants overexpressing γ-glutamyl-cysteine synthetase or transgenic plants expressing bacterial γ-glutamyl-cysteine synthetase evaluated its effect on metal tolerance based on the assumption that higher levels of GSH and PCs will lead to more efficient metal sequestration [45]. *Bacopa monnieri*, a wetland macrophyte, is well known for its accumulation potential of metals and metal tolerance and thus is suitable for phytoremediation. Aquatic plants respond to metal stress by increasing the production of PC as well as other antioxidants. The accumulation potential of *B. monnieri* for various metals warrants its evaluation for metal tolerance and detoxification mechanism and for its suitability

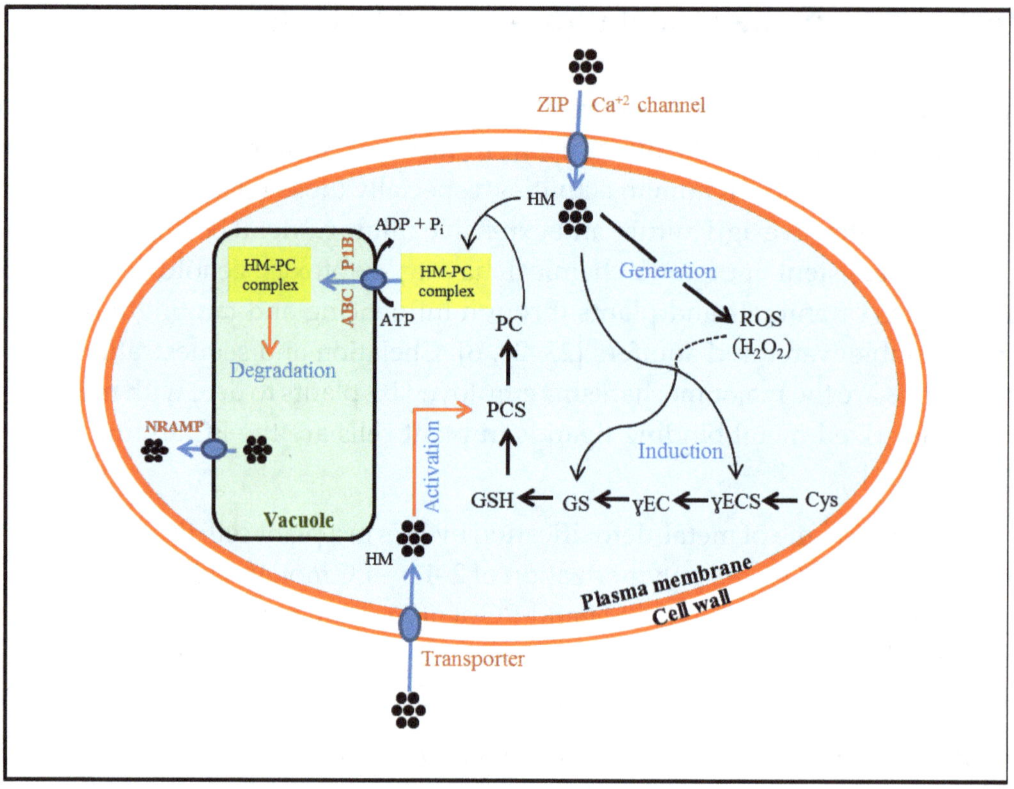

Figure 2. The scheme of heavy metal (HM) detoxification by phytochelatins (PC) in a plant cell. HM activates phytochelatin synthase (PCS) and the HM–PC complexes are established. These complexes are consequently transported through tonoplast to vacuole by ATP-binding-cassette and P1B-ATPase transporter (ABC-P1B). HM is chelated in the cytosol by ligands such as PC. Induction of PC synthesis by HM and a large flux of GSH is further achieved by increased activity of the GSH metabolic enzymes, γ-ECS and GS. It is possible that the enzyme activation is not directed through effects of HM but due to H2O2 produced as a result of HM-presence. Transport of HM through the plasma membrane (ZIP). Vacuolar transport of HM (NRAMP: natural resistance associated macrophage protein). Heavy metals are shown as black dots. Figure adapted and modified from [26].

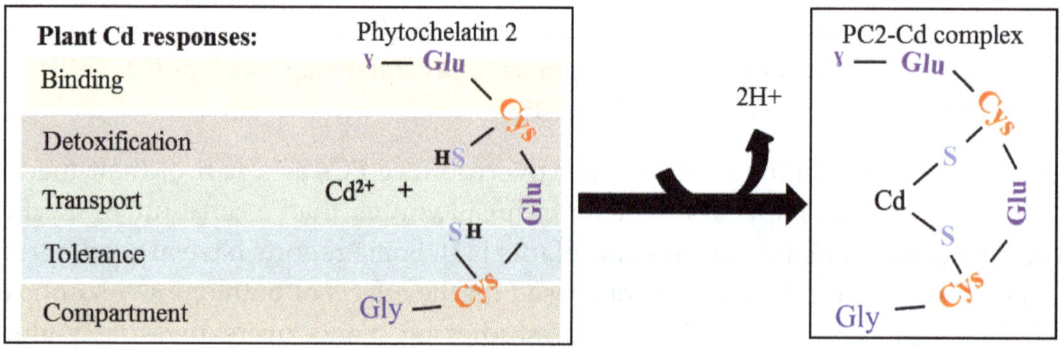

Figure 3. General functions of phytochelatins (PCs) and the model of complex between cadmium (Cd^{+2}) ion and one molecule of PC2. Cys, cysteine; Glu, glutamic acid; Gly, glycine; S, sulphur.

in phytoremediation [38]. In a study on *Arabidopsis thaliana* showed that Cd is immediately scavenged by thiols in root cells, in particular PC, at the expense of GSH. At the same time, a redox signal is suggested to be generated by a decreased GSH pool in combination with an altered GSH:GSSG ratio in order to increase the antioxidant capacity [46]. Overexpression of PCs synthetase in Arabidopsis led to 20–100 times more biomass on 250 and 300 μM arsenate than in the wild type. Also, the accumulation of thiol-peptides was 10 times higher after the exposure to Cd and arsenic, compared to the wild type. Gamma-glutamyl cysteine, which is a substrate for PC synthesis, increased rapidly after arsenate or Cd-exposure. Overexpression of PC synthase gene can be useful for phytoremediation [47]. Additionally, legumes are also capable of synthesizing homo-PCs in response to metal stress [45]. Citrus plants were also reported to synthesize PC in response to metal intoxication [48]. In wheat (*Triticum aestivum*), PC–metal complexes have been reported to accumulate in the vacuole. Retention of Cd in the root cell vacuoles might influence the symplastic radial Cd transport to the xylem and further transport to the shoot, resulting in genotypic differences in grain Cd accumulation [49].

3. Phytochelatins in metal/metalloid-exposed animals

As mentioned also above, PC proteins have been broadly described and characterized in plants, yeasts, algae, fungi and bacteria [22]. However, PC synthase genes are also present in animal species from several phyla. PC synthesis appears not to be transcriptionally regulated in animals [50]. Nevertheless, originally thought to be found only in plants and yeast, PC synthase genes have since been found in species that span almost the whole animal tree of life. Notably, PC synthase genes are found in species from several other metazoan phyla, including Annelida, Cnidaria, Echinodermata, Chordata and Mollusca (both Gastropoda and Bivalvia classes) [51, 52].

Several phyla of the Metazoa contain one or more species harbouring PC synthase homologous sequences: the Cnidaria (*Hydra magnipapillata*), the Chordata (*Molgula tectiformis*, as well the model chordate *Ciona intestinalis*), the Echinodermata (*Strongylocentrotus purpuratus*), the Annelida (*Lumbricus rubellus*) and the Platyhelminthes (*Schistosoma japonicum* and *Schistosoma mansoni*) [53, 51]. Biochemical studies have also shown that these PC synthase genes are functional. The *Caenorhabditis elegans* PC synthase produces PC when it is expressed in an appropriate host, and knocking out the gene increases the sensitivity of *C. elegans* to Cd [54]. Several studies have since measured PC by direct biochemical analysis of *C. elegans* tissue extracts, and found that Cd exposure did indeed increase PC levels in *C. elegans*. PC2, PC3 and PC4 have all been found, with PC2 in the highest concentration [55, 20, 56]. Therefore, these studies concluded that PCs production can play a major role in protecting *C. elegans* against Cd toxicity. PC2 and PC3 were increased in autochthonous *Lumbricus rubellus* populations sampled from contaminated sites [50]. The yeast (for example, *S. pombe*) possesses an ATP-binding cassette (ABC) transporter, Hmt1, which was originally thought to play a possible role in translocation of PCs–metal complexes to the vacuole. However, while knocking out the *C. elegans* HMT-1 (CeHMT-1) increases the sensitivity to Cd; the increase is greater than could be explained by a lack of PC synthase alone [57]. It is important to say that MTs are another widely established metal-binding ligand and a key metal detoxification system in animals. Additionally, MTs have many other important biological functions as well. Nevertheless, little is known about how MTs and PCs may complement each other for dealing with toxic metals [50].

The activation and function of PC synthase in animals came into light from studies on the nematode *C. elegans* [58], the flatworm *Schistosoma mansoni* [19, 59, 21], and Cionidae *Ciona intestinalis* [60]. The occurrence of PC synthase in animals suggests the occurrence, in these organisms, of a stress oxidative and metal detoxification system based on a class of molecules which was considered as the privilege of plants. The PC synthase gene has a wide phylogenetic distribution and can be found in species that cover almost all of the animal tree of life. But even though some members of particular taxonomic groups may contain PC synthase genes, there are also many species without these genes. Ron Elran et al. reported the regulation of GSH cycle genes in *Nematostella vectensis*, and an interesting finding was that PC synthase 1, which synthesizes the non-ribosomal formation of metal-binding PC, was upregulated after Hg and Cu treatments [15]. Phylogenetic analyses supported the hypothesis that PC synthase evolved independently in plants, cyanobacteria and green algae. Among the sequenced metazoan genomes, only a few contain a PC synthase gene. However, the reason for the scattered distribution of these genes remains unclear, considering that metazoans with PC synthase genes in their genomes do not share any physiological, behavioural or ecological features [60]. Just how (and if) PC in invertebrates complement the function of MTs remains to be elucidated, and the temporal, spatial and metal specificity of the two systems are still unknown [6].

4. Methods for the assays of phytochelatins and phytochelatin synthase enzyme

4.1. Determination of phytochelatins

We briefly discuss herein different methods for the detection and quantification of PC. Additionally, we are giving an overview of the methods used for determination of PC, comprising a broad range of electrochemical as well as spectrometric methods, which have been optimized and even hyphenated with different separation methods to detect PC. Recently, Wood et al. showed the analytical methodology for quantification of PC and their metal(loid) complexes [61]. The classical approach to the analysis of PC is reversed phase HPLC with post-column derivatization of the sulphydryl groups and spectrophotometric detection, but the detection is not specific to PC. The use of an analytical technique is able to detect compounds, specifically mass spectrometry. Independent studies showed a sensitive method for determining PCs by HPLC with fluorescence detection [62, 63]. A simple sensitive method for the identification, sequencing and quantitative determination of PCs in plants by electrospray tandem mass spectrometry (ESI MS-MS) was showed for different studies [64, 65]. Other study showed the combination of three processes for identification PC: (1) easy sample preparation including thiol reduction, (2) rapid and high-resolution separation using ultra-performance liquid chromatography (UPLC) and (3) specific and sensitive ESI-MS/MS detection using multi-reaction mode (MRM) transitions in alga's extract [66].

Nevertheless, in vitro formed Cd–PC2 complexes were characterized using ion exchange chromatography (IEC), flow injection analysis/high-performance liquid chromatography with CoulArray or Coulochem electrochemical detector and matrix-assisted laser desorption/ionization time-of-flight (MALDI-TOF) mass spectrometry [67, 68]. Zitka et al. optimized high-performance liquid chromatography coupled with electrochemical detector for determination of PC2 [69]. Many studies showed the determination of cysteine, reduced and oxidized glutathione and PC in different species of plants using high-performance liquid chromatography with electrochemical detection [70, 71].

4.2. Bioassays for phytochelatin synthase activity

The methods for identification and quantification of PC synthase are multidisciplinary, among themselves, comprising a broad range of molecular biology, electrochemical and spectrometric methods. HPLC coupled with electrochemical detector has been suggested as a new tool for the determination of PC synthase activity. The optimized procedure was subsequently used for studying PC synthase activity in the tobacco BY-2 cells treated with different concentrations of Cd(II) ions and the results were in good agreement with Nakazawa et al. [72]. Other study in animals showed that HPLC-LC system coupled to a single quadrupole LC–MS equipped with ESI was a sensitive method for PC synthase activity [22]. A highly sensitive assay for PC synthase activity was devised, where, the dequenching of Cu(I)-bathocuproinedisulphonate complexes was used in the detection system of a reversed-phase high-performance liquid chromatography. The present assay method is a sensitive tool that can be used to investigate this issue and would allow for the determination of PC synthase activity using 10–100-fold less protein [73]. Electrochemical methods such as differential pulse voltammetry and high-performance liquid chromatography with electrochemical detection were used for determination of Pt(IV) content, GSH levels, PC synthase activity in maize (*Zea mays*) and pea (*Pisum sativum*) plants treated with various doses of Pt(IV) [74].

Other methods required for the identification and characterization of PC synthase are, for example, the novel technology of molecular biology. Xu et al. showed a study that represents the first transcriptome-based analysis of miRNAs and their targets responsive to Cd stress in radish (*Raphanus sativus*) roots. Furthermore, a few target transcripts including PC synthase 1 (PCS1), iron transporter protein and ABC transporter protein were involved in plant response to Cd stress [75]. In 2009, Amaro et al. reported the identification and characterization of a cDNA encoding a PC synthase homologous sequence from the ciliated protozoan *T. thermophila,* the first to be described in ciliates. A quantitative real-time PCR (qRT-PCR) expression analysis of PC synthase has been carried out under different metal stress conditions. Several experimental evidences suggest that this enzyme is biosynthetically inactive in PC formation, which makes it the first pseudo-PC synthase to be described in eukaryotes [76].

5. Phytochelatins in connection with bionanoparticles

The connection of nanoparticles and PC has two faces: on one hand, the biosynthesis of nanoparticles and on the other hand, the protection of stress caused by the damage of any harmful nanoparticles. An *in vitro* study showed the enzyme-mediated synthesis of CdS nanocrystals by immobilized PC synthase, which converts GSH into the metal-binding peptide PC. Formation of CdS nanocrystals were observed upon the addition of CdCl2 and Na2S with PC as the capping agent [77]. This study is expected to help in designing a rational enzymatic strategy for the synthesis of nanoparticles of different chemical compositions, shapes and sizes. Also, an enzymatic synthesis route to peptide-capped gold nanoparticles was developed. Gold nanoparticles were synthesized using alpha-NADPH-dependent sulphite reductase and PC in vitro [78]. In Figure 4, we show the general structure of nanocrystal with cross-linked, PC-like coating (Figure modified from [79]). The microbiological production of inorganic nano-particles is an interesting and promising alternative to the known physical and chemical production methods. Extensive studies revealed the potential of bacteria, actinomycetes, algae, yeasts and fungi for biosynthesis of nanoparticles [80]. Few studies have discussed the possible synthesis of nanoparticles by algae. Particularly, *Phaeodactylum tricornutum* exposed to Cd, forms Cd–PC complexes, where sulphide ions (S^{2-}) can be incorporated to stabilize PC-coated CdS nanocrystallites [81, 82]. Metal is immobilized by an intracellular detoxification mechanism. Krumov et al. showed that Cd is associated to a protein fraction between 25 and 67 kDa which correspond to the theoretical molecular weight of CdS nanoparticles of 35 kDa coated with PC by size exclusion chromatography [83]. However, contingent to their types and concentrations, any nanoparticles can pose a risk to human health and to the environment [84]. Zinc oxide nanoparticles (ZnONPs) are used in large quantities by the cosmetic, food and textile industries. The harmful effects of ZnONPs are driven by their physicochemical properties and the resulting physical damage caused by the aggregation and agglomeration of nanoparticles. PC synthase may confer protection against ZnONPs-induced toxicity in *Caenorhabditis elegans* [24]. Effect of magnetic nanoparticles on tobacco BY-2 cell suspension culture showed induced PC biosynthesis. These trends were observable for almost all monitored PCs: PC2, PC3 and PC5 [85].

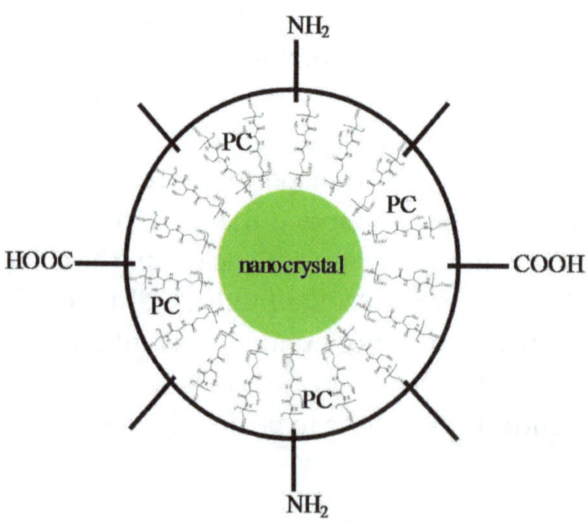

Figure 4. Nanocrystal with crosslinked, phytochelatin (PC) like coating, an effective strategy to make QDs as small with a crosslinked peptide sheath by mimicking PC-coated heavy metal nanoclusters. Figure adapted and modified from [79].

6. Conclusions

The concept of phytoremediation of contaminated soils has been increasingly supported by research in recent years. The identification of PC synthase genes from plants and other organisms is a significant breakthrough that will lead to a better understanding of the regulation of a critical step in PC biosynthesis. Many studies showed the mechanisms of chelation of metals–PC in plants in recent years. Chelation and sequestration of metals by particular ligands are also mechanisms used by plants to deal with metal stress. The two best-characterized metal-binding ligands in plant cells are the PCs and MTs. While the role played by PC synthase enzymes and PCs in animals still remains to be fully explored, there is increasing evidence that PC synthase genes are likely to be found in many important animal groups and that PCs may well turn out to be important players in metal ion detoxification in many of these species. It will be of interest in the future to see whether different animal species coordinate PC and MT responses to potentially toxic elements and if this is different for different metal ions.

Acknowledgements

Research of MAMR, ZH, OZ, VA and RK was financially supported by NANOSEMED KAN208130801. NAA and EP gratefully acknowledge the financial supports received from both FCT (Government of Portugal) through contract (SFRH/BPD/84671/2012) and from the Aveiro University Research Institute/CESAM (UID/AMB/50017/2013).

Author details

Miguel Angel Merlos Rodrigo[1,2], Naser A. Anjum[3], Zbynek Heger[1], Ondrej Zitka[1,2], Adam Vojtech[1,2], Eduarda Pereira[3] and Rene Kizek[1,2*]

*Address all correspondence to: kizek@sci.muni.cz

1 Department of Chemistry and Biochemistry, Laboratory of Metallomics and Nanotechnologies, Mendel University in Brno, Brno, Czech Republic

2 Central European Institute of Technology, Brno University of Technology, Brno, Czech Republic

3 CESAM-Centre for Environmental and Marine Studies & Department of Chemistry, University of Aveiro, Aveiro, Portugal

References

[1] Shi W-y, Shao H-b, Li H, Shao M-a, Du S. Progress in the remediation of hazardous heavy metal-polluted soils by natural zeolite. J Hazard Mater 2009;170:1–6.

[2] Wu G, Kang HB, Zhang XY, Shao HB, Chu LY, Ruan CJ. A critical review on the bio-removal of hazardous heavy metals from contaminated soils: issues, progress, eco-environmental concerns and opportunities. J Hazard Mater 2010;174:1–8.

[3] Vodyanitskii YN. Contamination of soils with heavy metals and metalloids and its ecological hazard (analytic review). Eurasian Soil Sci 2013;46:793–801.

[4] Sharma SS, Dietz K-J. The relationship between metal toxicity and cellular redox imbalance. Trend Plant Sci 2009;14:43–50.

[5] Yadav SK. Heavy metals toxicity in plants: an overview on the role of glutathione and phytochelatins in heavy metal stress tolerance of plants. S Afr J Bot 2010;76:167–79.

[6] Emamverdian A, Ding Y, Mokhberdoran F, Xie Y. Heavy metal stress and some mechanisms of plant defense response. Sci World J 2015;2015:1–18.

[7] Harris GK, Shi X. Signaling by carcinogenic metals and metal-induced reactive oxygen species. Mutat Res-Fundament Molecul Mechan Mutagen 2003;533:183–200.

[8] Valko M, Morris H, Cronin MTD. Metals, toxicity and oxidative stress. Curr Med Chem 2005;12:1161–208.

[9] Cobbett C, Goldsbrough P. Phytochelatins and metallothioneins: Roles in heavy metal detoxification and homeostasis. Ann Rev Plant Biol 2002;53:159–82.

[10] Tan W-N, Li Z-A, Zou B. Molecular mechanisms of plant tolerance to heavy metals. Zhiwu Shengtai Xuebao 2006;30:703–12.

[11] Rea PA. Phytochelatin synthase: of a protease a peptide polymerase made. Physiol Plant 2012;145:154–64.

[12] Mendoza-Cozatl DG, Butko E, Springer F, Torpey JW, Komives EA, Kehr J, Schroeder JI. Identification of high levels of phytochelatins, glutathione and cadmium in the phloem sap of Brassica napus. A role for thiol-peptides in the long-distance transport of cadmium and the effect of cadmium on iron translocation. Plant J 2008;54:249–59.

[13] Zenk MH. Heavy metal detoxification in higher plants - a review. Gene 1996;179:21–30.

[14] Rauser WE. Structure and function of metal chelators produced by plants - the case for organic acids, amino acids, phytin, and metallothioneins. Cell Biochem Biophys 1999;31:19–48.

[15] Pivato M, Fabrega-Prats M, Masi A. Low-molecular-weight thiols in plants: Functional and analytical implications. Arch Biochem Biophys 2014;560:83–99.

[16] Clemens S, Kim EJ, Neumann D, Schroeder JI. Tolerance to toxic metals by a gene family of phytochelatin synthases from plants and yeast. Embo J 1999;18:3325–33.

[17] Bolchi A, Ruotolo R, Marchini G, Vurro E, di Toppi LS, Kohler A, Tisserant E, Martin F, Ottonello S. Genome-wide inventory of metal homeostasis-related gene products including a functional phytochelatin synthase in the hypogeous mycorrhizal fungus Tuber melanosporum. Fungal Genet Biol 2011;48:573–84.

[18] Vatamaniuk OK, Bucher EA, Ward JT, Rea PA. A new pathway for heavy metal detoxification in animals - phytochelatin synthase is required for cadmium tolerance in Caenorhabditis elegans. J Biol Chem 2001;276:20817–20.

[19] Brulle F, Cocquerelle C, Wamalah AN, Morgan AJ, Kille P, Lepretre A, Vandenbuicke F. cDNA cloning and expression analysis of Eisenia fetida (Annelida : Oligochaeta) phytochelatin synthase under cadmium exposure. Ecotoxicol Environment Safety 2008;71:47–55.

[20] Schwartz MS, Benci JL, Selote DS, Sharma AK, Chen AGY, Dang H, Fares H, Vatamaniuk OK. Detoxification of multiple heavy metals by a half-molecule ABC transporter, HMT-1, and Coelomocytes of Caenorhabditis elegans. Plos One 2010;5.

[21] Ray D, Williams DL. Characterization of the phytochelatin synthase of Schistosoma mansoni. Plos Neglect Trop Dis 2011;5.

[22] Rigouin C, Vermeire JJ, Nylin E, Williams DL. Characterization of the phytochelatin synthase from the human parasitic nematode Ancylostoma ceylanicum. Molecul Biochem Parasitol 2013;191:1–6.

[23] Elran R, Raam M, Kraus R, Brekhman V, Sher N, Plaschkes I, Chalifa-Caspi V, Lotan T. Early and late response of Nematostella vectensis transcriptome to heavy metals. Molecul Ecol 2014;23:4722–36.

[24] Polak N, Read DS, Jurkschat K, Matzke M, Kelly FJ, Spurgeon DJ, Sturzenbaum SR. Metalloproteins and phytochelatin synthase may confer protection against zinc oxide nanoparticle induced toxicity in Caenorhabditis elegans. Compar Biochem Physiol C-Toxicol Pharmacol 2014;160:75–85.

[25] Yoon J, Cao X, Zhou Q, Ma LQ. Accumulation of Pb, Cu, and Zn in native plants growing on a contaminated Florida site. Sci Total Environ 2006;368:456–64.

[26] Anjum NA, Hasanuzzaman M, Hossein MA, Thangavel P, Roychoudhury A, Gill SS, Rodrigo MAM, Adam V, Fujita M, Kizek R, Duarte AC, Pereira E, Ahmed I. Jacks of metal/metalloid chelation trade in plants-an overview. Front Plant Sci 2015; 6:192. doi:10.3389/fpls. 2015.00192.

[27] Flora SJS, Mittal M, Mehta A. Heavy metal induced oxidative stress & its possible reversal by chelation therapy. Ind J Med Res 2008;128:501–23.

[28] Rascio N, Navari-Izzo F. Heavy metal hyperaccumulating plants: How and why do they do it? And what makes them so interesting? Plant Sci 2011;180:169–81.

[29] Raab A, Feldmann J, Meharg AA. The nature of arsenic-phytochelatin complexes in Holcus lanatus and Pteris cretica. Plant Physiol 2004;134:1113–22.

[30] Pagani MA, Tomas M, Carrillo J, Bofill R, Capdevila M, Atrian S, Andreo CS. The response of the different soybean metallothionein isoforms to cadmium intoxication. J Inorg Biochem 2012;117:306–15.

[31] Gupta DK, Huang HG, Corpas FJ. Lead tolerance in plants: strategies for phytoremediation. Environ Sci Pollut Res 2013;20:2150–61.

[32] Kim Y-O, Jung S, Kim K, Bae H-J. Role of pCeMT, a putative metallothionein from Colocasia esculenta, in response to metal stress. Plant Physiol Biochem 2013;64:25–32.

[33] Bashir H, Ibrahim MM, Bagheri R, Ahmad J, Arif IA, Baig MA, Qureshi MI. Influence of sulfur and cadmium on antioxidants, phytochelatins and growth in Indian mustard. AoB Plants 2015;7.

[34] Shahpiri A, Soleimanifard I, Asadollahi MA. Functional characterization of a type 3 metallothionein isoform (OsMTI-3a) from rice. Int J Biol Macromolecul 2015;73:154–9.

[35] Degola F, De Benedictis M, Petraglia A, Massimi A, Fattorini L, Sorbo S, Basile A, di Toppi LS. A Cd/Fe/Zn-responsive phytochelatin synthase is constitutively present in the ancient liverwort Lunularia cruciata (L.) Dumort. Plant Cell Physiol 2014;55:1884–91.

[36] Fischer S, Kuehnlenz T, Thieme M, Schmidt H, Clemens S. Analysis of plant Pb tolerance at realistic submicromolar concentrations demonstrates the role of phytochelatin synthesis for Pb detoxification. Environ Sci Technol 2014;48:7552–9.

[37] Garcia JD, Mendoza-Cozatl DG, Moreno-Sanchez R. An uncommon phytochelatin synthase gives hints on how to improve their catalytic efficiency on heavy metal hyperaccumulator organisms. Protein Sci 2014;23:185–6.

[38] Kuhnlenz T, Schmidt H, Uraguchi S, Clemens S. Arabidopsis thaliana phytochelatin synthase 2 is constitutively active in vivo and can rescue the growth defect of the PCS1-deficient cad1-3 mutant on Cd-contaminated soil. J Experiment Bot 2014;65:4241–53.

[39] Castro AV, de Almeida A-AF, Pirovani CP, Reis GSM, Almeida NM, Mangabeira PAO. Morphological, biochemical, molecular and ultrastructural changes induced by Cd toxicity in seedlings of Theobroma cacao L. Ecotoxicol Environment Safety 2015;115:174–86.

[40] Hazama K, Nagata S, Fujimori T, Yanagisawa S, Yoneyama T. Concentrations of metals and potential metal-binding compounds and speciation of Cd, Zn and Cu in

phloem and xylem saps from castor bean plants (Ricinus communis) treated with four levels of cadmium. Physiol Plant 2015;154:243–55.

[41] Lee BD, Hwang S. Tobacco phytochelatin synthase (NtPCS1) plays important roles in cadmium and arsenic tolerance and in early plant development in tobacco. Plant Biotechnol Rep 2015;9:107–14.

[42] Gonzalez-Mendoza D, Moreno AQ, Zapata-Perez O. Coordinated responses of phytochelatin synthase and metallothionein genes in black mangrove, Avicennia germinans, exposed to cadmium and copper. Aquatic Toxicol 2007;83:306–14.

[43] Freeman J, Gustin J, Salt D. Constitutively elevated salicylic acid signals glutathione mediated Ni tolerance in Thlaspi Ni hyperaccumulators. Plant Biol (Rockville) 2005;2005:111–2.

[44] Schat H, Llugany M, Vooijs R, Hartley-Whitaker J, Bleeker PM. The role of phytochelatins in constitutive and adaptive heavy metal tolerances in hyperaccumulator and non-hyperaccumulator metallophytes. J Experiment Bot 2002;53:2381–92.

[45] Zagorchev L, Seal CE, Kranner I, Odjakova M. A central role for thiols in plant tolerance to abiotic stress. Int J Molecul Sci 2013;14:7405–32.

[46] Jozefczak M, Keunen E, Schat H, Bliek M, Hernandez LE, Carleer R, Remans T, Bohler S, Vangronsveld J, Cuypers A. Differential response of Arabidopsis leaves and roots to cadmium: glutathione-related chelating capacity vs antioxidant capacity. Plant Physiol Biochem 2014;83:1–9.

[47] Li YJ, Dhankher OP, Carreira L, Lee D, Chen A, Schroeder JI, Balish RS, Meagher RB. Overexpression of phytochelatin synthase in Arabidopsis leads to enhanced arsenic tolerance and cadmium hypersensitivity (vol 45, pg 1787, 2004). Plant Cell Physiol 2005;46:387.

[48] Lopez-Climent MF, Arbona V, Perez-Clemente RM, Zandalinas SI, Gomez-Cadenas A. Effect of cadmium and calcium treatments on phytochelatin and glutathione levels in citrus plants. Plant Biol 2014;16:79–87.

[49] Stolt JP, Sneller FEC, Bryngelsson T, Lundborg T, Schat H. Phytochelatin and cadmium accumulation in wheat. Environment Experiment Bot 2003;49:21–8.

[50] Liebeke M, Garcia-Perez I, Anderson CJ, Lawlor AJ, Bennett MH, Morris CA, Kille P, Svendsen C, Spurgeon DJ, Bundy JG. Earthworms produce phytochelatins in response to arsenic. Plos One 2013;8.

[51] Clemens S. Evolution and function of phytochelatin synthases. J Plant Physiol 2006;163:319–32.

[52] Clemens S, Persoh D. Multi-tasking phytochelatin synthases. Plant Sci 2009;177:266–71.

[53] Vatamaniuk OK, Bucher EA, Ward JT, Rea PA. Worms take the 'phyto' out of 'phytochelatins'. Trend Biotechnol 2002;20:61–4.

[54] Bundy JG, Kille P. Metabolites and metals in Metazoa - what role do phytochelatins play in animals? Metallomics 2014;6:1576–82.

[55] Cui Y, McBride SJ, Boyd WA, Alper S, Freedman JH. Toxicogenomic analysis of Caenorhabditis elegans reveals novel genes and pathways involved in the resistance to cadmium toxicity. Genome Biol 2007;8.

[56] Hall J, Haas KL, Freedman JH. Role of MTL-1, MTL-2, and CDR-1 in mediating cadmium sensitivity in Caenorhabditis elegans. Toxicol Sci 2012;128:418–26.

[57] Vatamaniuk OK, Bucher EA, Sundaram MV, Rea PA. CeHMT-1, a putative phytochelatin transporter, is required for cadmium tolerance in Caenorhabditis elegans. J Biol Chem 2005;280:23684–90.

[58] Hughes SL, Bundy JG, Want EJ, Kille P, Sturzenbaum SR. The metabolomic responses of Caenorhabditis elegans to cadmium are largely independent of metallothionein status, but dominated by changes in cystathionine and phytochelatins. J Proteome Res 2009;8:3512–9.

[59] Bernard F, Brulle F, Douay F, Lemiere S, Demuynck S, Vandenbulcke F. Metallic trace element body burdens and gene expression analysis of biomarker candidates in Eisenia fetida, using an 'exposure/depuration' experimental scheme with field soils. Ecotoxicol Environ Safety 2010;73:1034–45.

[60] Franchi N, Piccinni E, Ferro D, Basso G, Spolaore B, Santovito G, Ballarin L. Characterization and transcription studies of a phytochelatin synthase gene from the solitary tunicate Ciona intestinalis exposed to cadmium. Aquatic Toxicol 2014;152:47–56.

[61] Wood BA, Feldmann J. Quantification of phytochelatins and their metal(loid) complexes: critical assessment of current analytical methodology. Anal Bioanal Chem 2012;402:3299–309.

[62] Kawakami SK, Gledhill M, Achterberg EP. Determination of phytochelatins and glutathione in phytoplankton from natural waters using HPLC with fluorescence detection. Trac-Trend Anal Chem 2006;25:133–142.

[63] Ju XH, Tang SR, Jia Y, Guo JK, Ding YZ, Song ZG, Zhao YJ. Determination and characterization of cysteine, glutathione and phytochelatins (PC2-6) in Lolium perenne L. exposed to Cd stress under ambient and elevated carbon dioxide using HPLC with fluorescence detection. J Chromatography B-Anal Technol Biomed Life Sci 2011;879:1717–24.

[64] Vacchina V, Chassaigne H, Oven M, Zenk MH, Lobinski R. Characterisation and determination of phytochelatins in plant extracts by electrospray tandem mass spectrometry. Analyst 1999;124:1425–30.

[65] Baralkiewicz D, Kozka M, Piechalak A, Tomaszewska B, Sobczak P. Determination of cadmium and lead species and phytochelatins in pea (Pisum sativum) by HPLC-ICP-MS and HPLC-ESI-MSn. Talanta 2009;79:493–8.

[66] Braeutigam A, Wesenberg D, Preud'homme H, Schaumloeffel D. Rapid and simple UPLC-MS/MS method for precise phytochelatin quantification in alga extracts. Anal Bioanal Chem 2010;398:877–83.

[67] Rodrigo MAM, Cernei N, Kominkova M, Zitka O, Beklova M, Zehnalek J, Kizek R, Adam V. Ion exchange chromatography and mass spectrometric methods for analysis of cadmium-phytochelatin (II) complexes. Int J Environ Res Public Health 2013;10:1304–11.

[68] Rodrigo MAM, Zitka O, Kominkova M, Adam V, Beklova M, Kizek R. Analysis of cadmium-phytochelatins 2 complexes using flow injection analysis coupled with electrochemical detection mass spectrometry. Int J Electrochem Sci 2013;8:4409–21.

[69] Zitka O, Skutkova H, Krystofova O, Sobrova P, Adam V, Zehnalek J, Havel L, Beklova M, Hubalek J, Provaznik I, Kizek R. Rapid and ultrasensitive method for determination of phytochelatin(2) using high performance liquid chromatography with electrochemical detection. Int J Electrochem Sci 2011;6:1367–81.

[70] Zitka O, Stejskal K, Kleckerova A, Adam V, Beklova M, Horna A, Supalkova V, Havel L, Kizek R. Utilizing electrochemical techniques for detection of biological samples. Chemicke Listy 2007;101:225–31.

[71] Skladanka J, Adam V, Zitka O, Krystofova O, Beklova M, Kizek R, Havlicek Z, Slama P, Nawrath A. Investigation into the effect of molds in grasses on their content of low molecular mass thiols. Int J Environ Res Public Health 2012;9:3789–805.

[72] Nakazawa R, Kato H, Kameda Y, Takenaga H. Optimum assay conditions of the activity of phytochelatin synthase from tobacco cells. Biol Plant 2002;45:311–3.

[73] Ogawa S, Yoshidomi T, Shirabe T, Yoshimura E. HPLC method for the determination of phytochelatin synthase activity specific for soft metal ion chelators. J Inorg Biochem 2010;104:442–5.

[74] Mikulaskova H, Merlos MAR, Zitka O, Kominkova M, Hynek D, Adam V, Beklova M, Kizek R. Employment of electrochemical methods for assessment of the maize (Zea mays L.) and pea (Pisum sativum L.) response to treatment with platinum(IV). Int J Electrochem Sci 2013;8:4505–19.

[75] Xu L, Wang Y, Zhai LL, Xu YY, Wang LJ, Zhu XW, Gong YQ, Yu RG, Limera C, Liu LW. Genome-wide identification and characterization of cadmium-responsive microRNAs and their target genes in radish (Raphanus sativus L.) roots. J Experiment Bot 2013;64:4271–87.

[76] Amaro F, Ruotolo R, Martin-Gonzalez A, Faccini A, Ottonello S, Gutierrez JC. A pseudo-phytochelatin synthase in the ciliated protozoan Tetrahymena thermophila. Compar Biochem Physiol C-Toxicol Pharmacol 2009;149:598–604.

[77] Liu F, Kang SH, Lee Y-I, Choa Y-h, Mulchandani A, Myung NV, Chen W. Enzyme mediated synthesis of phytochelatin-capped CdS nanocrystals. Appl Phys Lett 2010;97.

[78] Kumar SA, Abyaneh MK, Gosavi SW, Kulkarni SK, Ahmad A, Khan MI. Sulfite reductase-mediated synthesis of gold nanoparticles capped with phytochelatin. Biotechnol Appl Biochem 2007;47:191–5.

[79] Zheng Y, Yang Z, Li Y, Ying JY. From glutathione capping to a crosslinked, phytochelatin-like coating of quantum dots. Adv Mater 2008;20:3410–5.

[80] Krumov N, Perner-Nochta I, Oder S, Gotchev V, Angelov A, Posten C. Production of inorganic nanoparticles by microorganisms. Chem Eng Technol 2009;32:1026–35.

[81] Morelli E, Cruz BH, Somovigo S, Scarano G. Speciation of cadmium - gamma-glutamyl peptides complexes in cells of the marine microalga Phaeodactylum tricornutum. Plant Sci 2002;163:807–13.

[82] Gioacchino Scarano EM. Properties of phytochelatin-coated CdS nanochrystallites formed in a marine phytoplanktonic alga. Plant Sci 2003;165: 803–10.

[83] Krumov N, Oder S, Perner-Nochta I, Angelov A, Posten C. Accumulation of CdS nanoparticles by yeasts in a fed-batch bioprocess. J Biotechnol 2007;132:481–86.

[84] Nowack B, Brouwer C, Geertsma RE, Heugens EHW, Ross BL, Toufektsian M-C, Wijnhoven SWP, Aitken RJ. Analysis of the occupational, consumer and environmental exposure to engineered nanomaterials used in 10 technology sectors. Nanotoxicology 2013;7:1152–6.

[85] Krystofova O, Sochor J, Zitka O, Babula P, Kudrle V, Adam V, Kizek R. Effect of magnetic nanoparticles on tobacco BY-2 cell suspension culture. Int J Environ Res Public Health 2013;10:47–71.

6

Cowpea Breeding for Drought Tolerance — From Brazil to World

Maria Antonia Machado Barbosa, Allan Klynger da Silva Lobato, Milton Hélio Lima da Silva, Gabriel Mascarenhas Maciel and Douglas José Marques

Abstract

The aims of this chapter are to (i) present the importance of the cowpea crop, (ii) demonstrate problems related to drought, (iii) describe aspects related to flower structure and hybridization, and (iv) reveal how the genotype selection with tolerance to water deficit will promote increase of the yield in this culture. This chapter describes that Nigeria, Niger, and Brazil are the leading producers of cowpea crop worldwide, and this crop has a large influence on the economy of these countries. However, the drought problem can be frequently observed in areas with agricultural potential, with negative effects on production components and a consequent decrease of the yield in this culture. Breeding programs aimed at drought tolerance using selection strategies linked to genetic, biochemical, physiological, nutritional, and agronomic characteristics can help increase the yield and reduce losses promoted by the drought. In addition, flower structure and hybridization technique used in Núcleo de Pesquisa Vegetal Básica e Aplicada (NPVBA/UFRA) are presented, as well as populations are evaluated and plant management are explained in detail. This chapter describes the results obtained in other breeding programs aimed at drought tolerance and also explains the potential uses to increase the crop yield.

Keywords: Vigna unguiculata, production, drought, breeding

1. Introduction

The cowpea [*Vigna unguiculata* (L.) Walp.] is cultivated widely in several countries due to its social, economic, and nutritional importance [1]. The aim of growing cowpea crop in such a huge quantity is to market its grains. These species are abundant in the regions of Latin

America, Africa, and Asia, while Nigeria, Niger, and Brazil are emerging as the leading producers of cowpea [2].

In Brazil, the production of cowpea is concentrated in the North and Northeast regions, because it presents interesting metabolic characteristics, such as low nutrient requirements, less water consumption, and well adapted to tropical environments, compared to other legumes [3-4]. Until the 1980s, the cowpea was almost exclusively grown by small and medium farmers. Currently, this culture seems to occupy other agricultural scenarios in areas with high irrigation potential, or/and also during the off-season; next only to the soybean crop, cowpea crops are being exploited by large producers by adopting new technologies [5-6].

The plants of various species under field conditions are constantly exposed to abiotic or biotic stresses, and suffer interactions that can significantly affect their performance [7], including the culture of cowpea [8]. Lawlor [9] points out that the main limiting abiotic factors of productivity in various plant species are drought, flooding, low and high temperatures, salinity, excess radiation, toxic heavy metals, and excessive macro- or micronutrients [10]. Drought is considered a major cause of the reduction of global agricultural production [11-12].

Drought is a frequently observed event and recurrent in areas with agricultural potential, mainly in tropical regions, causing an increase in temperature and decrease in the relative humidity [13]. This stress can be caused by irregularities in the distribution of rainfall [14-15] and/or inappropriate supplement irrigation for crops [16-17].

Thus, the water deficit acts as a limiting factor for both quality and production in several species with agronomic potential [18-20], including the cowpea [21]. The occurrence of drought during the growing and development and reproductive and maturity stages usually results in lower growth rates and development [22], flower abortion [23], and reduced grain production (GP) [24], thus complicating the reproduction process in the cowpea crop.

The decrease in the growth and development in cowpea plants observed during vegetative stages is related to negative interference periods of drought, which is caused by the lower weight of the aerial part of the plant [25], reduction in the expansion rate of leaf area [26], and severe reductions in gas exchange [27].

Drought induces molecular, physiological, biochemical, and morphological changes in cowpea plants [28-31], which are considered as adaptation strategies and survival mechanisms of the species [32]. Another problem normally connected to drought is the rise in temperature in the plant tissues due to reduced transpiration rate and consequent loss of effectiveness of thermoregulatory mechanism [33, 34]. This is due to water stress, as water is an essential element for the growth and development of all species of plants and plays an important role in photosynthesis, transport of organic solutes, and temperature control [35].

The increase in the rate of growth of the root system is a possibility to overcome the water stress in plants [36]; for, under such conditions, the stimulation of root growth in depth and the wetter areas of the soil profile [18] adaptations can check and control drought [26]. The root system of common bean often lies in surface when water availability is adequate [37]. Guimarães et al. [38] observed that the genotypes more tolerant to water stress had more developed root systems compared with the susceptible genotypes. Additionally, Peña-Valdivia et al. [39], investigating common bean plants, describe negative changes related to the anatomy of the root, such as reductions in the xylem vessel and cross-sectional area of the root, in domesticated plants, while wild plants remained unchanged in structures and showed greater tolerance to drought.

The osmotic adjustment is another mechanism that can directly contribute to water retention in the plant tissue [40], for increasing the tolerance to dehydration [34]. This adjustment is accomplished by substances called osmoregulators such as carbohydrates and amino acids [41-42]. The accumulation of metabolites of plants provides a decrease in osmotic potential (ψ_s) via a net increase in intracellular solute [43], which induces higher water retention in the tissue [44], and thus can control the plant cell elongation and expansion in growth regions [45].

The type of substances used in osmotic adjustment depends on the species and tissue [46]. However, the degree of adjustment depends on the speed at which the water deficit is applied, and osmotic adjustment is directly proportional to the intensity of water deficit [47]. The proline and glycine betaine are usually amino acids that act as osmotic adjusters in cowpea plants exposed to water deficit [48-49]. In water stress conditions, an increase in the synthesis of abscisic acid (ABA) in the roots is also observed, which is then transported to the shoot via the xylem [50].

In cowpea plants, overexpression of genes *CPRD 8*, *CPRD12*, *CPRD14*, *CPRD22*, and *CPRD46* conferring tolerance to water stress [28], as well as *Vuc*APX genes, *Vup*APX, and *Vut*APX-connected *Vus*APX production of antioxidant enzymes [51], are reported, besides the high-level expression of *PvP5CS* gene associated with the production of proline, an amino acid that holds the osmotic adjusting function across species during drought.

The responses of cultivars and lines in cowpea are different in relation to water stress during the cycle [52]. In addition, there are factors such as the intensity and duration of drought that can intensify such stress [53-55].

The culture of cowpea requires basic and applied research related to the impact caused by water deficit, due to frequent and severe loss of production related to abiotic stress [56-58] and the high economic and nutritional importance to the producers and grain-consuming countries, such as Brazil [59-60].

The selection of genotypes of tolerance to drought has been carried out in several breeding programs [61-62]. However, the large number of genes involved in tolerance to drought [63], combined with the influence of the environment, hinders the selection of plants in segregating generations [53,57] and affects the evaluation lines/cultivars by virtue of the interaction, provided genotype environment often is significant [64].

The crosses between contrasting parenting have been widely used and allow to investigate the genetic control of quantitative traits such as drought tolerance [15,54]. Therefore, the characteristics described by Bastos et al. [8] are present in cultivars of cowpea, BRS Paraguaçu (drought tolerance) and Tracuateua-192 (sensitivity to drought), and are suitable for the purpose of this research and justify the hybridization between these cultivars, which are contrasting in relation to tolerance to drought.

Adequate selection methods used in breeding programs of a species can reduce time and investment and maintenance of evaluating plants that could be previously discarded [65]. Thus, selection strategies linked to genetic, biochemical, physiological, nutritional, and agronomic characteristics can help due to the high detection efficiency and low cost [66]. In addition, the knowledge of the genetic control related to these characteristics will establish breeding strategies and selection of agronomically superior plants, with genes of tolerance to drought.

2. Objectives

The aims of this chapter are to (i) present the importance of the cowpea crop, (ii) demonstrate problems related to drought, (iii) describe aspects related to flower structure and hybridization, and (iv) reveal how the genotype selection with tolerance to water deficit will promote increase of the yield in this culture.

3. Flower structure in cowpea

In relation to flowers, the cowpea bean flower is hermaphrodite, deciduous, cyclic, dichlamydeous, and heterochlamydeous with zygomorphic symmetry (Figure 1). The five sepals and petals present are in free condition, denominated polysepalous and polypetalous, respectively [67].

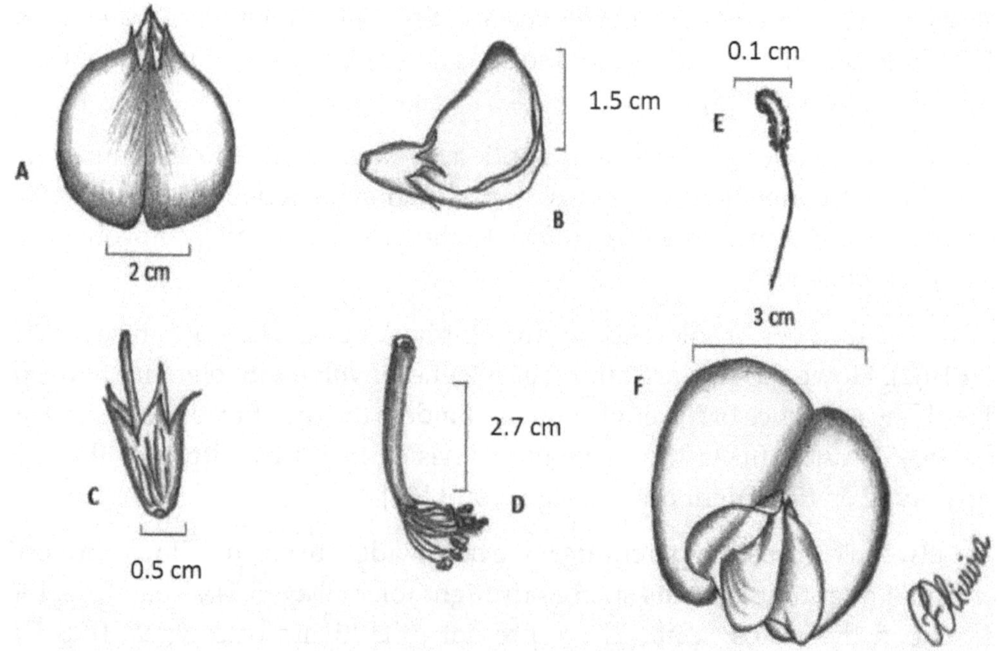

Figure 1. *V. unguiculata* floral pieces. (A) standard; (B) post-anthesis flower structure; (C) sepals; (D) stamen and pistil; (E) longer stamen; (F) keel formation [67].

Their corolla is papilionaceous; that is, it has an upper petal called standard, two side petals called wings and two lower, inner petals jointly called keel (Figure 1). One of the stamens is longer than the other nine (heterostemonous). Stamens are free, being diadelphous, with simple ramification; the anthers are free and basifixed, enclosed in relation to the corolla, longitudinally dehiscent and introrsed [67]. They are multicarpellary, syncarpous, with insertion in the terminal style. The hilum, chalaza, and micropyle are in the same straight line; that is, their ovule is classified as orthotropous [67].

4. Aspects linked to artificial hybridization

The results described by Ribeiro et al. on aspects of the floral biology of *Vigna unguiculata* reveals that anthesis occurred between 05:00 and 05:30 h [67]. In pre-anthesis, at 4:00 h, the greenish flowers remained. In the stigma receptivity test, peroxidase action was observed at all times. As noted by Rocha et al., the floral opening of cowpea bean flowers begins around 05:30 h and continues until 09:30 h, when the stigma is still receptive (Figure 2) [68].

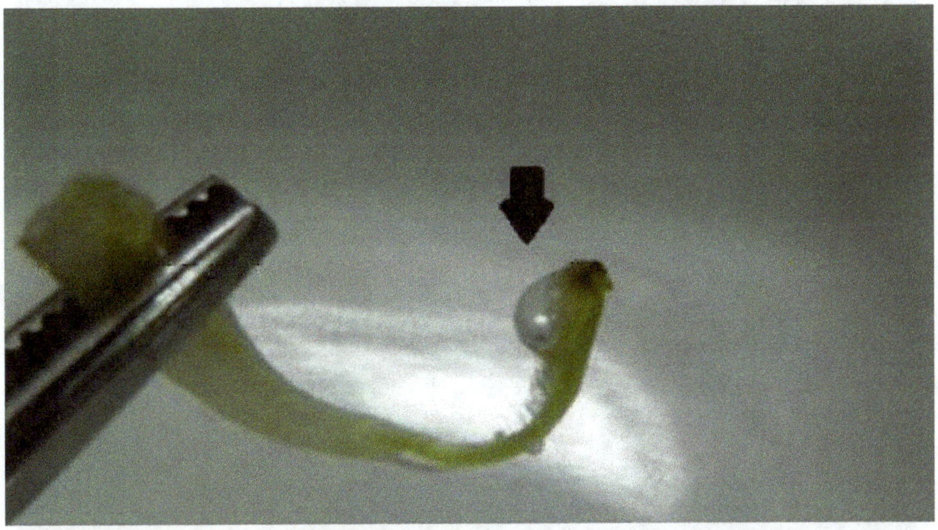

Figure 2. Peroxidase activity on *V. unguiculata* stigma, demonstrating receptiveness [67].

5. Adaptations to hybridization in Brazilian conditions

The method of artificial hybridization described by Rachie et al. [69], as well as Zary and Miller Jr [70], was adjusted to conditions of infrastructure from Núcleo de Pesquisa Vegetal Básica e Aplicada (NPVBA) of the Universidade Federal Rural da Amazônia (UFRA) and climatic conditions of Northern Brazil (Figure 3), which used flower pollen collected in the morning (between 06:00 and 08:00 h) and stored in the refrigerator until use.

In the evening, the flower buds were emasculated and pollinated (16:30 and 17:30 h) (Figure 3). This method provided a higher percentage of successful pollination. This result is in agreement with the Zary and Miller Jr [70] method, described by Teófilo et al. [71], who reported that the success of this method could be because the surface of the emasculated flowers are more receptive to pollen grains in the late afternoon, since, in this period, the temperature and moisture conditions are more appropriate.

Figure 3. Greenish flowers with flower from left in pre-anthesis stage (A); cut of standard in medium region (B); removal of the standard (C); removal of the anthers (D); application of pollen viable (E); successful pollination and pod formation (F).

6. Populations evaluated and plant management

After hybridizations (Figure 4), the hybrid seeds were multiplied to obtain the F_2, BC_1, and BC_2 generations. In each pot, only one seed was placed. In this study, containers with capacity of 3 L containing holes in bottom were used, aiming to drain the excess water. The substrate was composed of a mixture of Plantmax® and sand in 3:1 ratio (v/v). Fertilization was performed according to exigencies of cowpea crop and previous substrate analysis, with the fertilization applied at regular intervals of 15 days until the 45th day after implementation experiment.

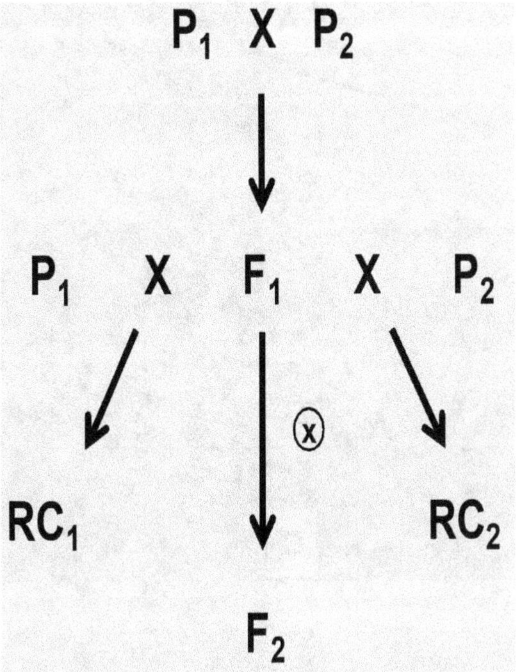

Figure 4. Scheme to obtain six populations (P_1, P_2, F_1, F_2, BC_1, and BC_2) from crosses between LP 97-28 (low tolerance to drought) × IPR-Uirapuru (high tolerance to drought).

7. Irrigation and water-deficit application

All plants were irrigated every day for 15 minutes at 10:00 and 13:00 h. The irrigation within the greenhouse was performed by a microsprinkler system, with a flow rate of 10 L h^{-1}. The three cycles of moderate stress were induced by irrigation suspension for four days at 25, 35, and 45 days after seedling emergence, whose periods coincided with the phenological stages V_3, R_6, and R_8, respectively. In the cowpea crop, these periods described are related to vegetative, flowering, and pod filling stages, respectively (Figure 5).

Figure 5. Seed placed into substrate (A); *V. unguiculata* seedling (B); plants with 14 days (C); trifoliate leaf (D); data obtained during experiment (E); infra-red gas analyzer (F).

8. Results obtained in breeding programs aimed at drought tolerance

The study was conducted by Lobato et al. [72] with six populations, with two parents P_1 and P_2, and F_1, F_2, BC_1, and BC_2 generations, derived from the cross between LP 97-28 (low tolerance to drought) × IPR-Uirapuru (high tolerance to drought). Regarding production components, the grain production (GP) results reveal that the evaluated plants had values between 0.01 and 9.78 g. The low and high means for all populations were 2.30 and 6.86 g, respectively (Table 1). The F_2 generation showed the best performance. The low and high values for variance were obtained in the parental P_1 (1.51) and F_2 (8.88) generations, respectively. In relation to the average weight of 100 seeds (W100s), the values ranged between 12.56 and 29.64 g. In addition, the low and high means were 18.49 and 21.83 g, with the greatest means observed in the F_1 and P_2 populations (Table 1). The greatest variance of 11.12 was obtained in the F_2 generation. For the number of pods per plant (NPP), the plants studied had values between 1 and 19, and the lowest and highest means were 2.70 and 7.02, respectively (Table 1). The best result was found in the F_2 generation, while the lowest value was observed in the BC_1 generation. The highest variance of 8.08 was observed in the F_2 generation (Table 1). For the number of seeds per pod (NSP), the plants collected in this study had values between 1.0 and 6.8. The low and high means were 4.17 and 4.64 in populations P_1 and BC_2, respectively (Table 1). Additionally, the best result was found in generation BC_2. The low and high variances were 0.38 and 1.54 (Table 1) and were obtained in the P_2 and F_2 generations, respectively [72].

The estimated means of the segregating generations and the existence of high genetic variation (σ^2) coupled with additivity indicated the presence of transgressive individuals. These findings enabled the selection of promising genotypes for drought tolerances higher than those of IPR-Uirapuru and LP 97-28, the parent lines in this study. In terms of the estimated variances in the study populations, the best performances were observed for the segregating generations (F_2, BC_1, and BC_2), which demonstrated higher values for all traits compared to the parent (P_1 and P_2) and F_1 generations (Table 1). These results can be attributed to the large segregation of genes and, consequently, the higher amplitude in the distribution of the drought stress tolerance values, indicating genetic variation for the evaluated traits [73]. Similar results to those found in this study in terms of the GP of the F_2 generation were reported by Szilagyi [74] for experiments with the common bean grown under adequate conditions (irrigation) and drought stress. This author studied production components in six populations derived from crosses between F332 and Ardeleana.

Higher variances for GP, W100s, NPP, and NSP were observed in the F_2 generation, revealing greater plant heterogeneity and suggesting great variability within this population. Genetic variability is extensively explored in breeding programs; it serves as the basis for selection and provides opportunities to establish a desired characteristic [75]. Smaller variances were obtained in the parents, confirming homozygosis in these populations due to the line and cultivar.

Regarding genetic control, the values of the phenotypic (σ_p^2), environmental (σ_e^2), genotypic (σ_g^2), additive (σ_a^2), and dominance variances (σ_d^2) for grain production were 8.88, 2.97, 5.91, 5.75, and 0.16, respectively (Table 2). The genotypic variance corresponded to 66.6% of the phenotypic variance (total), and the additive variance accounted for 97.3% of the genetic variance. In W100s, the phenotypic, environmental, genotypic, additive, and dominance variances were 11.12, 2.55, 8.57, 7.66, and 0.91, respectively (Table 2). For this characteristic, the genotypic variance accounted for 77.1% of the total variance, while the additive variance corresponded to 89.4% of the existing genetic variance. In relation to NPP, the phenotypic, environmental, genotypic, additive, and dominance variances were 8.08, 2.33, 5.75, 4.73, and 1.02, respectively (Table 2). The phenotypic variance accounted for 71.2% of the genetic variance. Additionally, the additive variance corresponded to 82.3% of the genotypic variance. For the NSP, the phenotypic, environmental, genotypic, additive, and dominance variances were 1.54, 0.56, 0.98, 0.76, and 0.22, respectively (Table 2). The genotypic variance represented 63.7% of the phenotypic variance in this characteristic. The additive and dominance variances contributed to 77.6 and 22.4% of the genotypic variance, respectively [72].

The results indicate high contributions of additive variances in relation to genotypic variance and intense additive allelic interactions on all the evaluated traits. The existence of high additive variance suggests the identification of superior genotypes [76]. Typically, breeding methods that take advantage of high additive variance to obtain genetic gains are more important for the improvement of autogamous species, such as *Phaseolus vulgaris* [73].

According to this research, the use of additive variance is recommended as an indicator when studying GP, W100s, NPP, and NSP in the cross (LP 97-28 × IPR-Uirapuru), because it accounts for a significant portion of genotypic variance.

The estimates of broad-sense heritability (H^2 %) ranged between 63.6 and 77.0% (Table 2), and the high and low values were found in the W100s and NSP characteristics, respectively. The estimates of narrow-sense heritability (h^2 %) oscillated between 49.2 and 68.9% (Table 2), and the high and low values also corresponded to the W100s and NSP characteristics.

The average degree of dominance (add) values were 0.22, 0.48, 0.65, and 0.76 for the GP, W100s, NPP, and NSP characteristics, respectively (Table 2). The minimum number of genes (mng) that controlled the GP, W100s, NPP, and NSP characteristics were 4.7, 4.4, 8.6, and 5.5, respectively (Table 2).

The results related to broad- and narrow-sense heritabilities described in this study are high because studies involving populations are normally conducted under field conditions and high levels of environmental interference reduce genetic variances and produce lower heritabilities. Higher heritability coefficients may be caused by greater additive genetic variance, lower environmental variance, or minor interactions between genotype and environment [77]. Additionally, similar results for broad- and narrow-sense heritabilities indicate that the dominance effect is null. However, if the broad-sense heritability is higher than the narrow-sense heritability, the dominance effect is present [78].

LP 97-28×IPR-Uirapuru		GP(g)		W100s(g)		NPP		NSP	
	n	m	σ^2	m	σ^2	m	σ^2	m	σ^2
LP 97-28 (P_1)	16	4.60	1.51	18.49	1.19	5.69	1.29	4.17	0.68
IPR-Uirapuru (P_2)	18	4.48	1.89	21.83	1.37	4.39	2.13	4.60	0.38
F_1	7	4.19	4.25	21.83	3.18	4.43	2.95	4.29	0.59
F_2	166	6.86	8.88	21.29	11.12	7.02	8.08	4.16	1.54
BC_1	10	2.30	4.92	19.29	6.44	2.70	5.34	4.54	1.43
BC_2	39	3.31	7.07	20.52	8.13	3.44	6.09	4.64	0.88

Table 1. Number of evaluated plants (*n*), means (*m*), and variances (σ^2) from grain production (GP), average weight of 100 seeds (W100s), number of pod per plant (NPP), and number of seeds per plant (NSP) obtained in six populations (P_1, P_2, F_1, F_2, BC_1, and BC_2), derived from cross between LP 97-28 × IPR-Uirapuru, Maringá-PR, Brazil, 2011 [72].

LP 97-28×IPR-Uirapuru	GP	W100s	NPP	NSP
Phenotypic variance ($\sigma_p{}^2$)	8.88	11.12	8.08	1.54
Enviromental variance ($\sigma_p{}^2$)	2.97	2.55	2.33	0.56
Genotypic variance ($\sigma_p{}^2$)	5.91	8.57	5.75	0.98
Additive variance ($\sigma_p{}^2$)	5.75	7.66	4.73	0.76
Dominance variance ($\sigma_p{}^2$)	0.16	0.91	1.02	0.22
Broad sense heritability (H^2%)	66.4	77.0	71.1	63.6
Narrow sense heritability (h^2%)	64.7	68.9	58.5	49.2
Average degree of dominance (*add*)	0.22	0.48	0.65	0.76
Minimum number of genes (*mng*)	4.7	4.4	8.6	5.5

Table 2. Estimates of phenotypic variance (σ_p^2), environmental variance (σ_e^2), genotypic variance (σ_g^2), additive variance (σ_a^2), dominance variance (σ_d^2), broad-sense heritability (H^2 %), narrow-sense heritability (h^2 %), average degree of dominance (add), and minimum number of genes (mng) related to grain production (GP), average weight of 100 seeds (W100s), number of pod per plant (NPP), and number of seeds per plant (NSP) obtained in six populations (P_1, P_2, F_1, F_2, BC_1, and BC_2), derived from cross between LP 97-28 × IPR-Uirapuru, Maringá-PR, Brazil, 2011 [72].In differential of selection (DS), gain by selection (GS) and predicted genetic gain, the characteristics of grain production, average weight of 100 seeds, number of pods per plant, and number of seeds per pod had differential of selection (DS) values ranging from 1.49 to 4.85 (Table 3).

Characteristic	Genetic component					
	Mi	Ms	Ds	GS	GS(%)	PGG
GP	6.86	11.45	4.59	2.93	42.7	9.79
W100s	21.29	26.14	4.85	3.29	15.5	24.58
NPP	7.02	11.12	4.10	2.37	33.8	9.39
NSP	4.61	6.10	1.49	0.73	15.9	5.34

Table 3. Mean initial in F_2 generation (Mi), mean of selected plants in F_2 generation (Ms), differential of selection (DS), gain by selection (GS), gain by selection expressed in percentage [(GS (%)], and predicted gain genetic (PGG) related to grain production (GP), average weight of 100 seeds (W100s), number of pod per plant (NPP), and number of seeds per plant (NSP) obtained in six populations (P_1, P_2, F_1, F_2, BC_1, and BC_2), derived from cross between LP 97-28 × IPR-Uirapuru, Maringá-PR, Brazil, 2011 [72].

The high and low values were obtained for the W100s and the NSP, respectively. In relation to gain by selection (GS), the GP, W100s, NPP, and NSP characteristics had values of 2.93, 3.29, 2.37, and 0.73, respectively (Table 3). When expressed as a percentage (% GS), the grain production trait had the highest value for gain by selection at 42.7%. The lowest value was found for the average weight of 100 seeds. The predicted genetic gain (PGG) values were 9.79, 24.58, 9.39, and 5.34 for the GP, W100s, NPP, and NSP characteristics, respectively (Table 3) [72].

Characteristic	GP	W100s	NPP	NSP
GP	-	0.36	0.96**	0.01
W100s		-	0.13	0.14
NPP			-	-0.20
NSP				-

Asterisks (**) indicate significance to 0.01 of probability by the *t*-test.

Table 4. Coefficients of phenotypic correlation between grain production (GP), average weight of 100 seeds (W100s), number of pod per plant (NPP), and number of seeds per plant (NSP) obtained in six populations (P_1, P_2, F_1, F_2, BC_1, and BC_2), derived from cross between LP 97-28 × IPR-Uirapuru, Maringá-PR, Brazil, 2011 [72].

Regarding correlations between characteristics, results indicated that all characteristics were directly proportional (Table 4), except between the NSP and NPP, which were inversely proportional. Additionally, the results show a high correlation (0.96) between the NPP and GP. Moderate associations were found between the GP and W100s and the W100s and NSP within six generations (P_1, P_2, F_1, F_2, BC_1, and BC_2) derived from crosses between LP 97-28 and IPR-Uirapuru [72].

9. Final considerations

This chapter described that the leading producers of chickpea worldwide are Nigeria, Niger, and Brazil, and this crop has a large influence on economy of these countries. However, the drought represents a problem frequently observed in areas with agricultural potential, with negative repercussion on production components and consequent decrease of the yield in this culture. Breeding programs aiming tolerance to drought using selection strategies linked to genetic, biochemical, physiological, nutritional, and agronomic characteristics can help increase the yield and reduce losses promoted by the drought. In addition, flower structure and hybridization technique used in Núcleo de Pesquisa Vegetal Básica e Aplicada (NPVBA/UFRA) were presented, as well as populations evaluated and plant management were explained in detail. This chapter described the results obtained in other breeding programs aimed at drought tolerance and also explained the potential uses to increase the crop yield.

Acknowledgements

This chapter had financial support from Fundação Amazônia Paraense de Amparo à Pesquisa (FAPESPA/Brazil), Universidade Federal Rural da Amazônia (UFRA/Brazil), and Conselho Nacional de Pesquisa (CNPq/Brazil) to Lobato AKS.

Author details

Maria Antonia Machado Barbosa[1], Allan Klynger da Silva Lobato[1*],
Milton Hélio Lima da Silva[2], Gabriel Mascarenhas Maciel[3] and Douglas José Marques[4]

*Address all correspondence to: allanllobato@yahoo.com.br

1 Núcleo de Pesquisa Vegetal Básica e Aplicada, Universidade Federal Rural da Amazônia, Paragominas, Brazil

2 Coordenação de Botânica, Museu Paraense Emílio Goeldi, Belém, Brazil

3 Instituto de Ciências Agrárias, Universidade Federal de Uberlândia, Monte Carmelo, Brazil

4 Setor de Olericultura e Experimentação, Universidade José do Rosário Vellano (UNIFENAS), Alfenas, Brazil

References

[1] Muchero W, Ehlers JD, Roberts PA. Seedling stage drought-induced phenotypes and drougth-responsive genes in diverse cowpea genotypes. Crop Science. 2008;48(10): 541-552.

[2] Fao. Faostat database gateway [Internet]. 2013. Available from: http://www.fao.org [Accessed: 11 de julho de 2013]

[3] Silveira JAG, Costa RCL, Oliveira JTA. Drough-induced effects and recovery of nitrate assimilation and nodule activity in cowpea plants inoculated with *Bradyrhizobium* spp. under moderate nitrate level. Brazilian Journal of Microbiology. 2001;32(3): 187-194. DOI: 10.1590/S1517-83822001000300005.

[4] Lobato AKS, Santos Filho BG, Costa RCL, Gonçalves-Vidigal MC, Moraes EC, Oliveira Neto CF, Rodrigues VLF, Cruz FJR, Ferreira AS, Pita JD, Barreto AGT. Morphological, physiological and biochemical responses during germination of the cowpea

(*Vigna unguiculata* cv. Pitiuba) seeds under salt stress.. World Journal of Agricultural Sciences. 2009a;5(5):590-596.

[5] Soares ALL. Agronomic efficiency of selected rhizobia strains and diversity of native nodulating populations in Perdões (MG - Brazil). I – cowpea. Revista Brasileira de Ciências do solo. 2006;30(5):795-802. DOI: 10.1590/S0100-06832006000500005

[6] Borges PRS, Saboya RCC, Saboya LMF, Santos ER, Souza SEA. Distribuição de massa seca e rendimento de feijão-caupi inoculadas com rizóbio em Gurupi, TO. Revista Caatinga. 2012;25:248-254.

[7] Atkinson NJ, Urwin PE. The interaction of plant biotic and abiotic stresses: from genes to the field. Journal of Experimental Botany. 2012;63(10):3523-3543. DOI: 10.1093/jxb/ers100

[8] Bastos EA, Nascimento SP, Silva EM, Freire filho FR, Gomide RL. Identification of cowpea genotypes for drought tolerance. Revista Ciência Agrônomica. 2011;42(1): 100-107.

[9] Lawlor DW. Limitation to photosynthesis in leaves water-stressed: stomata vs. metabolism and the role of ATP. Annals of Botany. 2002;89(7):871-885. DOI: 10.1093/aob/mcf110

[10] Cramer GR, Urano K, Delrote S, Pezzotti M, Shinozaki K. Effects of abiotic stress on plants: a systems biology perspective. BMC Plant Biology. 2011;11(163):1-14. DOI: 10.1186/1471-2229-11-163

[11] Kramer PJ, Boyer JS. Water relations of plant and soils. 1st ed. New York: Academic Press; 1995. 482 p.

[12] Reddy AR, Chaitanya KV, Vivekanandan M. Drought-induced responses of photosynthesis and antioxidant metabolism in higher plants.. Journal of Plant Physiology. 2004;161(11):1189-1202. DOI: 10.1016/j.jplph.2004.01.013

[13] Muñoz-Perea CG, Teran H, Allen RG, Wright JL, Westermann DT, Singh SP. Selection for drought resistance in dry bean landraces and cultivars.. Crop Science. 2006;46(5):2111-2120. DOI: 10.2135/cropsci2006.01.0029

[14] Asfaw A, Blair MW, Struik PC. Multienvironment quantitative trait loci analysis for photosynthate acquisition, accumulation, and remobilization traits in common bean under drought stress. G3: Genes, Genomes, Genetics. 2012;2(5):579-595. DOI: 10.1534/g3.112.002303.

[15] Blair MW, Galeano CH, Tovar E, Torres MCM, Castrillón AV, Beebe SE, Rao IM. Development of a Mesoamerican intra-genepool genetic map for quantitative trait loci detection in a drought tolerant x susceptible common bean (*Phaseolus vulgaris* L.) cross. Molecular Breeding. 2012;29(1):71-88. DOI: 10.1007/s11032-010-9527-9

[16] Antonino ACD, Sampaio EVSB, Dall'olio A, Salcedo IH. Balanço Hídrico em solo com cultivos de subsistencia no semi-árido do nordeste do Brasil. Revista Brasileira

de Engenharia Agrícola e Ambiental. 2000;4(1):29-34. DOI: 10.1590/S1415-43662000000100006.

[17] Fereres E, Soriano MA. Deficit irrigation for reducing agricultural water use.. Journal of Experimental Botany. 2007;58(2):147-159. DOI: 10.1093/jxb/erl165

[18] Santos RF, Carlesso R. Water deficit and morphologic and physiologic behavior of the plants. Revista Brasileira de Engenharia Agrícola e Ambiental. 1998;2(3):287-294.

[19] Acosta-Díaz E, Acosta-Gallegos J, Trejo-López C, Padilla-Ramírez J, Amador-Ramírez M. Adaptation traits in dry bean cultivars grown under drought stress. Agricultura Técnica en México. 2009;35(4):416-425.

[20] Andrezej S. Impact of drought stress on some physiological parameters of common bean (*Phaseolus vulgaris* L.). Annual Report of the Bean Improvement Cooperative. 2012;55(1):177-178.

[21] Leite ML, Rodrigues JD, Virgens Filho JS. Effects of water stress on the cowpea crop, cv. EMAPA-821. III - Production. Revista de Agricultura. 2000;75(1):9-20.

[22] Omae H, Kumar A, Kashiwaba K, Shono M. Assessing drought tolerance of snap bean (*Phaseolus vulgaris*) from genotypic differences in leaf water relations, shoot growth and photosynthetic parameters. Plant Production Science. 2007;10(1):28-35. DOI: 10.1626/pps.10.28.

[23] Omae H, Kumar A, Kashiwaba K, Shono M. Adaptation to high temperature and water deficit in the common bean (*Phaseolus vulgaris* L.) during the reproductive period.. Journal of Botany. 2012;2012(2012):1-6. DOI: 10.1155/2012/803413

[24] Leport L, Turner NC, French RJ, Tennant D, Thomson BD, Siddique KHM. Water relations, gas exchange and growth of cool-season grain legumes in a Mediterranean-type environment. European Journal of Agronomy. 1998;9(1):295-303. DOI: 10.1016/S1161-0301(98)00042-2

[25] Emam Y, Shekoofa A, Salehi F, Jalali AH. Water stress effects on two common bean cultivars with contrasting growth habits.. American-Eurasian Journal of Agriculture and Environment. 2010;9(1):495-499. DOI: 10.1080/03650340.2010.530256.

[26] Mohamed MF, Keutgen N, Tawfit AA, Noga G. Dehydration-avoidance responses of tepary bean lines differing in drought resistance.. Journal of Plant Physiology. 2002;159(1):31-38. DOI: 10.1078/0176-1617-00530

[27] Menuccini M, Mambelli S, Comstock J. Stomatal responses to leaf water status in common bean (*Phaseolus vulgaris* L.) is a function of time of day.. Plant Cell and Environment. 2000;23(10):1109-1118. DOI: 10.1046/j.1365-3040.2000.00617.x

[28] Iuchi S, Yamaguchi-Shinozaki K, Urao T, Shinozaki K. Novel drought inducible genes in the highly drought-tolerant cowpea: cloning of cDNA and analysis of their

gene expression.. Plant Cell Physiology. 1996a;37(8):1073-1082. DOI: 10.1093/oxfordjournals.pcp.a029056.

[29] Iuchi S, Yamaguchi-Shinozaki K, Urao T, Shinozaki K. Characterization of two cDNA for novel drought-inducible genes in the highly-tolerant cowpea. Journal of Plant Research. 1996b;109(4):415-424. DOI: 10.1007/BF02344557.

[30] Figueiredo MVB, Burity HA, França FP. Drought stress response in enzymatic activities of cowpea nodules.. Journal of Plant Physiology. 1999;155(2):262-268. DOI: 10.1016/S0176-1617(99)80017-0

[31] Gomes Filho RR, Tahin JF. Physiological responses of cowpea cultivars (*Vigna unguiculata*) erect and decumbent to different levels of irrigation.. Engenharia na Agricultura. 2002;10(1):56-60.

[32] Lobato AKS, Costa RCL, Oliveira Neto CF, Santos Filho BG, Gonçalves-Vidigal MC, Vidigal Filho PS, Silva CR, Cruz FJR, Carvalho PMP, Santos PCM, Gonela A. Consequences of the water deficit on water relations and symbiosis in Vigna unguiculata cultivars. Plant Soil and Environment. 2009b;55(4):139-145.

[33] Matsui T, Singh BB. Root characteristics in cowpea related to drought tolerance at the seedling stage. Experimental Agriculture. 2003;39(1):29-38. DOI: 10.1017/S0014479703001108

[34] Nascimento SP, Bastos EA, Araújo ECE, Freire Filho FR, Silva EM. Tolerance to water deficit of cowpea genotypes. Revista Brasileira de Engenharia Agrícola e Ambiental. 2011;15(8):853-860. DOI: 10.1590/S1415-43662011000800013

[35] Miranda JES, Tavera VM, Acosta-Gallegos JA, López JLA, Chavira MMG. Sequence and functional characterization of the dry bean gene STPP and its relationship to drought response. Annual Report of the Bean Improvement Cooperative. 2012;55(1): 175-176.

[36] Sharp RE, Poroyko V, Hejlek LG, Spollen WG, Springer GK, Bohnert HJ, Nguyen HT. Root growth maintenance during water deficits: physiology to functional genomics.. Journal of Experimental Botany. 2004;55(407):2343-2351. DOI: 10.1093/jxb/erh276

[37] Sponchiado BN, White JW, Castillo JA, Jones PG. Root growth of four common bean cultivars in relation to drought tolerance in environments with contrasting soil types.. Experimental Agriculture. 1989;25(2):249-257. DOI: 10.1017/S0014479700016756

[38] Guimarães CM, Brunini O, Stone LF. Common bean (*Phaseolus vulgaris* L.) adaptation to drought. Pesquisa Agropecuária Brasileira. 1996;31(2):393-399. DOI: 10.1590/S1415-43662006000100001

[39] Peña-Valdivia CB, Sanchez-Urdaneta AB, Rangel JM, Muñoz JJ, García-Nava R, Velázquez RC. Anatomical root variations in response to water deficit: wild and domes-

ticated common bean (*Phaseolus vulgaris* L).. Biological Reseach. 2010;43(4):417-427. DOI: S0716-97602010000400006

[40] Mitra J. Genetics and genetic improvement of drought resistance of crop plants.. Current Science. 2001;80(25):758-763.

[41] Nogueira RJMC, Moraes JAPV, Burity HA, Bezerra Neto E. Changes in diffusion resistance of vapor leaves and water relations in aceloreiras subjected to water deficit.. Revista Brasileira de Fisiologia Vegetal. 2002;13(1):75-87. DOI: 10.1590/S0103-31312001000100009.

[42] Vendruscolo ECG, Schuster I, Pileggi M, Scapim CA, Molinari HBC, Marur CJ, Vieira LGE. Stress-induced synthesis of proline confers tolerance to water deficit in transgenic wheat. Journal of Plant Physiology. 2007;164(10):1367-1376. DOI: 10.1016/j.jplph.2007.05.001

[43] Chaves MM, Depauw RM, Townley-Smith TF. How plants cope with water stress in the field. Photosynthesis and growth. Annals of Botany. 2002;89(7):907-916. DOI: 10.1093/aob/mcf105

[44] Muchow RC, Carberry PS. Designing improved plant types for the semi-arids tropics: Agronomist viewpoints. In: Penning de Vries FWT, Teng P, Metselaar k, editors. Systems appraches for agricultural development; 1993; Dordrecht: Kluver. Dordrecht: Kluver:1993. p. 37-61. DOI: 10.1007/978-94-011-2840-7

[45] Spollen WG, Nelson CJ. Response of fructan to water deficit in growing leaves of tall fescue.. Plant Physiology. 1994;106(1):329-336. DOI: 10.1104/pp.106.1.329

[46] Barrios-Gómez EJ, López-Castañeda C, Kohashi-Shibata J, Acosta-Gallegos JA, Miranda-Colín S, Mayek-Pérez N. Osmotic adjustment and seed yield of dry bean under drought. Annual Report of the Bean Improvement Cooperative. 2009;52(1):100-101.

[47] Kozlowski TT, Pallardy SG. Acclimation and adaptive responses of woody plants to environmental stresses. The Botanical Review. 2002;68(2):270-334. DOI: 10.1663/0006-8101

[48] Maestri M, DaMatta FM, Regazzi A, Barros RS. Accumulation of proline and quaternary ammonium compounds in mature leaves of water stressed coffee plants (*Coffea arabica* and *Coffea canephora*). Journal Horticultural Science. 1995;70(1):229-233.

[49] Barbosa MAM, Lobato AKS, Tan DKY, Viana GDM, Coelho KNN, Barbosa JRS, Moraes MCHS, Costa RCL, Santos Filho BG, Oliveira Neto CF. Bradyrhizobium improves nitrogen assimilation, osmotic adjustment and growth in contrasting cowpea cultivars under drought.. Australian Journal of Crop Science. 2013;7(13):1983-1989.

[50] Hartung W, Sauter A, Hose E. Abscisic acid in the xylem: where does it come from, where does it go to?. Journal of Experimental Botany. 2002;53(366):27-32. DOI: 10.1093/jexbot/53.366.27

[51] D'arcy-Lameta A, Ferrari-Iliou R, Contour-Ansel D, Pham-Thi AT, Zuily-Fodil Y. Isolation and characterization of four ascorbate peroxidase cDNA responsive to water deficit in cowpea leaves. Annals of Botany. 2006;97(1):133-140. DOI: 10.1093/aob/mcj010

[52] Leite ML, Virgens Filho JS. Dry matter production of cowpea (*Vigna unguiculata* (L.) Walp.) plants submitted to water deficits.. Publication UEPG/Ciências Agrárias e Engenharias. 2004;10(1):43-51.

[53] Singh SP. Selection for water-stress tolerance in interracial populations of commom bean.. Crop Science. 1995;35(1):118-124. DOI: 10.2135/cropsci1995.0011183X003500010022x

[54] Schneider KA, Brothers ME, Kelly JD. Marker assisted selection to improve drought resistance in common bean.. Crop Science. 1997;37(1):51-60. DOI: 10.2135/cropsci1997.0011183X003700010008x

[55] Karamanos AJ, Papatheohari AY. Assessment of drought resistance of crop cultivars and lines by means of the water potential index. Crop Science. 1999;39(1):1792-1797.

[56] Dubetz S, Mahlle PS. Effect of soil water stress on bush beans (*Phaseolus vulgaris* L.), at three stages of growth.. Journal of American Society for Horticultural Science. 1969;94:479-481.

[57] White JW, Ochoa MR, Ibarra PF, Singh SP. Inheritance of seed yield, maturity and seed weight of common bean (*Phaseolus vulgaris*) under semi-arid rainfed conditions.. Journal of Agricultural Science. 1994;122(2):265-273. DOI: 10.1017/S0021859600087451

[58] Beebe SE, Rao IM, Blair MW, Acosta-Gallegos JA. Phenotyping common beans for adaptation to drought. Fronters in Physiology. 2013;4(35):1-20. DOI: 10.3389/fphys.2013.00035

[59] Broughton WJ, Hernandez G, Blair M, Beebe S, Gepts P, Vanderleyden J. Beans (*Phaseolus* spp.) – model food legumes. Plant and Soil. 2003;252(1):407-415. DOI: 10.1023/A:1024146710611

[60] Londero PMG, Ribeiro ND, Cargnelutti Filho A. Fiber content and yield grain in common bean populations. Ciência e Agrotecnologia. 2008;32(1):167-173. DOI: 10.1590/S1413-70542008000100024

[61] Freire Filho FR, Lima JAA, Ribeiro VQ. Cowpea: Technological advances. Embrapa Informação Tecnológica. 2005;1:519.

[62] Hall AE. Phenotyping cowpeas for adaptation to drought.. Frontiers in Physiology. 2012;3(155):1-8. DOI: 10.3389/fphys.2012.00155.

[63] Torres GAM, Pflieger S, Corre-Menguy F, Mazubert C, Hartmann C, Lelandais-Briere C. Identification of novel drought related mRNAs in common bean roots by dif

ferential display RT-PCR. Plant Science. 2006;171(3):300-307. DOI: 10.1016/j.plantsci.2006.03.008.

[64] Cooper M, DeLacy IH. Relationship among analytical methods used to study genotypic variation and genotype-by-environment interaction in plant breeding multi-environment experiments. Theorical and Applied Genetics. 1994;88(5):561-572. DOI: 10.1007/BF01240919.

[65] Clarke JM, Depauw RM, Townley-Smith TF. Evaluation of methods for quantification of drought tolerance in wheat.. Crop Science. 1992;32(3):723-728. DOI: 10.2135/cropsci1992.0011183X003200030029x

[66] Subbarao GV, Johansen C, Slinkard AE, Nageswara Rao RC, Saxena NP, Chauhan YS. Strategies for improving drought tolerance in grain legumes. Critical Reviews in Plant Sciences. 1995;14(6):469-523. DOI: 10.1080/07352689509701933.

[67] Ribeiro GS, Ferreira AF, Neves CML, Sousa FSM, Oliveira C, Alves EM, Sodré GS, Carvalho CAL. Aspects of the floral biology and pollen properties of *Vigna unguiculata* L. Walp (Fabaceae). African Journal of Plant Science. 2013;7(5):149-154. DOI: 10.5897/AJPS13.1014.

[68] Rocha MM, Filho FRF, Silva KSD, Ribeiro VQ. Feijão-caupi: Biologia floral. Embrapa Meio-Norte. Teresina – PI.

[69] Rachie K, Rawal KM, Franckowial JD. A rapid method of hand crossing cowpeas. 1th ed. Nigeria:1975. 5p.

[70] Zary KW, Miller Junior JC. Comparison of two methods of hand-crossing *Vigna unguiculata* (L.) Walp. HortScience. 1982;17(2):246-248.

[71] Teófilo EM, Paiva JB, Medeiros Filho S. Arficial pollination in cowpea (*Vigna unguiculata* (L.) Walp. Ciência e Agrotecnologia. 2001;25(1):220-223.

[72] Lobato AKS, Gonçalves-Vidigal MC, Vidigal Filho PS, Ramos VMS, Poletine JP, Andrade CAB. Genetic parameters of grain production and its components in common bean (*Phaseolus vulgaris* L.) under drought stress. Australian Journal of Crop Science. 2014;8(8):1152-1159.

[73] Gravina GA, Martins Filho S, Sediyama CS, Cruz CD. Genetic parameters of soybean resistance to Cercospora sojina. Pesquisa Agropecuária Brasileira. 2004;39(7):653-659. DOI: 10.1590/S0100-204X2004000700006.

[74] Szilagyi L. Influence of grought on seed yield components in common bean. Bulgarian Journal Plant Physiology. 2003;9(1):320-330.

[75] Franco MC, Cassini STA, Oliveira VR, Tsai SM. Characterization of the genetic diversity of common beans by RAPD markers.. Pesquisa Agropecuária Brasileira. 2001;36(2):381-385. DOI: 10.1590/S0100-204X2001000200023

[76] Cruz CD, Regazzi AJ, Carneiro PCS. Modelos Biométricos Aplicados ao Melhoramento Genético. 4th ed. Brazil:2012. 668 p.

[77] Acquaah G. Principles of plant genetics and breeding.. 1st ed. Australia: Blackwell; 2007. 584 p.

[78] Abney M, Mcpeek MS, Ober C. Broad and narrow heritabilities of quantitative traits in a founder population. American Journal of Human Genetics. 2001;68(5):1302-1307. DOI: 10.1086/320112.

7

Tolerance of Plants to Toxicity Induced by Micronutrients

Allan Klynger da Silva Lobato, Emily Juliane Alvino Lima, Elaine Maria Silva Guedes Lobato, Gabriel Mascarenhas Maciel and Douglas José Marques

Abstract

Micronutrient elements such as zinc, boron copper, iron, manganese, molybdenum, and chlorine are frequently responsible by the regulatory activity of the cell organelles, being nutrients that are absorbed and found in lower concentrations in plant tissues, they also contribute to supply the nutritional exigency of the plant. Study with Zea mays plants exposed to Zn toxicity + Si presented significant increases in stomatal conductance, net photosynthetic rate, transpiration rate, and water use efficiency, respectively, in comparison with treatment only with Zn. In relation to chlorophylls a, b and total and carotenoids presented non-significant increases, when compared to plants exposed to Zn toxicity. This study revealed the positive contribution of the Si on gas exchange and reduction of the negative effects provoked on chlorophylls and carotenoids in maize plants under Zn toxicity. Other results described that prolonged exposure to excessive Cu resulted in serious toxic effects on the rice seedlings. In contrast, Tre pretreatment has been shown to be beneficial in alleviating Cu toxicity, which was mainly attributed to the ability of Tre to restrict Cu uptake and accumulation to maintain Cu homeostasis, and to induce production of antioxidant and Gly enzymes to alleviate excessive Cu-triggered oxidative stress. Stress caused by the excessive supply of micronutrients to plants frequently promotes repercussions on oxidant system, inducing the overproduction of reactive oxygen species. The oxidative damage is a situation characterized by the large ROS accumulation and insufficient detoxification promoted by antioxidant enzymes, such as catalase and glutathione peroxidase. Different mechanisms have been proposed to explain the tolerance of plants to toxicity induced by micronutrients, as uses of other elements and substances, in which it can positively act with specific transporters, metal ion homeostasis and compartmentalization of micronutrients into the vacuole.

Keywords: gas exchange, antioxidant metabolism, metals, photosynthetic pigments

1. Introduction

Large amounts of minerals and organic materials can be found in soil [1]; and some of these materials have great significance in the anatomical development and plant morphology. The soil fertility is an important factor of agronomic order. It is related to provision of essential elements in the soil, where the plant will absorb adequate levels of important substances to work with its metabolism [2]. The nutrients are arranged in ionic form, which can interact with the root system of plants via three mechanisms: interception root, diffusion, or mass flow [3].

The plants require basic conditions to develop and supply the essential elements to aid with the metabolism and produce simple and/or complex compounds [4]. These are divided into non-minerals and minerals, which are sub-divided into macro and micronutrients.

2. Non-mineral nutrients

The non-mineral nutrients are hydrogen (H), oxygen (O), and carbon (C). These elements help in the formation of tissues. They are responsible for the composition of lipids, carbohydrates, proteins, nucleic acids, and hormones in plants [5]. Hydrogen (H) is the element that contributes to nitrogen fixation and photosynthesis; this is due to their reducing capacity molecules, being a nutrient assimilated from the water. Oxygen (O) is extracted from atmosphere and water, it is used during the process of cell respiration, where sugar transferred by the photosynthetic process is transformed into energy. The plant captures the Carbon (C) in the form of CO_2 from atmosphere. It is an organic nutrient with function to ensure the occurrence of physical, chemical, and biological cycles, ensuring the maintenance of energy reserves throughout vegetative growth.

3. Nutrients minerals

These are elements essential for plant to complete its cycle [6]. They are divided into macro and micronutrients [7].

3.1. Macronutrients

The macronutrients are composed by nitrogen (N), phosphorus (P), potassium (K), calcium (Ca), magnesium (Mg), and sulphur (S). Large amount of these elements are required for plants to develop and meet its physiological activity. The macronutrients play a vital role in plant structure [8].

3.2. Micronutrients

These elements are responsible for the regulatory activity of the cell organelles. These nutrients are absorbed and found in lower concentrations in plant tissues and supply the nutritional exigency of the plant [9].

3.2.1. Zinc (Zn)

Zinc is an essential component in thousands of proteins in plants, in which it performs the role of enzymatic activator in various processes related to plant metabolism, such as dehydrogenases, aldolases, isomerases and transphorylases [10]. Its deficiency affects plant development, with stems and leaf expansion directly affected, and also reported to have consequences on chlorophyll levels and possibly necrosis [11].

3.2.2. Boron (B)

Boron is a nutrient that works with the carbohydrate metabolism and transport of sugars through membranes. This element acts in the formation of the cell wall, sap movement, cell division, development of leaves and inflorescence, and confer resistance to tissues [12]. Its deficiency causes disorder in meristematic activities and reserve mechanisms can be directly affected, with the stem apex, flowers and fruits suffer deformations [13].

3.2.3. Copper (Cu)

The copper activity is related to photosynthetic and respiration processes, fixation, and distribution of the nitrogen, all being necessary during plant cycle [14]. When the plant does not absorb adequate amount of this element for their development, it may present negative characteristics, such as changes of colour and structure in leaf [15].

3.2.4. Iron (Fe)

Iron is an essential micronutrient used in photosynthesis, cell division, nitrogen fixation, and formation and morphology of plants [16-17]. The excess of this element affects the leaf, and their growth is strongly reduced [18]. The deficiency induces chlorosis [19].

3.2.5. Manganese (Mn)

Manganese is a micronutrient of essential character. It plays the metabolic activities of direct or indirect orders, with capacity to activate enzymes responsible for the control of oxidation-reduction processes, such as the production of chlorophylls [20]. The low concentration of this micronutrient causes loss of leaf due to appearance of small spots on leaves and chlorosis, and structural development of the plant normally is compromised [21].

3.2.6. Molybdenum (Mo)

Molybdenum is a micronutrient that helps in nitrogen fixation and nitrate assimilation [22]. With the deficiency of this nutrient, plants are affected by chlorotic spots and marginal necrosis, thus the plant yield is negatively affected [23].

3.2.7. Chlorine (Cl)

Chlorine acts in the photosynthetic process of plants, aside from interacting with water and during transpiration process [24]. The lack of chlorine causes chlorosis and necrosis, roots suffer atrophy, affecting the development of the plant [25]. The excess of chlorine in the soil is more common than the deficiency. Indication of its excess is signalled by the burn of the leaf edges [26].

4. Toxicity of micronutrients

Despite micronutrients be required in higher plants, in higher concentrations frequently is toxic and provokes negative effects [27], as reduction in photosynthetic pigments [28], minor integrity and permeability of membranes [29], increase of the oxidative stress related with production and accumulation of reactive oxygen species (ROS), besides to increase the activities of antioxidant enzymes [30], and in levels more extremes to induce cell death [31].

Stress caused by the excessive supply of nutrients to plants promotes repercussion on oxidant system [32-33], inducing the overproduction of reactive oxygen species (ROS) as superoxide radical (O_2^-) and hydrogen peroxide (H_2O_2) [34]. The oxidative damage is a situation characterized by the large ROS accumulation and insufficient detoxification promoted by antioxidant enzymes, such as catalase and glutathione peroxidase [35].

Different mechanisms have been proposed to explain the tolerance of plants to toxicity induced by heavy metals and nutrients. Two specific transporters are metal ion homeostasis and compartmentalization of metals into the vacuole [36-37]. However, responses linked to contribution of Si in plants submitted to Zn excess, more specifically on gas exchanges and photosynthetic pigments, are unknown.

Beneficial repercussions related to Si uses in higher plants are intensively found [38-40]. Isa et al. [41] reported that Si is largely accumulated in leaves. Silva et al. [42] described increases in chlorophylls produced by exogenous Si application. Si also induces higher mechanical resistance from cell wall [43]. Chen et al. [44] found better light reception and increasing net photosynthesis rate and CO_2 capitation after Si treatment.

5. Objectives

This chapter aim to: (i) define what nutrient toxicity is; (ii) present the modifications produced in the biochemical and physiological levels; (iii) explain the consequences to plant induced by the nutrient toxicity.

6. Effects of zinc toxicity in gas exchange

Study conducted by Paula et al. [45] with *Zea mays* plants under Zn toxicity and treated with Si (silicon) evaluating the gas exchanges and photosynthetic pigments was shown in Figure 1. Zinc toxicity produced a negative interaction promoting stomatal closing, and consequently, reduction in stomatal conductance. This effect can be attributed to Zn excess, in which it will induce a minor density and size of these structures, with structural differences in adaxial and abaxial sides, besides minor stomatal aperture sizes [46]. Similar results were obtained by Pavlíková et al. [47] in *Nicotiana tabacum* plants submitted to stress by Zn.

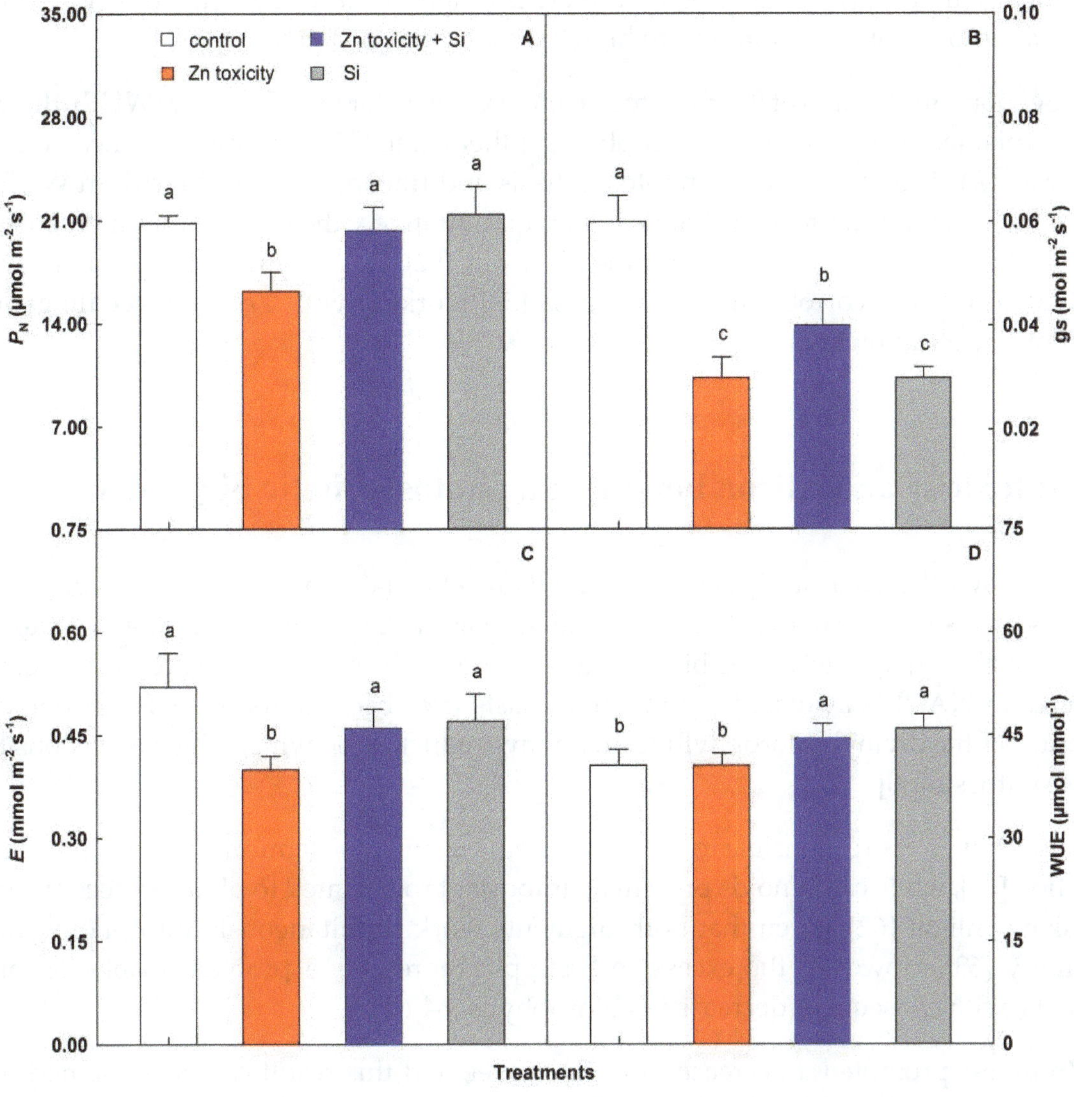

Figure 1. Stomatal Conductance (A), Photosynthesis Rate (B), Transpiration Rate (C), and Water Use Efficiency (D) in *Zea mays* plants subjected to silicon and zinc toxicity. Different letters to treatments indicate significant differences from the Skott-Knott test (P < 0.05). Columns represent the mean values from four repetitions, and bars represent the standard deviations [45].

Zn induced a reduction in net photosynthetic rate, as explained by the stomatal limitation, arising of minor stomatal conductance, and consequent decrease of the CO_2 assimilation to photosynthetic process [48-49]. Similar results were found by Shi and Cai [50] working with *Arachis hypogaea* plants submitted to Zn stress, corroborating with results obtained in this research.

The reduction of the transpiration in plants under exogenous application of Zn was possibly attributed to decrease in stomatal conductance. This stomatal limitation reduces the transpiration rate, promoting minor water loss from plant to atmosphere, and consequently limited nutrients reposition, in form of adsorbed ions into substrate with water, using the via root system [51]. In other words, the transpiration is responsible with the dynamic of nutrient transport form substrate in direction root and leaf [52], thus avoid the cavitation in xylem [53]. Fernàndez et al. It was also described thatthere is a significant reduction in transpiration rate in *Populus deltoides* plants submitted to high Zn concentrations [54].

The exogenous application of Si promoted an increase in water use efficiency (WUE), this result can be explained by the increase in net photosynthetic rate (*PN*) and maintenance in transpiration rate (*E*). The ratio between photosynthesis and transpiration will result in WUE [55], being a physiological parameter that describes quantitatively the behavior momentaneous of the gas exchanges in leaf, it also reveals the efficiency that the plant utilizes the water resource [56]. Our results are corroborated by Moussa [57] working with *Zea mays* seedlings under exogenous application of Si.

7. Zinc toxicity and silicon benefits on photosynthetic pigments

Figure 2 shows the same study conducted by Paula et al. [45] on photosynthetic pigments in *Zea mays* plants under Zn toxicity and Si. The decrease in CHL *a* occurred due to Zn toxicity, being probably related to minor biosynthesis rate of CHL *a* [58]. The Zn excess negatively interferes in NADPH availability into chloroplasts [59], because the NADPH is one of the substrates of the divinyl chlorophyllide an 8-vinyl-reductase enzyme, which is responsible to CHL *a* synthesis [60].

The chlorophylls are responsible to the photochemical and biochemical reactions during light capitation [61], while carotenoids present an important role related to photoprotection against excessive sunlight [62], given that both pigments work simultaneously into photosynthetic machinery [63]. However, the excessive B supply represents a problem to photosynthetic pigments, with consequent decrease in chlorophylls [64-65].

The Zn excess promoted a decrease in CHL b level, and this result can be explained by the oxidative stress induced by the overproduction of reactive species oxygen (ROS), such as H_2O_2 [66-67]. The H_2O_2 aside from being toxic in chloroplasts, is considered an inhibitor of the carbon metabolism, [68], resulting in acceleration of leaf senescence through of the lipid peroxidation and oxidative damages [69]. Similar results were reported by Bettaieb [70] evaluating *Solanum tuberosum* cultivars.

The CHL total levels were reduced after Zn toxicity, which is related to magnesium (Mg) substitution in molecule of chlorophyll by the Zn. It will result to the inadequate work of the light-harvesting complex (LHCII), and consequently the photosynthesis limitation [71-73]. Our results on reduction in CHL total were corroborated by Bassi and Sarma [74] in *Triticum aestivum* seedlings.

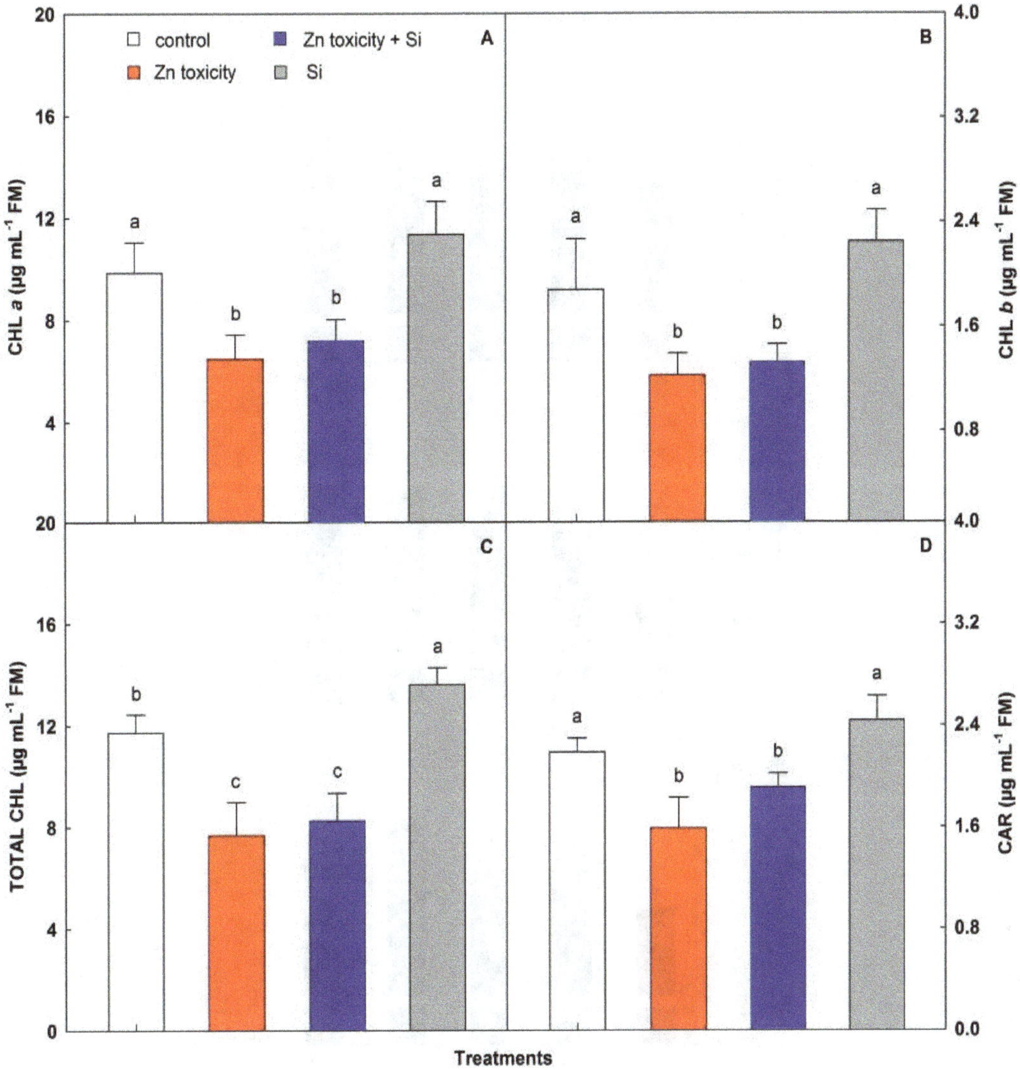

Figure 2. Chlorophyll *a* (A), Chlorophyll *b* (B), Total Chlorophyll (C) and Carotenoids (D) in *Zea mays* plants subjected to silicon and zinc toxicity. Different letters to treatments indicate significant differences from the Skott-Knott test ($P < 0.05$). Columns represent the mean values from four repetitions, and bars represent the standard deviations [45].

8. Boron toxicity induces modifications in nitrogen metabolism

Silva et al. [75] working with *Schizolobium parahyba* plants exposed to progressive boron levels determined amino acids total and proline is depicted in Figure 2. The total amino acids accumulated in leaf must be associated to proteases enzyme activity and concomitantly to inhibition of the protein biosynthesis rate, due to the high B levels. According to Paula et al. [76], the increases in proteases activities promote the breakdown and consequent decrease of proteins, increasing the total AA amount, which will be used in synthesis of other AA, like as PRO. Cervilla et al. [77] evaluating the effects of the B toxicity in *Solanum lycopersicum* plants found similar results.

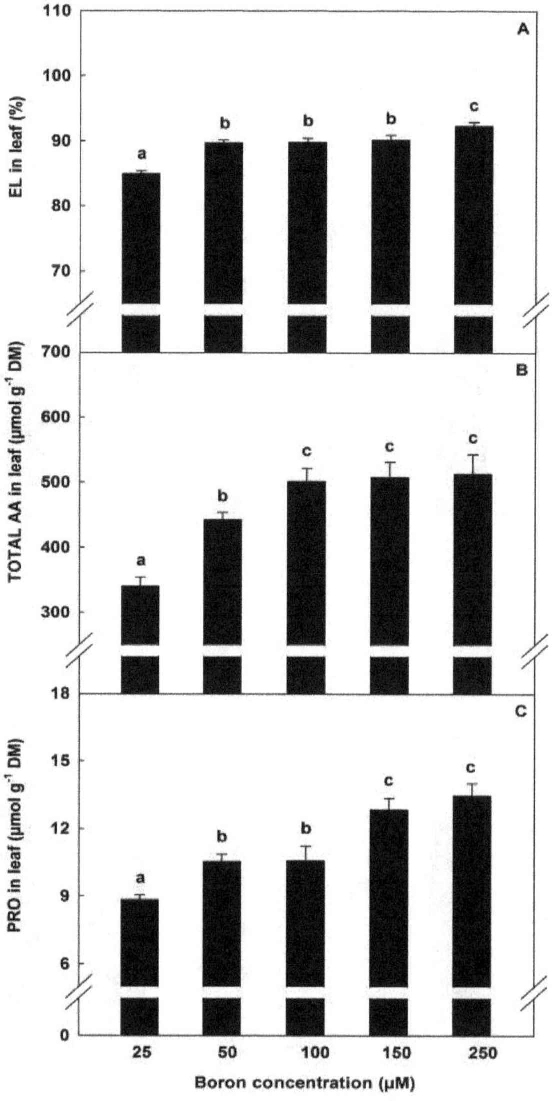

Figure 3. Electrolyte Leakage (A), Total Amino Acids (B), Proline (C) of young *Schizolobium parahyba* var. *amazonicum* plants subjected to boron toxicity. Different letters for boron levels indicate significant differences from the Skott-Knott test ($P<0.05$). Columns represent the mean values from 5 repetitions, and bars represent the standard deviations [75].

The proline concentration was maximized and this result is related to the amino acid acting in detoxification process of ROS and membrane protection against lipid peroxidation [78-79]; associated with ROS antagonist, it aims to attenuate the oxidative stress and to avoid the cell death. In other activities, the PRO protects the protein structure against denaturation and it will stabilize the cell membranes during interaction with phospholipids [80]. Contreras et al. [81] evaluating the B and NaCl effects in *Solanum lycopersicum* plants verified similar results of this research.

Proline is an amino acid synthetized into nitrogen metabolism with functions related to osmoprotection [82] and cellular homeostasis [83], which can contribute to improve the plant tolerance under situations of abiotic stress, as B toxicity.

9. Effects of the copper toxicity and exogenous trehalose on antioxidant enzymes

In agreement with Mostofa et al. [84], the regulating Cu homeostasis is crucial in maintaining the intracellular Cu level to avoid toxicity. Plants have developed various mechanisms to restrict Cu toxicity, such as inhibition of Cu uptake by binding with root exudates like organic acids, intracellular sequestration by strong ligands like cysteine-rich compounds and phytochelatins, and exclusion of excessive Cu from the cells by sugar alcohols like trehalose (Tre) [26–28]. Tre, a non-reducing disaccharide of glucose, protects plant cells against long-term desiccation by stabilizing enzymes, proteins, and biological membranes under dehydration [29].

In relation to antioxidant enzymes superoxide dismutase (SOD), catalase (CAT), glutathione peroxidase (GPX), and glutathione reductase (GST), the pretreatment with non-reducing disaccharide Tre induces increases in enzyme activities when compared with control or Cu stress treatments (Figure 4).

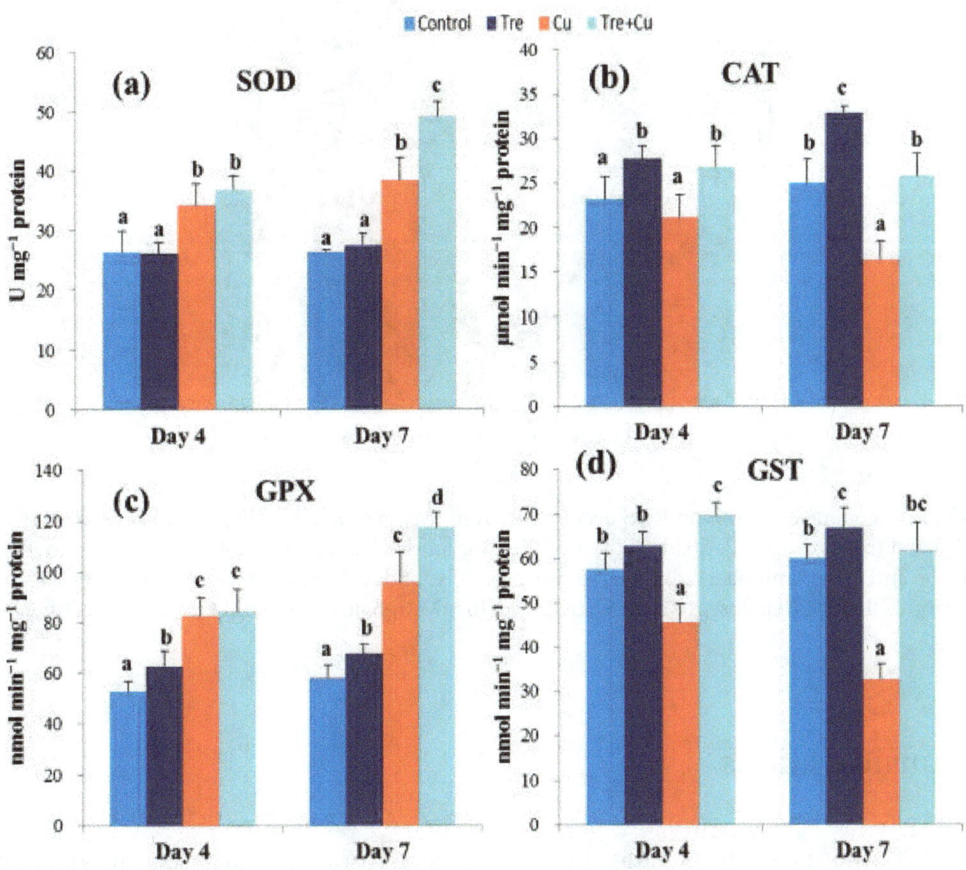

Figure 4. Effect of exogenous trehalose on the activities of antioxidant enzymes in rice seedlings with or without Cu stress. (a) Superoxide Dismutase (SOD), (b) Catalase (CAT), (c) Glutathione Peroxidase (GPX), and (d) Glutathione Reductase (GST). Control, Tre, Cu, and Tre + Cu correspond to control, 10 mM trehalose, 100 µM $CuSO_4$, and 10 mM trehalose + 100 µM $CuSO_4$, respectively. Bars represent standard deviation (SD) of the mean (n = 3). Different letters indicate significant differences among treatments at $p < 0.05$, according to Duncan's multiple range test [84].

NBT staining indicated an increased amount of O_2^- as scattered dark blue spots in the leaf plate of the Cu-stressed seedlings compared with the non-treated control (Figure 5a). Similarly, DAB staining confirmed a marked increase in brown polymerization products, which indicated the over-accumulation of H_2O_2 in the leaves of the Cu-stressed seedlings relative to control (Figure 5b) [84].

Results described by Mostofa et al. [84] indicate that prolonged exposure to excessive Cu resulted in serious toxic effects on the rice seedlings. In contrast, Tre pretreatment has been shown to be beneficial in alleviating Cu toxicity, which was mainly attributed to the ability of Tre (i) to restrict Cu uptake and accumulation to maintain Cu homeostasis, and (ii) to induce production of antioxidant and Gly enzymes to alleviate excessive Cu-triggered oxidative stress [84].

Figure 5. Effect of exogenous trehalose on ROS accumulation in leaves of rice seedlings with or without Cu stress. (a) Superoxide (O_2^-) and (b) Hydrogen Peroxide (H_2O_2) production in rice leaves were detected using nitro-blue tetrazolium (NBT) solution and 3,3'-diaminobenzidine (DAB), respectively, at day 7 of Cu stress. Control, Tre, Cu, and Tre + Cu correspond to control, 10 mM trehalose, 100 μM $CuSO_4$, and 10 mM trehalose + 100 μM $CuSO_4$, respectively [84].

10. Final considerations

Micronutrient elements such as zinc, boron copper, iron, manganese, molybdenum, and chlorine are responsible to the regulatory activity of the cell organelles. These nutrients are absorbed and found in lower concentrations in plant tissues, and contribute to supply the nutritional exigency of the plant.

Study with *Zea mays* plants exposed to Zn toxicity + Si presented more significant increases in stomatal conductance, net photosynthetic rate, transpiration rate, and water use efficiency compared with the treatment only with Zn. On the contrary, chlorophylls a, b and total and carotenoids presented non-significant increases when compared with plants exposed to Zn toxicity.

This study also revealed the positive contribution of the Si on gas exchange and reduction of the negative effects provoked on chlorophylls and carotenoids in *Zea mays* plants under Zn toxicity. Other result described that prolonged exposure to excessive Cu resulted in serious toxic effects on the rice seedlings. In contrast, Tre pretreatment has been shown to be beneficial in alleviating Cu toxicity, which was mainly attributed to the ability of Tre to restrict Cu uptake and accumulation to maintain Cu homeostasis. It induced production of antioxidant and Gly enzymes to alleviate excessive Cu-triggered oxidative stress.

Stress caused by the excessive supply of micronutrients to plants promotes repercussion on oxidant system, inducing the overproduction of reactive oxygen species. The oxidative damage is a situation characterized by the large ROS accumulation and insufficient detoxification promoted by antioxidant enzymes, such as catalase and glutathione peroxidase. Different mechanisms have been proposed to explain the tolerance of plants to toxicity induced by micronutrients, as uses of other elements and substances, in which it can positively act with specific transporters, metal ion homeostasis and compartmentalization of micronutrients into the vacuole.

Acknowledgements

This chapter had financial support from Fundação Amazônia Paraense de Amparo à Pesquisa (FAPESPA/Brazil), Universidade Federal Rural da Amazônia (UFRA/Brazil) and Conselho Nacional de Pesquisa (CNPq/Brazil) to Lobato AKS.

Author details

Allan Klynger da Silva Lobato[1*], Emily Juliane Alvino Lima[1], Elaine Maria Silva Guedes Lobato[1], Gabriel Mascarenhas Maciel[2] and Douglas José Marques[3]

*Address all correspondence to: allanllobato@yahoo.com.br

1 Núcleo de Pesquisa Vegetal Básica e Aplicada, Universidade Federal Rural da Amazônia, Paragominas, Brazil

2 Instituto de Ciências Agrárias, Universidade Federal de Uberlândia, Monte Carmelo, Brazil

3 Setor de Olericultura e Experimentação, Universidade José do Rosário Vellano (UNIFENAS), Alfenas, Brazil

References

[1] Pavinato, P.S. and Rosolem, C.A. Disponibilidade de nutrientes no solo - Decomposição e liberação de compostos orgânicos de resíduos vegetais. Revista Brasileira de Ciência do Solo. 2008;32:911-920. DOI: 10.1590/S0100-06832008000300001.

[2] Arnon, D.I. and Stout, P.R.. The essentiality of certain elements in minute quantity for plants with special reference to copper. Plant Physiology. 1939;14(2):371-375.

[3] Epstein, E. and Hagen, C.E. A kinetic study of the absorption of alkali cations by barley roots. Plant Physiology. 1952;27:457-474. DOI: 10.1104/pp.27.3.457.

[4] Laviola, B.G and Dias, A.S. Teor e acúmulo de nutrientes em folhas e frutos de pinhão-manso. Revista Brasileira de Ciencia do Solo. 2008;32:1969-1975.

[5] Malavolta, E. Manual de nutrição mineral de plantas.1ed. São Paulo: Agronômica Ceres; 2006.638p.

[6] Borkert, C.M., Gaudêncio, C.A., Pereira, J.E., Pereira, L.R., Oliveira, A.J. Junior. Nutrientes minerais na biomassa da parte aérea em culturas de cobertura de solo. Pesquisa Agropecuária Brasileira. 2003;38:143-153. DOI: 10.1590/S0100-204X2003000100019.

[7] Mengel, K., Kirkby, E. Principles of plant nutrition. 5ed. Dordrecht/Boston/London: Kluwer Academic Publishers; 2001.849p.

[8] Coelho, V.A.T., Rodas, C.L., Coelho, L.C, Carvalho, J.G., Almeida, E.F.A., Figueiredo, M.A. Caracterização de sintomas visuais de deficiências de macronutrientes e boro em plantas de gengibre. Revista Brasileira de Horticultura Ornamental. 2012;18: 47-55.

[9] Welch, R.M and Shuman, R. Micronutrient Nutrition of Plants. Plant Sciences. 1995;14(1):49-82. DOI: 10.1080/07352689509701922.

[10] Andreotti, M., Souza, E.C.A., Crusciol, C.A.C. Componentes morfológicos e produção de matéria seca de milho em função da aplicação de calcário e zinco. Scientia Agricola. 2001;58:321-327.

[11] Cakmak, I.. Possible role of zinc in protecting plant cells from damage by reactive oxygen species. New Phytologist. 2000;146:185–205.

[12] Bastos, A.R.R. and Carvalho, J.G. Absorção radicular e redistribuição do boro pelas plantas, e seu papel na parede celular. Revista Univiversidade Rural, Série Ciencias da Vida. 2004.;24:47-66.

[13] Viégas, I.J.M., Thomaz, M.A.A., Silva, J.F., Conceição, H.E.O., Naiff, A.P.M. Efeito da omissão de macronutrientes e boro no crescimento, nos sintomas de deficiências nutricionais e na composição mineral de plantas de camucamuzeiro. Revista Brasileira de Fruticultura. 2004;26:315-319.

[14] Luchese, A.V., Gonçalves, A.C.J., Luchese, E.B., Braccini, M.C.L. Emergência e absorção de cobre por plantas de milho (*Zea mays*) em resposta ao tratamento de sementes com cobre. Ciência Rural. 2004;34:1949-1952.

[15] Yruela,I.. Copper in plants. Brazilian Journal Plant Physiology. 2005;17(1):145-156.

[16] Marenco, R.A., Lopes, N.F.. Fisiologia Vegetal: Fotossíntese, respiração, relações hídricas e nutrição mineral.Viçosa: Editora UFV; 2005. 451p.

[17] Alexander J.R., Oliveira, M.L.F., Santos, T.C.S., Canton, G.C., Conceição, J.M, Eutrópio, F.J., Cruz, Z.M.A., Dobbss, L.B., Ramos, A.C. Zinco e ferro: de micronutrientes a contaminantes do solo. ESFA [online] http://www.naturezaonline.com.br. 2012 ;10(1): 23-28.

[18] Kuki, K.N., Oliva, M.A., Costa, A.C. The simulated effects of iron dust and acidity during the early stages of establishment of two coastal plant Species. Water Air Soil Pollution. 2009;196:287–295.

[19] Hellín, E., Hernández-Cortés, J.A., Piqueras, A., Olmos, E., Sevilla, F. The influence of the iron content on the superoxide dismutase activity and chloroplast ultrastructure of *Citrus limon*. Developments in Plant and Soil Sciences. 1995;59:247-254. DOI: 10.1007/978-94-011-0503-3_36.

[20] Hoagland, D.R. and Arnon, D.L. The water culture methods for growing plants without soil. California Agriculture Experiment Station. 1950.;347:1-32.

[21] Fernandes, D.S., Soratto, R.P., Kulczynski, S.M., Biscaro, G.A., Reis, C.J. Produtividade e qualidade fisiológica de sementes de feijão em conseqüência da aplicação foliar de manganês. Pesquisa Agropecuária Brasileira. 2007;42:419-426.

[22] Dechen, A.R., Haag, H.P., Carmello, Q.A.C. Mecanismos de absorção e de translocação de micronutrientes In: FERREIRA, M.E., CRUZ, M.C.P. (Ed.) Micronutrientes na agricultura. Piracicaba: Potafós/CNPq, 1991. 134p, p.79-97.

[23] Yuri, J.E., Resende, G.M., Mota, J.H., Gonçalves, L.D.,Souza, R.J. Doses e épocas de aplicação de molibdênio na produção e qualidade de alface americana. Horticultura Brasileira. 2004;22:589-592.

[24] Broyer, T.C., Carlton, A.B., Johson, C.M., Stout, P.R. Chlorine: a micronutrient element for higher plants. Plant Physiology. 1954;29:526-532.DOI 10.1104/pp.29.6.526.

[25] Xu, S., Huang, H.K., Kaiser, P., Latterich, M., Hunter, T. Phosphorylation and spindle pole body localization of the Cdc15p mitotic regulatory protein kinase in budding yeast. Current Biology. 2000;10(6):329-32.

[26] Mascarenhas, H.A.A., Neptune, A.M.L., Muraoka, T., Bulisani, E.A., Hiroce, R. Absorção de nutrientes por cultivares de soja (*Glicine max* (L.) Merril). Revista Brasileira de Ciência do Solo. 1980;4:92-96.

[27] Garbisu, C. and Alkorta, I. Basic concepts on heavy metal soil bioremediation. European Journal of Mineral Processing and Environmental Protection. 2003;3:58-66.

[28] Radic, S., Babic, M., Skobic, D., Roje, V. Ecotoxicological effects of aluminum and zinc on growth and antioxidants in *Lemna minor* L. Ecotoxicology and Environmental Safety. 2010;73(3):336-342. DOI: 10.1016/j.ecoenv.2009.10.014.

[29] Stoyanova, Z. and Doncheva S. The effect of zinc supply and succinate treatment on plant growth and mineral uptake in pea plant. Brazilian Journal of Plant Physiology. 2002;14:111-116. DOI: 10.1590/S1677-04202002000200005.

[30] Jain, R., Srivastava, S., Solomon, S., Shrivastava, A.K., Chandra A. Impact of excess zinc on growth parameters, cell division, nutrient accumulation, photosynthetic pigments and oxidative stress of sugarcane (*Saccharum* spp.). Acta Physiologiae Plantarum. 2010;32:979-986. DOI: 10.1007/s11738-010-0487-9.

[31] Chang, H.B., Lin, C.W., Huang, H.J. Zinc-induced cell death in rice (*Oryza sativa* L.) roots. Plant Growth Regulation. 2005;46(3):261-266. DOI: 10.1007/s10725-005-0162-0.

[32] Molassiotis, A., Sotiropoulos, T., Tanou, G., Diamantidis, G., Therios, I. Boron-induced oxidative damage and antioxidant and nucleolytic responses in shoot tips culture of the apple rootstock EM 9 (*Malus domestica* Borkh). Environmental and Experimental Botany. 2006;56(1):54-62. DOI: 10.1016/j.envexpbot.2005.01.002.

[33] Wang, J.Z., Tao, S.T., Qi, K.J., Wu, J., Wu, H.Q., Zhang, G.S.L. Changes in photosynthetic properties and antioxidative system of pear leaves to boron toxicity. African Journal of Biotechnology. 2011;10:19693-19700.

[34] Landi, M., Remorini, D., Pardossi A., Guidi, L. Boron excess affects photosynthesis and antioxidant apparatus of greenhouse *Cucurbita pepo* and *Cucumis sativus*. Journal of Plant Research. 2013;126(6):775-786. DOI: 10.1007/s10265-013-0575-1.

[35] Gill, S.S. and Tuteja, N. Reactive oxygen species and antioxidant machinery in abiotic stress tolerance in crop plants. Plant Physiology and Biochemistry. 2010;48(12): 909-930. DOI: 10.1016 / j.plaphy.2010.08.016.

[36] Verkleij, J.A.C., Koevoets, P.L.M., Mechteld, M.A., Blake-Kalff, M.M.A., Chardonnens A. Evidence for an important role of the tonoplast in the mechanism of naturally selected zinc tolerance in *Silene vulgaris*. Journal of Plant Physiology. 1998;153(1-2): 188-191. DOI: 10.1016/S0176-1617(98)80064-3.

[37] Williams, G.M., Kroes, R., Munro, I.C. Safety Evaluation and Risk Assessment of the Herbicide Roundup and Its Active Ingredient, Glyphosate, for Humans. Elsevier. 2000;31(2):117-165. DOI: doi:10.1006/rtph.1999.1371.

[38] Lobato, A.K.S., Coimbra, G.K., Neto, M.A.M., Costa, R.C.L., Santos Filho, B.C., Oliveira Neto, C.F., Luz, L.M., Barreto, A.G.T., Pereira, B.W.F., Alves, G.A.R., Monteiro, B.S., Marochio, C.A. Protective action of silicon on relations and photosynthetic pig-

ments in pepper plants induced to water deficit. Research Journal of Biological Sciences. 2009;4(5):617-623.

[39] Pereira, T.S., Lobato, A.K.S., Tan, D.K.Y., Costa, D.V., Uchôa, E.B., Ferreira, R.N., Pereira, E.S., Avila, F.W., Marques, D.J., Guedes, E.M.S. Positive interference of silicon on water relations, nitrogen metabolism, and osmotic adjustment in two pepper (*Capsicum annuum*) cultivars under water deficit. Australian Journal of Crop Science. 2013;7(8):1064-1071.

[40] Marques, D.J., Ferreira, M.M., Lobato, A.K.S., Carvalho, J.G., Carvalho, J.Á., Freitas, W.A., Bastos, A.R.R., Pereira, F.J., Castro, E.M. CaSiO3 improves water potential and gas exchange but not contribute to the production parameters of maize plants exposed to different irrigation depths. Australian Journal of Crop Science. 2014;8(8): 1257-1265.

[41] Isa, M., Bai S., Yokoyama, T., Ma, J.F., Ishibashi, Y., Yuasa, T., Iwaya-Inoue, M. Silicon enhances growth independent of silica deposition in a low-silica rice mutant, lsi1. Plant and Soil. 2010;331(1-2):361-375. DOI: 10.1007/s11104-009-0258-9.

[42] Silva, O.N., Lobato, A.K.S., Avila, F.W., Costa, R.C.L., Oliveira Neto, C.F., Santos Filho, B.G., Martins Filho, A.P., Lemos, R.P., Pinho, J.M., Medeiros, M.B.C.L., Cardoso, M.S., Andrade, I.P. Silicon-induced increase in chlorophyll is modulated by the leaf water potential in two water-deficient tomato cultivars. Plant Soil And Environment. 2012;58(11):481-486.

[43] Kim, S.G., Kim, K.W., Park, E.W., Choi, D. Silicon-induced cell wall fortification of rice leaves: A possible cellular mechanism of enhanced host resistance to blast. Phytopathology. 2002;92(10):1095-1103. DOI: 10.1094/PHYTO.2002.92.10.1095.

[44] Chen, W., Yao, X., Cai, K., Chen, J. Silicon alleviates drought stress of rice plants by improving plant water status, photosynthesis and mineral nutrient absorption. Biological Trace Element Research. 2011;142(1):67–76. DOI: 10.1007/s12011-010-8742-x.

[45] Paula, L.S., Silva, B.C., Pinho, W.C.S., Barbosa, M.A.M., Lobato, E.M.S.G., Batista, B.L., Barbosa Júnior, F., Lobato, A.K.S. Silicon (Si) ameliorates the gas exchange and reduces negative impacts on photosynthetic pigments in maize plants under Zn toxicity. Australian Journal of Crop Science. 2015;9: paper in press.

[46] Sagardoy, R., Vázquez, S., Florez-Sarasa, I.D., Albacete, A., Ribas-Carbó, M., Flexas, J., Abadía, J., Morales, F. Stomatal and mesophyll conductances to CO2 are the main limitations to photosynthesis in sugar beet (*Beta vulgaris*) plants grown with excess zinc. New Phytologist. 2010;187(1):145-158. DOI: 10.1111/j.1469-8137.2010.03241.x.

[47] Pavlíková, D., Pavlík, M., Procházková, D., Zemanová, V., Hnilicka, F., Wilhelmová, N. Nitrogen metabolism and gas exchange parameters associated with zinc stress in tobacco expressing an ipt gene for cytokinin synthesis. Journal of Plant Physiology. 2014;171:559-564.

[48] Ripley, B.S., Gilbert, M.E., Ibrahim, D.G., Osborne, C.P. Drought constraints on C4 photosynthesis: stomatal and metabolic limitations in C3 and C4 subspecies of *Alloteropsis semialata*. Journal of Experimental Botany. 2007;58(6):1351-1363.

[49] Flexas, J., Ribas-Carbó, M., Bota, J., Galmés, J., Henkle, M., Martínez-Cañellas, S., Medrano, H. Decreased rubisco activity during water stress is not induced by decreased relative water content but related to conditions of low stomatal conductance and chloroplast CO_2 concentration. New Phytologist. 2006;172(1):73-82.

[50] Shi, G.R. and Cai, Q.S. Photosynthetic and anatomic responses of peanut leaves to zinc stress. Biologia Plantarum. 2009;53(2):391-394.

[51] Mott, K.A. and Franks, P.J. The role of epidermal turgor in stomatal interactions following a local perturbation in humidity. Plant Cell and Environment. 2001;24(6): 657-662. DOI: 10.1046/j.0016-8025.2001.00705.x.

[52] Ronda, R.J., Bruin, H.A.R., Holtslag, A.A.M. Representation of the canopy conductance in modelling the surface energy budget for low vegetation. Journal of Applied Meteorology. 2001;40(8):1431-1444. DOI: http://dx.doi.org/10.1175/1520-0450(2001)040<1431:ROTCCI>2.0.CO;2.

[53] Sperry, J.S. Hydraulic constraints on plant gas exchange. Agricultural and Forest Meteorology. 2000;104:13-23.

[54] Fernàndez, J., Zacchini, M., Fleck, I. Photosynthetic and growth responses of Populus clones Eridano and I-214 submitted to elevated Zn concentrations. Journal of Geochemical Exploration. 2012;123:77-86. DOI: 10.1016/j.gexplo.2012.01.010.

[55] Ma, C.C., Gao, Y.B., Guo, H.Y., Wang, J.L. Photosynthesis, transpiration and water use efficiency of *Caragana microphylla*, C. intermedia and C. korshinskii. Photosynthetica. 2004;42(1):65-70.

[56] Silva, C.E.M., Gonçalves, J.F.C., Feldpausch, T.R. Water-use efficiency of tree species following calcium and phosphorus application on an abandoned pasture, central Amazonia, Brazil. Environmental and Experimental Botany. 2008;64(2):189-195. DOI: 10.1016/j.envexpbot.2008.03.001.

[57] Moussa, H.R. Influence of exogenous application of silicon on physiological response of salt-stressed maize (*Zea mays* L.). International Journal of Agriculture and Biology. 2006;2:293-297.

[58] Manivasagaperumal, R., Balamurugan, S., Thiyagarajan, G., Sekar, J. Effect of zinc on germination, seedling growth and biochemical content of cluster bean (*Cyamopsistetragonoloba* (L.) Taub). Current Botany. 2011;2:11-15.

[59] Mousavi, S.R. Zinc in crop production and interaction with phosphorus. Australian Journal of Basic and Applied Sciences. 2011;5(9):1503-1509.

[60] Parham, R. And Rebeiz, C.A. Chloroplast Biogenesis 72: A [4-vinyl] chlorophyllide a reductase assay using divinyl chlorophyllide a as an exogenous substrate. Analytical Biochemistry. 1995;231(1):164-169.

[61] Streit, N.M., Canterle, L.P., Canto, M.W., Hecktheuer, L.H.H. The chlorophylls. Ciencia Rural. 2005;35:748-755. DOI: 10.1590/S0103-84782005000300043.

[62] Cazzonelli, C.I. Carotenoids in nature: insights from plants and beyond. Functional Plant Biology. 2011;38(11):833–847. DOI: 10.1071/FP11192.

[63] Croce, R., Müller, M.G., Bassi, R., Holzwarth, A.R. Carotenoid-to-chlorophyll energy transfer in recombinant major light-harvesting complex (LHCII) of higher plants. I. Femtosecond transient absorption measurements. Biophysical Journal. 2001;80(2): 901-915. DOI: 10.1016/S0006-3495(01)76069-9.

[64] Papadakis, I. E., Dimassi, K.N., Bosabalidis, A.M., Therios, I.N., Patakas, A., Giannakoula, A. Boron toxicity in 'Clementine' mandarin plants grafted on two rootstocks. Plant Science. 2004;166(2):539-547. DOI: 10.1016/j.plantsci.2003.10.027.

[65] Tepe, M. And Aydemir, T. Antioxidant responses of lentil and barley plants to boron toxicity under different nitrogen sources. African Journal of Biotechnology. 2011;10:10882-10891. DOI: 10.5897/AJB10.1076.

[66] Madhava, R.K.V. and Sresty, T.V.S. Antioxidative parameters in the seedlings of pigeon-pea (*Cajanus cajan* (L.) Millspaugh) in response to Zn and Ni stresses. Plant Science. 2000;157(1):113-128.

[67] López, M.A.F., Ellis D.R., Grusak M.A. Effect of zinc and manganese supply on the activities of superoxide dismutase and carbonic anhydrase in *Medicago truncatula* wild type and raz mutant plants. Plant Science. 2005;168:1015-1022. DOI: 10.1016/ j.plantsci.2004.11.018.

[68] Hung, S.H., Yu, C.W., Lin, C.H. Hydrogen peroxide functions as a stress signal in plants. Botanical Bulletin of Academia Sinica. 2005;46:1-10.

[69] Sairam, R.K. and Srivastava, G.C. Induction of oxidative stress and antioxidant activity by hydrogen peroxide treatment in tolerant and susceptible wheat genotypes. Biologia Plantarum. 2000;43(2):381-386. DOI: 10.1023/A:1026730008917.

[70] Bettaieb, T., Denden, M., Mhamdi, M. In vitro regeneration and physiological characterization of Gladiolus (*Gladiolus grandiflorus* Hort.) Somaclonal variants tolerant to low temperatures. Tropicultura. 2008;26:10-16.

[71] Kowalewska, G., Lotocka, M., Latala, A. Formation of the copper-chlorophyll complexes in cells of phytoplankton from the Baltic Sea. Polskie Archiwum Hydrobiologii. 1992;39:41-49.

[72] Wettstein, D.V., Gough, S., Kannangara, C.G. Chlorophyll biosynthesis. Plant Cell. 1995;7(7):1039-1057. DOI: 10.1105/tpc.7.71039.

[73] Küpper, H., Küpper, F., Spiller, M. Environmental relevance of heavy metal substituted chlorophylls using the example of water plants. Journal of Experimental Botany. 1996;47(2):259–266. DOI: 10.1093/jxb/47.2.259.

[74] Bassi, R. and Sharma, S.S. Proline accumulation in wheat seedlings exposed to zinc and copper. Phytochemistry. 1993; 33:1339-1342.

[75] Silva, P.F.N., Lobato, E.M.S.G., Souza, P.R., Santos, H.J.M., Braga, R.O., Lobato, A.K.S.

[76] Proline and total glutathione are accumulated in leaf of young *Schizolobium parahyba* var. *amazonicum* plants exposed to boron toxicity. Brazilian Journal of Botany. 2015;38: paper in press.

[77] Paula, M.T., Filho, B.G.S., Cordeiro, Y.E.M., Conde, R.A., Neves, P.A.P.F.G. Ecophysiology the brazilian mahogany (*Swietenia macrophylla* King) agroforestry systems in Santa Barbara-Pa. Enciclopedia Biosfera. 2013;9:813-824.

[78] Cervilla, L.M., Blasco, B., Ríos, J.J., Rosales, M.A., Rubio-Wilhelmi, M.M., Rodríguez, E.S., Romero, L., Ruiz, J.M. Response of nitrogen metabolism to boron toxicity in tomato plants. Plant Biology. 2009;11(5):671-677. DOI: 10.1111/j.1438-8677.2008.00167.x.

[79] Hong, Z., Lakkineni, K., Zhang, Z., Verma, D.P.S. Removal of feedback inhibition of Δ1-pyrroline-5- carboxylate synthetase results in increased proline accumulation and protection of plants from osmotic stress. Plant Physiology. 2000;122:1129-1136. DOI: http://dx.doi.org/10.1104/pp.122.4.1129.

[80] Cervilla, L.M., Blasco, B., Rios, J.J., Rosales, M.A., Sánchez-Rodríguez, E., Rubio-Wilhelmi, M.M., Romero, L., Ruiz, J.M. Parameters symptomatic for boron toxicity in leaves of tomato plants. Journal Botany. 2012;2012:1-17. DOI: 10.1155/2012/726206

[81] Cervilla, L.M., Blasco, B., Ríos, J.J., Romero, L., Ruiz, J.M. Oxidative stress and antioxidants in tomato (*Solanum lycopersicum*) plants subjected to boron toxicity. Annals of Botany. 2007;100(4):747-756. DOI: 10.1093/aob/mcm156.

[82] Contreras, C., Montoya, A., Pacheco, P., Martinez-Ballesta, M.C., Carvajal, M., Bastias, E. The effects of the combination of salinity and excess boron on the water relations of tolerant tomato (*Solanum lycopersicum* L.) cv. Poncho Negro, in relation to aquaporin functionality. Spanish Journal of Agricultural Research. 2011;9:494-503.

[83] Costa, R.C.L., Lobato, A.K.S., Silveira, J.A.G., Laughinhouse, I.V.H.D. ABA-mediated proline synthesis in cowpea leaves exposed to water deficiency and rehydration. Turkish Journal of Agriculture and Forestry. 2011;35:309-317.

[84] Szabados, L. and Savouré, A. Proline: a multifunctional amino acid. Trends Plant Science. 2010;15(2):89-97. DOI: 10.1016/j.tplants.

[85] Mostofa, M.G., Hossain, M.A., Fujita, M., Tran, L.S. Physiological and biochemical mechanisms associated with trehalose-induced copper-stress tolerance. Scientific Repost. 2014;5:11433. DOI: 10.1038/srep11433.

8

Effect of Salinity Stress on Gene Expression in Black Tiger Shrimp *Penaeus monodon*

Shekhar S. Mudagandur, Gopikrishna Gopalapillay and Koyadan K. Vijayan

Abstract

Penaeid shrimps are euryhaline in nature and have the ability to survive and adapt in a wide range of salinities (3–50 ppt). The shrimps are cultured under a variety of conditions in many tropical and subtropical countries. Osmotic and ionic regulation is an important mechanism of environmental adaptation in crustaceans. However, drastic changes in abiotic and biotic conditions result in stress to the shrimps during the culture period. Salinity and temperature are the two major environmental factors that have huge impact on culture shrimp, affecting their physiological and metabolic parameters, which in turn affect shrimp growth, molting, and survival. Changes in the abiotic factors, chemical and biotic factors result in reduced immunity of shrimp and vulnerability to bacterial and viral diseases. This chapter describes the effects of low and high salinity on the gene profile changes of black tiger shrimp *Penaeus monodon*, and the functional role of these genes in shrimp salinity stress is discussed.

Keywords: *Penaeus monodon*, Salinity stress, Differentially expressed genes

1. Introduction

Penaeid shrimps being euryhaline can adapt to thrive and survive in a wide range of salinity conditions. Shrimps are, therefore, cultured and reared under different farming conditions in tropical and subtropical countries. Water quality management is an important criterion in shrimp farming for survival and growth of the shrimp. The optimal salinity conditions for penaeid shrimp ranges differently for different species. *Penaeus monodon,* which can tolerate low salinity of 5 ppt to high salinity conditions of 40 ppt has optimal range of salinity (15–25 ppt) for optimal growth [1]; juvenile *Penaeus chinensis* grows best at 20–30 ppt salinity range [2]; the optimal conditions of salinity for growth was estimated to be in the range of 22–34 ppt for *Penaeus latisulcatus* [3]. *Penaeus semisulcatus*, an Indo-Pacific species requires higher salinity (30–35 ppt) for growth [4]. Best performance (growth, survival, total biomass) for *Penaeus indicus* PL20 and PL60 at salinities between 20 and 30 ppt was observed after acclimation period [5]. Highest increase in biomass and production was observed at 25 ppt for *Penaeus merguiensis* [6]. *Litopenaeus vannamei*, a native species of the Pacific coast is a very important shrimp species, which is cultured under semi-intensive and intensive conditions in many parts of world. The juveniles of this species have optimal growth and survival in salinity range of 33–40 ppt [7]. It is also preferred for culture in low salinity water as it can tolerate very low salinities of 1–2 ppt.

However, the variable climate conditions result in drastic changes in abiotic factors causing stress to the shrimps during the culture period, which influences culture of euryhaline penaeids. In summer months, there is increase in salinity in ponds due to high evaporation rates and in rainy season the salinity decreases. The marine crustaceans are generally osmotic and ionic conformers in nature. The shrimps hypo-osmoregulate above the iso-osmotic point and hyper-osmoregulate below the iso-osmotic point through a osmoregulation mechanism.

In crustaceans, gills which are highly permeable external surfaces are the primary sites involved in osmoregulation. In the larval stages of penaeids, the typical features of osmoregulatory epithelia are present in pleurae and branchiostegites [8]. The overall ion-transport and osmotic regulation process involves ions absorbtion or excretion between the external and internal medium through osmoregulatory organs, such as gills, the antennal glands that mainly function in urine production, and the gut in decapod crustaceans [9-10]

The crustaceans have two well-known important enzymes which are central to osmotic and ionic regulation and ion uptake. The transepithelial movement of monovalent ions requires the action of Na^+/K^+-ATPase or the sodium pump utilizing ATP as energy source. The other major enzyme involved in osmoregulation is transport-related enzyme carbonic anhydrase, which plays a role in producing H^+ and HCO^{3-} through catalysis of respiratory CO_2, for counterions in Na^+ and Cl^- uptake. In penaeids, carbonic anhydrase is reported to be involved in both hyper- and hypo-osmotic regulation and is induced against low and high salinity exposure, indicating its role in ion uptake and excretion process [11].

In our study, we have constructed suppression subtractive hybridization (SSH) cDNA libraries to identify differentially expressed genes in shrimp *P. monodon*, in response to salinity stress. The SSH clones obtained from the forward SSH cDNA libraries (Figure 1) and reverse SSH cDNA libraries (Figure 2) constructed from gut tissues of shrimp exposed to low (3 ppt) and high (55 ppt) salinity on BLAST analysis, revealed several functional categories.

Similarly, we obtained several functional categories of genes from the forward SSH cDNA libraries (Figure 3) and reverse SSH cDNA libraries (Figure 4) constructed from gill tissues of shrimp exposed to low (3 ppt) and high (55 ppt) salinity conditions. These differentially expressed genes were subjected to RT-qPCR for gene expression analysis. Based on our study, we discuss herein in this chapter some of the important genes identified as differentially expressed in response to salinity stress in shrimp.

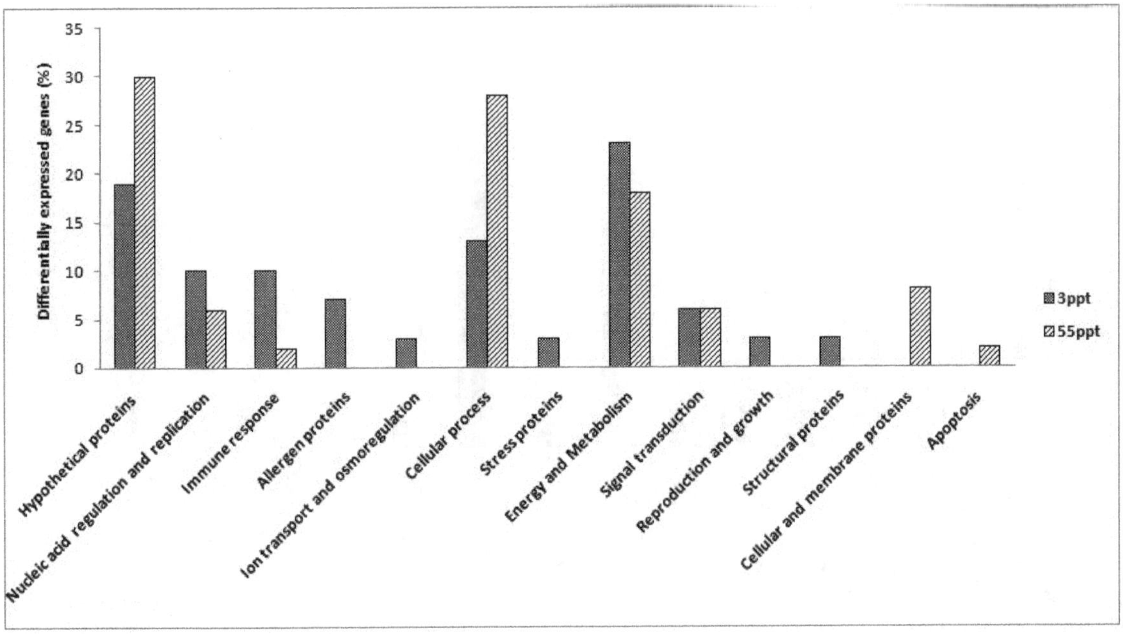

Figure 1. Differentially expressed genes from the forward SSH library of gut tissues of *P. monodon* under low (3 ppt) and high (55 ppt) salinity conditions

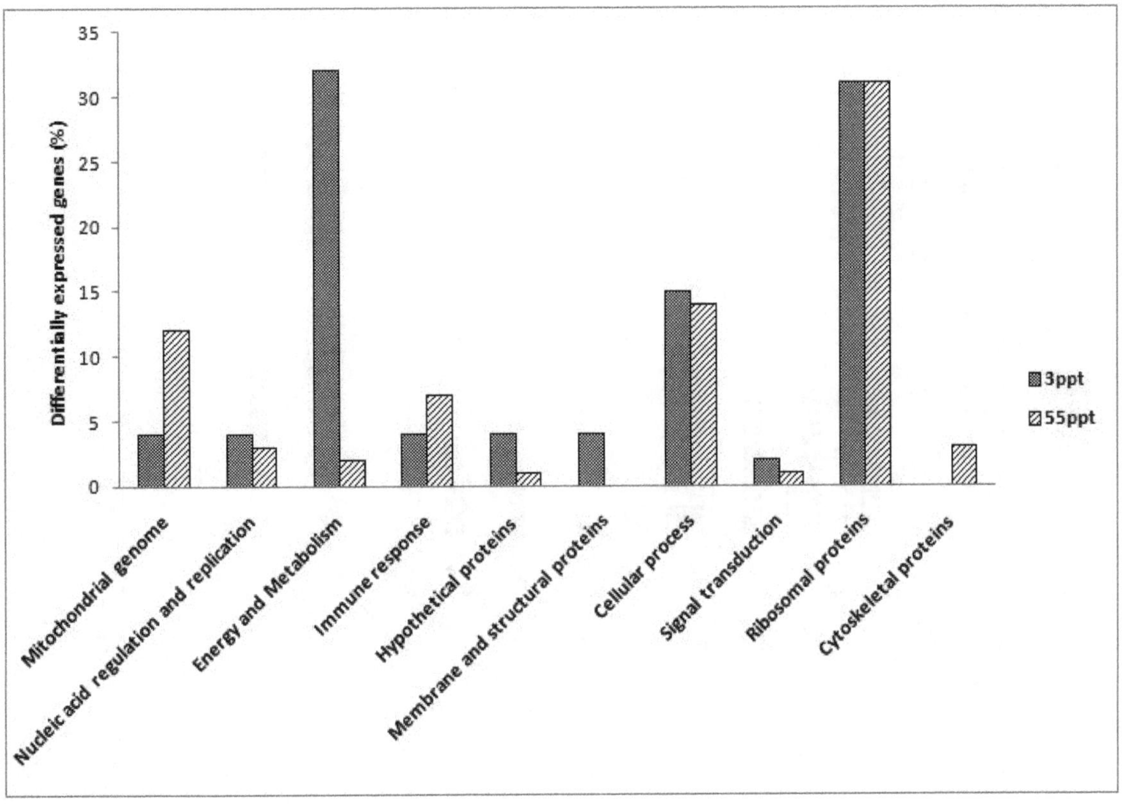

Figure 2. Differentially expressed genes from the reverse SSH library of gut tissues of *P. monodon* under low (3 ppt) and high (55 ppt) salinity conditions

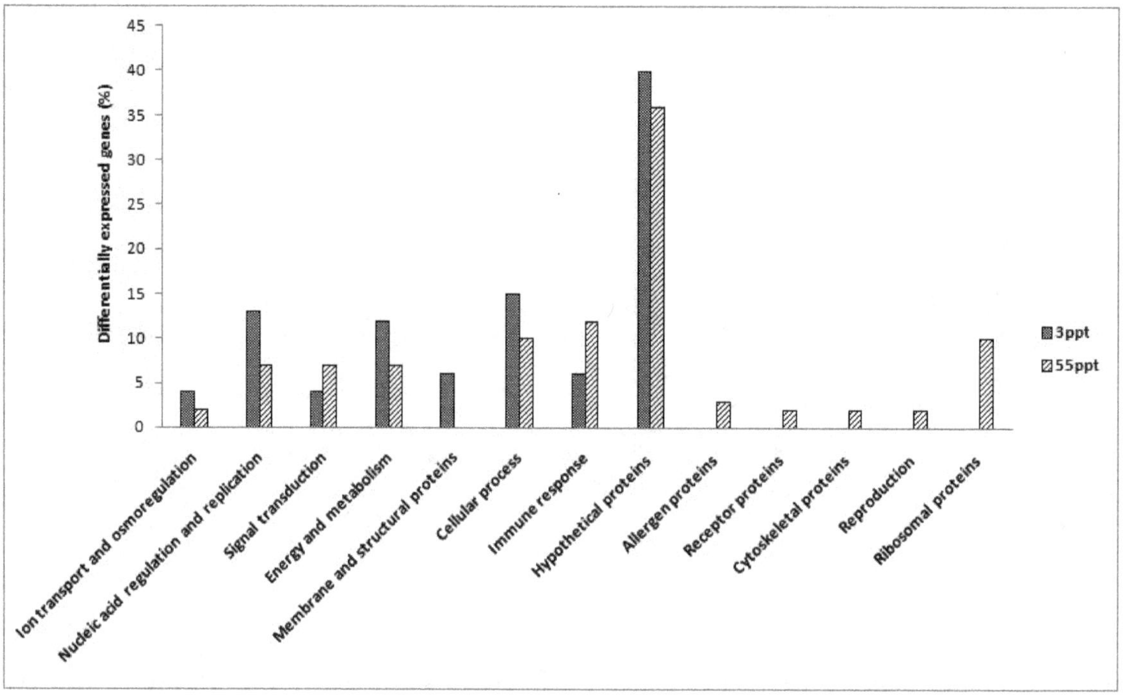

Figure 3. Differentially expressed genes from forward SSH library of gill tissues of *P. monodon* under low (3 ppt) and high (55 ppt) salinity conditions

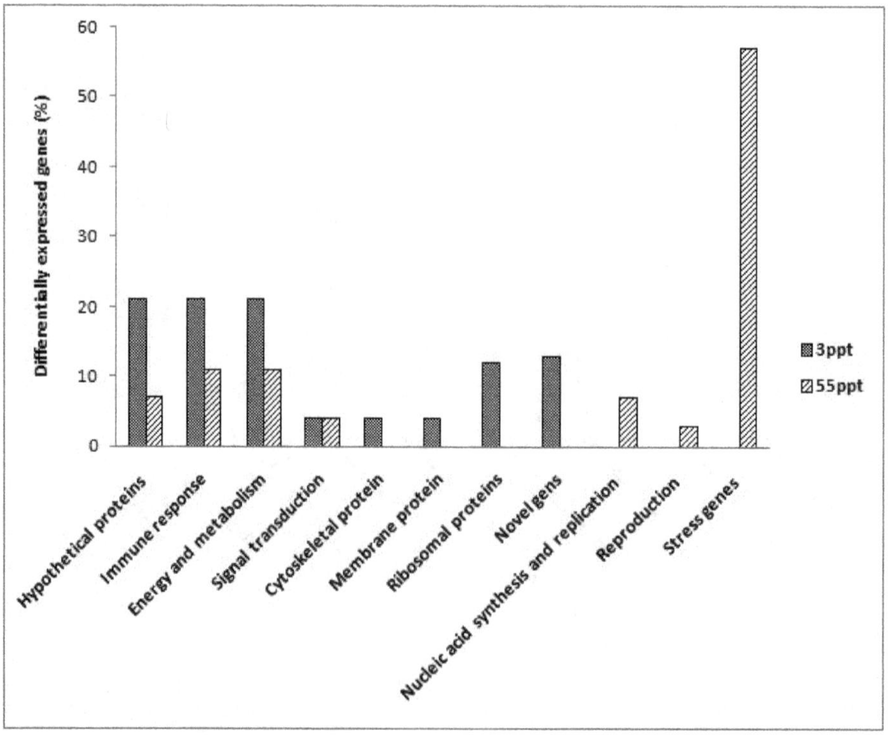

Figure 4. Differentially expressed genes from the reverse SSH library of gill tissues of *P. monodon* under low (3 ppt) and high (55 ppt) salinity conditions

2. Osmoregulatory genes

2.1. Na^+/K^+-ATPase

Na^+/K^+-ATPase, a transmembrane protein, contains three subunits, α-, β, and γ- subunit, which are involved in exporting three Na^+ from cytosol in exchange for two K^+ or NH_4^+ from extracellular fluid for each ATP hydrolyzed. The crustacean α-subunit, which is 71–74% identical in amino acid sequence to those of vertebrate α-subunit sequences, binds to ATP and functions for the catalytic action of the enzyme [12]. The binding of Na^+/K^+-ATPase complex to basolateral membrane requires participation of β-subunit and the γ-subunits of enzyme [13]. The activity of Na^++K^+-ATPase in gill tisues of crustaceans depends on the osmoconcentration gradient occurring between hemolymph and the external medium. In crustaceans, there is an increase in Na^+/K^+-ATPase activity when transferred from natural seawater to dilute seawater [14-15]. With the lowering of salinity when compared to that of normal seawater, the euryhaline crustacea undergo hyperosmoregulation. Increased enzymatic activity of Na^++K^+-ATPase and increased α-subunit gene expression has been observed in the gill tissues of crabs [12]. Substantial increase in Na^+/K^+-ATPase specific activity (300%), Na^+/K^+-ATPase protein levels (200%), and gene expression level of α-subunit (150%) has been observed in blue crab *Callinectes sapidus* crabs during acclimitization to dilute seawater of 10 ppt salinity [16]. *P. monodon* reared in 7 ppt seawater showed drastic morphological alterations of the antennal glands. The shrimps also showed higher expression and activity of the enzyme Na^+/K^+-ATPase in the antennal glands under low salinity conditions [17]. *L. vannamei* when transferred to different low salinity conditions ranging from 15 ppt to 1 ppt revealed no significant difference within 3 h for Na^+/K^+-ATPase α-subunit gene expression and enzyme activity. However, there was a rapid increase at 6 h followed by decrease in the expression level from 12 h to 24 h suggesting Na^+/K^+-ATPase is stimulated by salinity stress [18]. The study involving V-H ATPase α -subunit and Na^+/K^+-ATPase β-subunit response to environmental stress (bacteria, pH, Cd, salinity, and low temperature) revealed both the genes to be responsive to these environmental stress conditions. However, the V-H ATPase α -subunit and the Na^+/K^+-ATPase β-subunit, which is involved in proper folding and transport of Na^+/K^+-ATPase enzyme, were found to be more sensitive to salinity stress when compared to other stress factors. The exposure of *L. vannamei* to salinity stress resulted in significant changes in the expression of V-H ATPase α-subunit and Na^+/K^+-ATPase β-subunit gene expression levels in the hepatopancreas and gills of the shrimp. Na^+/K^+-ATPase β-subunit gene expression after exposure to 5 ppt increased to a highest level (17-fold) at 12 h in the gill tissues, whereas, in hepatopancreas the maximum gene expression levels (4.4-fold) were observed 6 h after exposure to 10 ppt salinity conditions [19].

At low (3 ppt) salinity conditions stress conditions, significant increase in the Na^+/K^+-ATPase α-subunit gene expression levels was observed in gill (34.28-fold) tissues of *P. monodon* [20]. At higher salinity stress of 55 ppt, *P. monodon* Na^+/K^+-ATPase gene responded to salinity stress conditions with significant expression levels in gill (15.23-fold) tissues [21]. These results suggests that *P. monodon* Na^+/K^+-ATPase gene is involved in osmoregulatory process in shrimp and responds significantly under salinity stress.

3. Immune genes

Crustins, belonging to family of antimicrobial peptides (AMPs) are shown to be differentially expressed in response to various immunostimulants and microbes [22]. Different isoforms of crustins have been isolated from a variety of penaeid shrimps [23-25]. The up-regulation of crustin-like AMP in shrimps suggests functional role of AMPs primarily in the shrimp immune response [26]. Under hyperosmotic stress, crustinPm5, crustinPm1, and crustin-like Pm gene levels were found to be up-regulated [27]. In our study, we observed crustin gene was up-regulated in all the three shrimp tissues of gills, gut, and muscle analyzed at low salinity stress [20]. Hence, crustins having antimicrobial activity also functionally respond to salinity stress in shrimps.

Another class of AMPs, the penaeidin gene was found to be down-regulated in gill, gut, and up-regulated in antennal gland tissues of shrimp exposed to high salinity stress (55 ppt) conditions [21]. These changes in the gene expression levels of penaeidin may be regulated through variations in hemocyte numbers in salinity-stressed shrimp. Antibacterial proteins such as lysozymes are involved in nonspecific innate immunity in shrimps [28]. Significant up-regulation of lysozyme gene expression occurred in gill (16.25-fold) tissues of shrimp exposed to low salinity stress [20]. In these shrimps, the anti-lipopolysaccharide factor, which is considered to play an important role in shrimp immune response [29], was significantly down-regulated in gill and gut tissues as compared to up-regulation in the muscle tissues of shrimp.

In general, as a result of environmental stress, the expression of immune-related shrimp genes transcripts gets affected [30]. The shrimp exhibits reduction in immune parameters when exposed to salinity stress conditions, leading to decreased resistance against pathogens [31-32]. White shrimp *L. vannamei*, when transferred to low salinity conditions showed significant decrease in hemocyte count, phenoloxidase activity, respiratory burst, and superoxide dismutase activity, which further reduced on challenge with *Vibrio alginolyticus*. Hence, innate immunity in shrimps reduces with combined effect of low salinity and bacterial challenge [32].

4. Cellular-process-related genes

4.1. Ubiquitin-conjugating enzyme

The ubiquitin proteolytic system is involved in various cellular processes such as cell cycle regulation, cellular response to stress, and immune response [33-34]. In shrimps, the ubiquitin-conjugating enzyme E2r (UBE2r) gene isolated from *Marsupenaeus japonicus* showed that the gene expression level changed significantly in the developing testis and ovary with higher level in the testis than in the ovary, which indicates the functional role of UBE2r in oogenesis and spermatogenesis of shrimp [35]. The gene expression pattern of ubiquitin in abdominal muscle of 3-month-old shrimps *L. vannamei* at different molt stages on examination did not increase significantly in premolt stages and was found to be relatively stable at all stages of the molt cycle [36]. Antiviral function has been demonstrated with ubiquitin-conjugating enzyme, E2, isolated from *F. chinensis*, which could inhibit WSSV replication and ubiquitinate WSSV RING domain-containing proteins [37]. In *P. monodon* ubiquitin-conjugating enzyme E2, gene was up-regulated in muscle and gill tissues under low salinity stress conditions [20]. In American lobster, *Homarus americanus* the polyubiquitin gene expression revealed significant changes in abdominal muscle and hepatopancreas during osmotic stress, indicating that proteins may be more susceptible to ionic fluctuations [38].

4.2. Cathepsins

The proteases are distributed into four major classes: aspartyl proteases, metalloproteases, serine proteases, and cysteine proteases [39]. The cathepsins belonging to cysteine proteases in addition to cellular protein degradation and turnover are also involved in numerous other physiological processes. In fishes, the cathepsin La isoform has been implicated in yolk-processing mechanism during oogenesis and embryogenesis [40]. In shrimps (*Metapenaeus ensis*), cathepsin L, with predominant expression in hepatopancreas, is suggested to have a role in food digestion. The immunolocalization of cathepsin L in the nucleus of oocyte suggests its specific physiological role in shrimp oocytes [41]. The cathepsin L-like proteinases isolated from *L. vannamei* showed that specific activity and mRNA expression of cysteine protease were associated during the molt cycle [42]. The expression of the cathepsin C gene expression in hemocytes of *L. vannamei* could be induced after *V. alginolyticus* challenge, which reached a maximum level at 4 h post challenge period, indicating that the cathepsins in shrimp may be involved in immune response [43]. The aspartic protease cathepsin D obtained from *M. japonicus* and *P. monodon* were characterized for physical and chemical properties and were shown to have identical subunits and similar optimal pH [44]. The analysis of cathepsin B gene expression in *L. vannamei* tissues revealed high expression levels in midgut gland and gut, which are involved in food digestion process. The changes in the gene expression levels and enzymatic activity were induced by starvation in the midgut gland of the white shrimp [45]. Significant increase in the cathepsin B gene expression levels in gills, gut, and muscle tissues of shrimp exposed to low (3 ppt) salinity conditions suggest that *P. monodon* cathepsin gene responds to the shrimp adaptive mechanism to salinity stress [20].

5. Signal transduction genes

5.1. 14-3-3 protein

The members of the 14-3-3 protein family are dimeric proteins that are expressed in a wide range of organisms and tissues. They are involved in modulation of protein interactions through phosphorylation process. The other diverse functional roles include interaction with a large number of protein kinases, DNA, Raf kinase, and regulation of cell cycle progression [46]. In plants, 14-3-3 protein activates and regulates plasma membrane H^+-ATPase through fusicoccin responsive system [47]. Two Na^+/K^+-ATPase α -subunit forms were detected in the gill transcripts of crab *Pachygrapsus marmoratus*, which differed in presence of an 81-nucleotide sequence near the translation start site in the D form when compared to the C form. The extended D form was found to contain the binding motif for the regulatory protein 14-3-3, suggesting that Na^+/K^+-ATPase may be stimulated by this regulatory protein binding [48]. The two isoforms of 14-3-3 proteins identified from the *P. monodon* were shown to have varied gene expression profiles during salinity adaptation in response to hypo-osmotic (3 ppt) or hyper-osmotic (40 ppt) salinity stress conditions. Induction of 14-3-3B gene expression in gills of shrimp acclimated to low salinity water suggests that it is likely to be involved in controlling osmoregulation in *P. monodon* under hyperosmotic conditions [49]. The up-regulation of the 14-3-3 gene expression in the tissues of shrimp exposed to low (3 ppt) or high (55 ppt) salinity conditions suggests that *P. monodon* 14-3-3 gene may have a potential role in shrimp response to salinity stress [20-21].

5.2. Calreticulin

Calreticulin, a versatile lectin-like chaperone and important endoplasmic reticulum luminal resident protein, is involved in Ca^{2+} homeostasis and molecular chaperoning. Calreticulins, which are highly conserved in most of the eukaryotes, are involved in the synthesis of various molecules and in many other biological and physiological processes of an organism. The highest expression of calreticulin was detected in ovary of *F. chinensis.* The gene expression varied at different molting stages and could be induced by heat shock and WSSV challenge, indicating multifunctional role of calreticulin [50]. The *P. monodon* calreticulin showed changes in expression profile in response to high-temperature stress, indicating its potential as a biomarker for stress responses in shrimps [51]. Calreticulin also responds to salinity stress with significant increase in the gene expression levels in gills and muscle tissues of *P. monodon* [20].

5.3. Innexins

Innexins, which are members of large multigene families, are involved in formation of gap junctions for cell-to-cell communication [52]. In crustaceans, innexin expression has been associated with developing lobster stomatogastric nervous system [53]. In *P. monodon,* the transcripts of innexin-2-like protein showed increased expression in response to yellow head viral disease [54]. Innexin-2 was found to be more abundantly expressed in testes than ovaries of *P. monodon,* indicating functional role of innexin-2 in spermatogenesis but not in oogenesis [55]. The high gene expression observed for innexin 2 with 14.43-fold in muscle tissue of shrimp under low salinity stress of 3 ppt indicates gap junctions regulation during salinity stress in shrimps [20].

6. Energy and metabolism genes

6.1. Arginine kinase

Arginine kinase plays a major role in energy metabolism and is a phosphotransferase that catalyzes the reversible transfer of phosphate from phosphoguanidine to ADP, resulting in generation of ATP [56]. In shrimps, Penm2 allergen gene having conserved guanidino specificity region and showing very high sequence similarity with crustacean arginine kinase has been isolated and characterized from *P. monodon* [57]. In crustaceans, such as in blue crab *Callinectes sapidus*, the arginine kinase gene expression is associated with salinity changes. The arginine kinase flux reduced under hyperosmotic treatments and increased with the hypo-osmotic treatments [58]. *M. japonicus* when subjected to severe hypoxic stress revealed up-regulation of arginine kinase indicating metabolic response of arginine kinase under hypoxic stress [59]. Arginine kinase was found to be differentially expressed and up-regulated in WSSV-infected blue shrimp (*Penaeus stylirostris*) [60] and the gene expression could be induced against LPS immunostimulation in *L. vannamei* indicating its correlation with immune response in shrimps [61]. The differential expression of arginine kinase in gills, gut, and muscle tissues of shrimp exposed to low (3 ppt) salinity conditions indicates that arginine kinase plays an important role in metabolic process associated with salinity stress in crustaceans [20].

6.2. Ferritin

Ferritin plays a functional role in iron storage and metabolism. In shrimps, this large multi-functional and multisubunit protein gene has been isolated and characterized from *L. vannamei*, which revealed that ferritin is expressed in most of the tissues of shrimp with major expression in hemocytes [62]. The administration of ferritin resulted in increased immune response in *L. vannamei* with enhanced survival rate in WSSV-challenged shrimps and maintained physiological homeostasis of shrimps [63]. In *M. rosenbergii*, the isolated ferritin gene showed conserved domain for the ferroxidase center and the administration of iron enhanced expression of ferritin gene in a tissue-specific manner [64]. The recombinant ferritin was shown to confer protection in *P. monodon* infected with *Vibrio harveyi* [65]. The expression of ferritin mRNA could be induced with heavy metal ions Cu^{2+} and Zn^{2+} and WSSV challenge in *F. chinensis* [66]. The gill tissues of *P. monodon* when subjected to low (3 ppt) salinity stress revealed significant increase in ferritin gene expression (8.79-fold), indicating its functional role in salinity stress in shrimps [20].

6.3. Intracellular fatty-acid-binding proteins

Intracellular fatty-acid-binding proteins (FABPs) are lipid-binding proteins that help in transport of fatty acids across extra- and intracellular membranes and are involved in various other biological processes such as modulation of signal transduction; gene transcription, especially of lipid metabolism; and cell growth and differentiation [67]. FABPs have been well-characterized in vertebrates as compared to that in invertebrates. In crustaceans, the FABP cDNA having fatty-acid-binding motifs has been cloned and characterized from the freshwater crayfish *Pacifastacus leniusculus* and *P. monodon* [68]. The activity of specific Na^+, K^+, Ca^{2+}, and Cl^- ion channels are regulated by various fatty acids. These ion channel regulations may be carried out directly through fatty acid interactions [69]. The shrimp (*P. monodon*) gut tissue revealed highest gene expression level of FABP (14.05-fold) at high salinity stress conditions (55 ppt) and at low salinity stress conditions (3 ppt) with 13.30-fold; the osmoregulatory process may therefore involve the FABPs in shrimps [20-21].

6.4. Acyl-CoA binding protein

Acyl-CoA binding protein (ACBP) is a highly conserved protein. In yeast it is involved in transportation of acyl-CoA esters from the fatty acid synthetase to acyl-CoA-consuming process [70]. The protein, which was first identified in mammals, acts as a neuropeptide that prevents binding of diazepam/endozepine to GABA receptor system [71-72] and is also involved in regulation of several acyl-CoA-dependent processes [73]. In addition, ACBP is involved in many other functions, which include regulating biosynthesis of fatty acid, functional regulation of enzymes and genes, intracellular acyl-CoA pool regulation, acyl-CoA esters donation required for β-oxidation and vesicular trafficking [74], and in regulation of m-calpain [75]. In plants *Arabidopsis thaliana*, different types of ACBPs are encoded [76], such as the ACBP1 and ACBP2, which are membrane-associated proteins [77-80]; ACBP3, which is the extracellularly targeted protein [81]; and ACBP4, ACBP5, and ACBP6, which are the cytosolic proteins [82-83]. ACBPs are involved in abiotic stress tolerance in plants. The ACBP2 in *Arabidopsis* was shown to be involved in heavy metal (Cd) tolerance [84-85]. ACBP6 and ACBP1 are functionally involved in increased freezing tolerance [78],[86]. In shrimp, ACBP functions in antibacterial [87] and antiviral response [88]. The ACBP gene in *P. monodon* was identified to be differentially expressed in the SSH libraries constructed from the gut tissues of both low (3ppt) and high (55ppt) salinity stressed shrimps. The complete cDNA sequence of ACBP consisted of 273 bp ORF coding for 90 amino and showed ligand-binding conserved domains similar to the other members of ACBP family. At 2 weeks post 3 ppt salinity stress conditions, a significant increase in the ACBP transcripts expression was observed in gills and muscle tissues with highest levels in the gut tissues (28.08-fold). Similar increase in the gene expression levels was observed in shrimps exposed to high salinity stress conditions of 55 ppt in gills and muscle tissues with gut tissues revealing high (11.95-fold) levels of gene expression [89]. These results suggest a functional role of ACBP gene during salinity stress in shrimps.

6.5. Catechol-O-methyltransferase

O-methyltransferase (OMT) is an enzyme found in a wide range of organisms such as microbes [90], where it is involved in antibiotic biosynthesis [91], and in fungi, where it is involved in biosynthesis of aflatoxins [92]. The OMT found in plants are well characterized for their functional role in O-methylation during biosynthesis of lignin, stress resistance, and disease tolerance [93]. In crustaceans, farnesoic acid O-methyltransferase (FAMeT) catalyzes farnesoic acid methylation resulting in production of isoprenoid methyl farnesoate, which is involved in metabolic and physiological regulation [94]. Catechol-O-methyltransferase (COMT), which is a type of O-methyltransferases, helps in catalyzing the transfer of methyl group to the hydroxyl group of catechol compounds from S-adenosyl-L-methionine. In higher animals, the COMT helps in catalysis of methylation of various macromolecules that are involved in different functional and regulatory purposes and is present in soluble and membrane-bound forms [95]. In shrimps, the construction of suppression subtractive hybridization (SSH) libraries from *P. monodon* gill tissues resulted in identification of COMT gene as one of the differentially expressed genes subjected to salinity stress. The COMT gene was differentially regulated in both the SSH libraries generated from low and high salinity conditions. The ORF of COMT gene of 666 bp size revealed the coding protein with 221 amino acids [96]. The *P. monodon* COMT showed the conserved domains present in superfamily of S-adenosylmethionine-dependent methyltransferases, which includes COMT, CCoAOMT family of indolethylamine N-methyltransferase from humans, and OMT from *Bacillus subtilis* [97]. The *P. monodon* COMT was found to be up-regulated in low and high salinity stress conditions at different time intervals in shrimp tissues (gills, guts, and muscles), suggesting a functional role of this gene in salinity stress tolerance in shrimps [96].

7. Stress genes

7.1. Heat shock proteins

Heat shock proteins (HSPs) initially discovered in *Drosophila melanogaster* are a highly conserved set of polypeptides present in both prokaryotic and eukaryotic cells. They are generally involved in conferring thermotolerance as molecular chaperones by refolding the denatured proteins and also respond against various other stresses. They play a crucial role in organisms' response to heat shock and cellular stress. In addition, the HSPs are also important for cellular damage protection and in maintaining cellular homeostasis [98]. In aquatic animals, the HSPs respond to environmental pollutants, abiotic stress, and are involved in disease resistance against viral and bacterial pathogens. The functional role and significance of HSP in farmed aquatic animals is demonstrated in stimulating the immune response [99]. In shrimp (*P. monodon*), some of the HSPs such as HSP21 have been characterized for gene expression against WSSV infection [100]; HSP70 gene expression was found to increase in the shrimp hemocytes after heat shock treatment [101]; and HSP90 gene expression has been related with the ovarian maturation [102]. The transcriptional levels of HSP21, HSP70, and HSP90 were inducible under the heat shock and responded upon bacterial exposure in *P. monodon* [103]. Expression of HSP70, which is one of the widely studied HSP in aquatic organisms, was high during short-term hyperthermic treatment when compared to hypoxic and osmotic stress in *P. monodon* [104]. The significant increase in gene expression level HSP21 in the gut and muscle tissues of *P. monodon* exposed to low salinity stress (3 ppt) conditions, indicates its functional role in osmotic stress in shrimps [20].

In conclusion, the construction of SSH cDNA library in response to low (3 ppt) and high (55 ppt) salinity stress in shrimp (*P. monodon*) led to identification of various differentially expressed genes. The significant up-regulation expression of several genes at transcription level in gills, gut, antennal gland, and muscle tissues of shrimp in response to two-week post-salinity stress condition indicates their functional role in gene pathways and regulatory mechanism of osmotic stress at the molecular level.

Acknowledgements

The authors wish to thank NAIP (ICAR) for the financial support provided under the NAIP project "Bioprospecting of genes and allele mining for abiotic stress tolerance."

Author details

Shekhar S. Mudagandur*, Gopikrishna Gopalapillay and Koyadan K. Vijayan

*Address all correspondence to: msshekhar@ciba.res.in

Central Institute of Brackishwater Aquaculture, Chennai, India

References

[1] Ferraris RP, Parado-Estepa FD, DeJesus EG, Ladja JM. Osmotic and chloride regulation in the haemolymph of the tiger prawn Penaeus monodon during molting in various salinities. *Marine Biol.* 1986;95:377-385.

[2] Jiann-Chu C, Jin-Nien L, Chung-Tin C, Min-Nan L. Survival, growth and intermolt period of juvenile *Penaeus chinensis* (Osbeck) reared at different combinations of salinity and temperature. *J* Exper *Marine Biol Ecol.* 1996;204:169-178.

[3] Sang HM, Fotedar R. Growth, survival, haemolymph osmolality and organosomatic indices of the western king prawn (*Penaeus latisulcatus* Kishinouye, 1896) reared at different salinity. *Aquaculture.* 2004;234:601-614.

[4] Kumlu M, Eroldogan OT, Aktas M. The effect of salinity on larval growth, survival and development of *Penaeus semisulcatus* (Decapoda: Penaeidae). *Israeli J* Aquacult-Bamidgeh. 1999;51:114-121.

[5] Kumlu M, Jones DA. Salinity tolerance of hatchery-reared postlarvae of *Penaeus indicus* H. Milne Edwards originating from India. *Aquaculture.* 1995;130:287-296.

[6] Staples DJ, Heales DS. Temperature and salinity optima for growth and survival of juvenile banana prawns *Penaeus merguiensis. J* Exper *Marine Biol Ecol.* 1991;154:251-274.

[7] Ponce-Palafox J, Martinez-Palacios CA, Ross LG. The effects of salinity and temperature on the growth and survival rates of juvenile white shrimp, *Penaeus vannamei*, Boone, 1931. *Aquaculture.* 1997;157:107-115.

[8] Bouaricha N, Charmantier-Daures M, Thuet P, Trilles JP, Charmantier G. Ontogeny of osmoregulatory structures in the shrimp *Penaeus japonicus* (Crustacea Decapoda). *Biologic Bull.* 1994;186:29-40.

[9] ChungKF, LinHC. Osmoregulation and Na,K-ATPase expression in osmoregulatory organs of *Scylla. Comp Biochem Physiol Part A: Mol Integr Physiol.* 2006;144:48-57.

[10] Chu KH. Sodium transport across the perfused midgut and hindgut of the blue crab, *Callinectes sapidus*: the possible role of the gut in crustacean osmoregulation. *Comp Biochem Physiol Part A: Physiol.* 1987;87:21-25.

[11] Roy LA, Davis DA, Saoud IP, Henry RP. Branchial carbonic anhydrase activity and ninhydrin positive substances in the Pacific white shrimp, *Litopenaeus vannamei*, acclimated to low and high salinities. *Comp Biochem Physiol Part A: MolIntegr Physiol*. 2007;147:404-411.

[12] Lucu C, Towle DW. $Na^+ + K^+$-ATPase in gills of aquatic crustacea. *Comp Biochem Physiol Part A: Mol Integr Physiol*. 2003;135: 195-214.

[13] Therien AG, Blostein R. Mechanisms of sodium pump regulation. *Am J Physiol*. 2000;279:C541-C566.

[14] Furriel RPM, McNamara JC, Leone FA. Characterization of (Na^+, K^+)-ATPase in gill microsomes of the freshwater shrimp *Macrobrachium olfersii*. *Comp Biochem Physiol – Part B: Biochem Mol Biol*. 2000; 126:303-315.

[15] Furriel RPM, McNamara JC, Leone FA. Nitrophenylphosphate as a tool to characterize gill Na^+, K^+-ATPase activity in hyperegulating Crustacea. *Comp Biochem Physiol Part A: Mol Integr Physiol*. 2001;130:665-676.

[16] Lovett DL, Verzi MP, Burgents JE, Tanner CA, Glomski K, Lee JJ, Towle DW. Expression profiles of Na+,K+-ATPase during acute and chronic hypo-osmotic stress in the blue crab *Callinectes sapidus*. *Biologic Bull*. 2006;211:58-65.

[17] Buranajitpirom D, Asuvapongpatana S, Weerachatyanukul W, Wongprasert K, Namwong W, Poltana P. et al. Adaptation of the black tiger shrimp, *Penaeus monodon*, to different salinities through an excretory function of the antennal gland. *Cell Tissue Res*. 2010;340:481-489.

[18] Sun H, Zhang L, Ren C, Chen C, Fan S, Xia JJ. et al. The expression of Na, KATPase in *Litopenaeus vannamei* under salinity stress. *Marine Biol Res*. 2011;7:623-628.

[19] Wang L, Wang WN, Liu Y, Cai DX, Li JZ, Wang AL. Two types of ATPases from the Pacific white shrimp, *Litopenaeus vannamei* in response to environmental stress. *Mol Biol* Rep. 2012;39:6427-6438.

[20] Shekhar MS, Kiruthika J, Ponniah AG. Identification and expression analysis of differentially expressed genes from shrimp (*Penaeus monodon*) in response to low salinity stress. *Fish Shellfish Immunol*. 2013;35:1957-1968.

[21] Shekhar MS, Kiruthika J, Rajesh S, Ponniah AG. High salinity induced expression profiling of differentially expressed genes in shrimp (*Penaeus monodon*). *Mol Biol Rep*. 2014;41:6275-6289.

[22] Tassanakajon A, Somboonwiwat K. Antimicrobial peptides from the black tiger shrimp *Penaeus monodon*: a review. In: Bondad-Reantaso MG, Jones JB, Corsin F, Aoki T, editors. *Diseases in Asian Aquaculture VII*. Fish health section. Selangor, Malaysia: Asian Fisheries Society, 2011. p. 229-240. 385 pp.

[23] Supungul P, Klinbunga S, Pichyangkura R, Hirono I, Aoki T, Tassanakajon A. Antimicrobial peptides discovered in the black tiger shrimp *Penaeus monodon* using the EST approach. *Dis Aquatic Organ*. 2004;61:123-135.

[24] Chen JY, Pan CY, Kou CM. cDNA sequence encoding an 11.5 kDa antibacterial peptide of the shrimp *Penaeus monodon*. *Fish Shellfish Immunol*. 2004;16:659-664.

[25] Amparyup P, Kondo H, Hirono I, Aoki T, Tassanakajon A. Molecular cloning, genomic organization and recombinant expression of a crustin-like antimicrobial peptide from black tiger shrimp *Penaeus monodon*. *Mol Immunol*. 2008;45:1085-1093.

[26] Antony SP, Bright IS, Sudheer NS, Vrinda S, Priyaja P, Philip R. Molecular characterization of a crustin-like antimicrobial peptide in the giant tiger shrimp, *Penaeus monodon*, and its expression profile in response to various immunostimulants and challenge with WSSV. *Immunobiology*. 2011;216;184-194.

[27] Vatanavicharn T, Supungul P, Puanglarp N, Yingvilasprasert W, Tassanakajon A. Genomic structure, expression pattern and functional characterization of crustinPm5, a unique isoform of crustin from *Penaeus monodon*. *Comp Biochem Physiol – Part B: Biochem Mol Biol*. 2009;153:244-252.

[28] Tyagi A, Khushiramani R, Karunasagar I, Karunasagar I. Antivibrio activity of recombinant lysozyme expressed from black tiger shrimp, *Penaeus monodon*. *Aquaculture*. 2007;272:246-253.

[29] Tharntada S, Somboonwiwat K, Rimphanitchayakit V, Tassanakajon A. Antilipopolysaccharide factors from the black tiger shrimp, *Penaeus monodon*, are encoded by two genomic loci. *Fish Shellfish Immunol*. 2008;24:46-54.

[30] Aoki T, Wang H, Unajak S, Santos MD, Kondo H, Hirono I. Microarray analyses of shrimp immune responses. *Marine Biotechnol*. 2011;13:629-638.

[31] Wang LU, Chen JC. The immune response of white shrimp *Litopenaeus vannamei* and its susceptibility to *Vibrio alginolyticus* at different salinity levels. *Fish Shellfish Immunol*. 2005;8:269-278.

[32] Li CC, Yeh ST, Chen JC. Innate immunity of the white shrimp *Litopenaeus vannamei* weakened by the combination of a *Vibrio alginolyticus* injection and low-salinity stress. *Fish Shellfish Immunol*. 2010;28:121-127.

[33] Ciechanover A, Orian A, Schwartz AL. Ubiquitin-mediated proteolysis: biological regulation via destruction. *Bioessays*. 2000;22:442-451.

[34] Hershko A, Ciechanover A. The ubiquitin system. *Annu* Rev *Biochem*. 1998;67:425-479.

[35] Shen B, Zhang Z, Wang Y, Wang G, Chen Y, Lin P. et al. Differential expression of ubiquitin-conjugating enzyme E2r in the developing ovary and testis of penaeid shrimp *Marsupenaeus japonicus*. *Mol Biol Rep*. 2009;36:1149-1157.

[36] Cesar JRO, Yang J. Expression patterns of ubiquitin, heat shock protein 70, α actin and β-actin over the molt cycle in the abdominal muscle of marine shrimp *Litopenaeusvannamei*. *Mol Reprod Dev*. 2007;74:554-559.

[37] Chen AJ, Wang S, Zhao XF, Yu XQ, Wang JX. Enzyme E2 from Chinese white shrimp inhibits replication of white spot syndrome virus and ubiquitinates its RING domain proteins. *J Virol*. 2011;85:8069-8079.

[38] Spees JL, Chang SA, Snyder MJ, Chang ES. Osmotic induction of stress responsive gene expression in the lobster *Homarus americanus*. *Biologic Bull*. 2002;203:331–337.

[39] McGrath ME. The lysosomal cysteine proteases. *Annu RevBiophys Biomol Struct*. 1999;28:181-204.

[40] Tingaud-Sequeira A, Cerdà J. Phylogenetic relationships and gene expression pattern of three different cathepsin L (Ctsl) isoforms in zebrafish: Ctsla is the putative yolk processing enzyme. *Gene*. 2007;386:98-106.

[41] Hu KJ, Leung PC. Shrimp cathepsin L encoded by an intronless gene has predominant expression in hepatopancreas, and occurs in the nucleus of oocyte. *Comp Biochem Physiol – Part B: Biochem Mol Biol*. 2004;137:21-33.

[42] Le Boulay C, Van Wormhoudt A, Sellos D.. Cloning and expression of cathepsin L-like proteinases in the hepatopancreas of the shrimp *Penaeus vannamei* during the intermolt cycle. *J* Comp *Physiol B*. 1996;166:310-318.

[43] Yishan L, Shitian L, Zaohe W, Jichang J. Molecular cloning and mRNA expression of cathepsin C in white shrimp, *Litopenaeus vannamei*. *Aquaculture Res*. 2011;42: 1569-1576.

[44] Jiang S, Nei FP, Chen HC, Wang JH. Comparative study on the cathepsin D from banded shrimp (*Penaeus japonicus*) and grass shrimp (*Penaeus monodon*). *J Agri* Food Chem. 1992;40:961-966.

[45] Stephens A, Rojo L, Araujo-Bernal S, Garcia-Carreño F, Muhlia-Almazan A. Cathepsin B from the white shrimp *Litopenaeus vannamei*: cDNA sequence analysis, tissues-specific expression and biological activity. *Comp Biochem Physiol – Part B: Biochem Mol Biol*. 2012;161:32-40.

[46] Aitken A. 14-3-3 proteins: a historic overview. *Semin Canc Biol*. 2006;16: 162-172.

[47] Baunsgaard L, Fuglsang AT, Jahn T, Korthout HA, Deboer AH, Palmgren, MG. The 14-3-3 proteins associate with the plant plasma membrane H+-ATPase to generate a fusicoccin binding complex and a fusicoccin responsive system. *Plant J*. 1998; 13:661-671.

[48] Jayasundara N, Towle DW, Weihrauch D, Spanings-Pierrot C. Gill-specific transcriptional regulation of Na+/K+ -ATPase a-subunit in the euryhaline shore crab *Pachy-*

grapsus marmoratus: sequence variants and promoter structure. *J Exper Biol.* 2007;210:2070-2081.

[49] Kaeodee M, Pongsomboon S, Tassanakajon A. Expression analysis and response of *Penaeus monodon* 14-3-3 genes to salinity stress. *Comp Biochem Physiol – Part B: Biochem Mol Biol.* 2011;159:244-251.

[50] Luana W, Li F, Wang B, Zhang X, Liu Y, Xiang J. Molecular characteristics and expression analysis of calreticulin in Chinese shrimp *Fenneropenaeus chinensis. Comp Biochem Physiol – Part B: Biochem Mol Biol.* 2007;147:482-491.

[51] Visudtiphole V, Watthanasurorot A, Klinbunga S, Menasveta P, Kirtikara K. Molecular characterization of Calreticulin: a biomarker for temperature stress responses of the giant tiger shrimp *Penaeus monodon. Aquaculture.* 2010;308:100-108.

[52] Phelan P. Innexins: members of an evolutionarily conserved family of gapjunction proteins. *Biochim* et *Biophys Acta.* 2005;1711:225-245.

[53] Ducret E, Alexopoulos H, Le Feuvre Y, Davies JA, Meyrand P, Bacon JP, Fénelon VS. Innexins in the lobster stomatogastric nervous system: cloning, phylogenetic analysis, developmental changes and expression within adult identified dye and electrically coupled neurons. *Eur J Neurosci.* 2006;24:3119-3133.

[54] Chintapitaksakul L, Udomkit A, Smith DR, Panyim S, Sonthayanon B. Expression analysis of selected haemocyte transcripts from black tiger shrimp infected with yellow head virus. *Sci Asia.* 2008;34:327-333.

[55] Leelatanawit R, Klinbunga S, Aoki T, Hirono I, Valyasevi R, Menasveta P. Suppression subtractive hybridization (SSH) for isolation and characterization of genes related to testicular development in the giant tiger shrimp *Penaeus monodon. BMB Reports.* 2008;41:796-802.

[56] Strong SJ, Ellington WR. Isolation and sequence analysis of the gene for arginine kinase from the chelicerate arthropod, *Limulus polyphemus*: insights into catalytically important residues. *Biochim Biophysic Acta.* 1995;1246:197-200.

[57] Yu CJ, Lin YF, Chiang BL, Chow LP. Proteomics and immunological analysis of a novel shrimp allergen, Pen m2. *J Immunol.* 2003;170:445-453.

[58] Holt SM, Kinsey ST. Osmotic effects on arginine kinase function in living muscle of the blue crab *Callinectes sapidus. J Exper Biol.* 2002;205:1775-1785.

[59] Abe H, Hirai S, Okada S. Metabolic responses and arginine kinase expression under hypoxic stress of the kuruma prawn *Marsupenaeus japonicus. Comp Biochem Physiol Part A: Mol Integr Physiol.* 2007;146:40-46.

[60] Astrofsky KM, Roux MM, Klimpel KR, Fox JG, Dhar AK. Isolation of differentially expressed genes from white spot virus (WSV) infected Pacific blue shrimp (*Penaeus stylirostris*). *Arch Virol.* 2002;147:1782-1799.

[61] Yao CL, Ji PF, Kong P, Wang ZY, Xiang JH. Arginine kinase from *Litopenaeus vannamei*: cloning, expression and catalytic properties. *Fish Shellfish Immunol.* 2009;26:553-558.

[62] Hsieh SL, Chiu Y.C, Kuo C.M. Molecular cloning and tissue distribution of ferritin in Pacific white shrimp (*Litopenaeus vannamei*). *Fish Shellfish Immunol.* 2006;21:279-283.

[63] Ruan YH, Kuo CM, Lo CF, Lee MH, Lian JL, Hsieh SL. Ferritin administration effectively enhances immunity, physiological responses, and survival of pacific white shrimp (*Litopenaeus vannamei*) challenged with white spot syndrome virus. *Fish Shellfish Immunol.* 2010;28:542-548.

[64] Qiu GF, Zheng L, Liu P. Transcriptional regulation of ferritin mRNA levels by iron in the freshwater giant prawn, *Macrobrachium rosenbergii. Comp Biochem Physiol – Part B: Biochem Mol Biol.* 2008;150:320-325.

[65] Maiti B, Khushiramani R, Tyagi A, Karunasagar I, Karunasagar I. Recombinant ferritin protein protects *Penaeus monodon* infected by pathogenic *Vibrio harveyi. Dis Aquatic Organ.* 2010;88:99-105.

[66] Zhang J, Li F, Wang Z, Zhang X, Zhou Q, Xiang J. Cloning, expression and identification of ferritin from Chinese shrimp, *Fenneropenaeus chinensis. J Biotechnol.* 2006;125:173-184.

[67] Zimmerman AW, Veerkamp JH. New insights into the structure and function of fatty acid-binding proteins. *Cell Mol Life* Sci. 2002;59:1096–1116.

[68] Söderhäll I, Tangprasittipap A, Liu H, Sritunyalucksana K, Prasertsan P, Jiravanichpaisal P. et al. Characterization of a hemocyte intracellular fatty acid-binding protein from crayfish (*Pacifastacus leniusculus*) and shrimp (*Penaeus monodon*). *FEBS J.* 2006;273:2902-2912.

[69] Ordway RW, Singer JJ, Walsh JV. Direct regulation of ion channels by fatty acids. *Trends Neurosci.* 1991;14:96-100.

[70] Schjerling CK, Hummel R, Hansen JK, Børsting C, Mikkelsen JM, Kristiansen K, Knudsen J. Disruption of the gene encoding the acyl-CoA-binding protein (ACB1) perturbs acyl-CoA metabolism in *Saccharomyces cerevisiae. J Biologic Chem.* 1996;271:22514-22521.

[71] Guidotti A, Forchetti CM, Corda MG, Konkel D, Bennett CD, Costa E. Isolation, characterization, and purification to homogeneity of an endogenous polypeptide with agonistic action on benzodiazepine receptors. Proc Natl Acad Sci. 1983;80:3531-3535.

[72] Kragelund BB, Knudsen J, Poulsen FM. Acyl-Coenzyme A binding protein (ACBP). *Biochim Biophys Acta.* 1999;1441:150-161.

[73] Elholm M, Garras A, Neve S, Tornehave D, Lund TB, Skorve J, Flatmark T, Kristiansen K, Berge R.K. Long-chain acyl-CoA esters and acyl-CoA binding protein are present in the nucleus of rat liver cells. *J Lipid Res*. 2000;41:538-545.

[74] Burton M, Rose TM, Færgeman NJ, Knudsen J. Evolution of the acyl-CoA binding protein (ACBP). *Biochem J*. 2005;392:299-307.

[75] Melloni E, Averna M, Salamino F, Sparatore B, Minafra R, Pontremoli S. Acyl-CoA-binding protein is a potent m-calpain activator. *J Biologic Chem*. 2000;275:82-86.

[76] Xiao S, Chye ML. An *Arabidopsis* family of six acyl-CoA-binding proteins has three cytosolic members. *Plant Physiol Biochem*. 2009;47:479-484.

[77] Chye ML. *Arabidopsis* cDNA encoding a membrane associated protein with an acyl-CoA binding domain. *Plant Mol Biol*. 1998;38:827-838.

[78] Chye ML, Huang BQ, Zee SY. Isolation of a gene encoding *Arabidopsis* membrane associated acyl-CoA binding protein and immunolocalization of its gene product. *Plant J*. 1999;18:205-214.

[79] Chye ML, Li HY, Yung MH. Single amino acid substitutions at the acyl-CoA-binding domain interrupt ^{14}[C] palmitoyl-CoA binding of ACBP2, an *Arabidopsis* acyl-CoA-binding protein with ankyrin repeats. *Plant Mol Biol*. 2000;44:711-721.

[80] Li HY, Chye ML. Membrane localization of *Arabidopsis* acyl-CoA binding protein ACBP2. *Plant Mol Biol*. 2003;51:483-492.

[81] Leung KC, Li HY, Xiao S, Tse MH, Chye ML. *Arabidopsis* ACBP3 is an extracellularly targeted acyl-CoA-binding protein. *Planta*. 2006;223:871-881.

[82] Chen QF, Xiao S, Chye ML. Overexpression of the *Arabidopsis* 10-kilodalton acyl-Coenzyme A-binding protein, ACBP6, enhances freezing tolerance. *Plant Physiol*. 2008;148:304-315.

[83] Xiao S, Li HY, Zhang JP, Chan SW, Chye ML. *Arabidopsis* acyl-CoAbinding proteins ACBP4 and ACBP5 are subcellularly localized to the cytosol and ACBP4 depletion affects membrane lipid composition. *Plant Mol Biol*. 2008;68:571-583.

[84] Gao W, Xiao S, Li HY, Tsao SW, Chye ML. *Arabidopsis thaliana* acyl-CoA binding protein ACBP2 interacts with a heavy-metal-binding farnesylated protein AtFP6. *New Phytologist*. 2008;181:89-102.

[85] Gao W, Li HY, Xiao S, Chye ML. Acyl-CoA-binding protein 2 binds lysophospholipase 2 and lysoPC to promote tolerance to cadmium-induced oxidative stress in transgenic Arabidopsis. *Plant J*. 2010;62:989-1003.

[86] Du ZY, Xiao S, Chen QF, Chye ML. *Arabidopsis* acyl-CoA-binding proteins ACBP1 and ACBP2 show different roles in freezing stress. *Plant* Signal *Behav*. 2010;5:5607-5609.

[87] Ren Q, Du ZQ, Zhao XF, Wang JX. An acyl-CoA-binding protein (FcACBP) and a fatty acid binding protein (FcFABP) respond to microbial infection in Chinese white shrimp, *Fenneropenaeus chinensis*. *Fish Shellfish Immunol*. 2009;27:739-747.

[88] Zhao ZY, Yin ZX, Weng SP, Guan HJ, Li SD, Xing K, Chen SM, He JG. Profiling of differentially expressed genes in hepatopancreas of white spot syndrome virus-resistant shrimp (*Litopenaeus vannamei*) by suppression subtractive hybridization. *Fish Shellfish Immunol*. 2007;22:520-534.

[89] Kiruthika J, Rajesh S, Ponniah AG, Shekhar MS. Molecular cloning and characterization of acyl-CoA binding protein (ACBP) gene from shrimp *Penaeus monodon* exposed to salinity stress. *Dev Comp* Immunol. 2013;40:78-82.

[90] Yoon Y, Park Y, Yi YS, Lee Y, Jo G, Park JC, et al. Characterization of an O-methyltransferase from Streptomyces avermitilis. MA-4680. *J Microbiol Biotechnol*. 2010;20:1359-1366.

[91] Connors NC, Strohl WR. Partial purification and properties of carminomycin 4-O-methyltransferase from *Streptomyces* sp. strain C5. *J Gen Microbiol*. 1993;139:1353-1362.

[92] Yu J, Cary JW, Bhatnagar D, Clevel TE, Keller NP, Chu FS. Cloning and characterization of a cDNA from *Aspergillus parasiticus* encoding an O-methyltransferase involved in aflatoxin biosynthesis. *Appl Environ Microbiol*. 1993;59:3564-3571.

[93] Lam KC, Ibrahim RK, Behdad B, Dayanandan S. Structure, function, and evolution of plant O-methyltransferases. *Genome*. 2007;50:1001-1013.

[94] Li DX, Du XJ, Zhao XF, Wang JX. Cloning and expression analysis of an OMethyltransferase (OMT) gene from Chinese shrimp, *Fenneropenaeus chinensis*. *Fish Shellfish Immunol*. 2006;21:284-292.

[95] Bonifacio MJ, Vieira-Coelho MA, Borges N, Soares-da-Silva P. Kinetics of rat brain and liver solubilized membrane-bound catechol-O methyltransferase. *Arch Biochem Biophys*. 2000;384:361-367.

[96] Rajesh S, Kiruthika J, Ponniah AG, Shekhar MS. Identification, cloning and expression analysis of Catechol-O-methyltransferase (COMT) gene from shrimp, *Penaeus monodon* and its relevance to salinity stress. *Fish Shellfish Immunol*. 2012;32:693-699.

[97] Kodama KI, Nakabeppu Y, Sekiguchi M. Cloning and expression of the *Bacillus subtilis* methyltransferase gene in *Escherichia coli* ada-cells. *Mut Res*. 1989;218:153-163.

[98] Kregel KC. Heat shock proteins: modifying factors in physiological stress responses and acquired thermotolerance. *J Appl Physiol*. 2002;92:2177-2186.

[99] Roberts RJ, Agius C, Saliba C, Bossier P, Sung YY. Heat shock proteins (chaperones) in fish and shellfish and their potential role in relation to fish health: a review. *J Fish Dis*. 2010;33:789-801.

[100] Huang P, Kang S, Chen W, Hsu T, Lo C, Liu K et al. Identification of the small heat shock proteins HSP21, of shrimp *Penaeus monodon* and the gene expression of HSP21 is inactivated after white spot syndrome virus (WSSV) infection. *Fish Shellfish Immunol.* 2008;25:250-257.

[101] Lo W, Liu K, Song Y. Cloning and molecular characterization of heat shock cognate 70 from tiger shrimp (*Penaeus monodon*). *Cell Stress Chaper.* 2004;9:332-343.

[102] Jiang S, Qiu L, Zhou F, Huang J, Guo Y, Yang K. Molecular cloning and expression analysis of a heat shock protein (Hsp90) gene from black tiger shrimp (*Penaeus monodon*). Mol BiolReport. 2009;36:127-134.

[103] Rungrassamee W, Leelatanawit R, Jiravanichpaisal P, Klinbunga S, Karoonuthaisiri N. Expression and distribution of three heat shock protein genes under heat shock stress and under exposure to *Vibrio harveyi* in *Penaeus monodon. Dev Comp* Immunol. 2010;34:1082-1089.

[104] de la Vega E, Hall MR, Degnan BM, Wilson KJ. Short-term hyperthermic treatment of *Penaeus monodon* increases expression of heat shock protein 70 (HSP70) and reduces replication of gill associated virus (GAV). =*Aquaculture.* 2006;253:82-90.

9

Sunflower Breeding for Resistance to Abiotic and Biotic Stresses

Dragan Škorić

Abstract

Due to a specific structure of its main organs (root, stem, leaves, and head), sunflower can be successfully grown on marginal soils and in semiarid conditions, and it is more resistant to abiotic stresses, than other field crops. Unfortunately, it is very sensitive to biotic stresses.

In sunflower breeding for resistance to abiotic stresses, the greatest progress has been made in selection for drought resistance. Breeders use over 30 different parameters in sunflower screening for drought resistance, with physiological ones being the predominant type. The best breeding results have been achieved using the phenomenon of stay-green, with the added bonus that this method incorporates into the cultivated sunflower not only drought resistance but resistance to *Macrophomina* and *Phomopsis* as well. The diversity of the wild *Helianthus* species offers great possibilities for increasing the genetic resistance of the cultivated sunflower toward abiotic stresses. In using wild sunflower species in sunflower breeding for drought resistance and resistance to salinity, best results have so far been achieved with H. *argophyllus* and H. *paradoxus*, respectively. In addition to the use of wild *Helianthus* species, sunflower breeding for abiotic stress resistance should also make more use of molecular breeding techniques. More progress has been made in sunflower breeding for heat resistance than in that for cold resistance. Specific breeding programs dealing with sunflower resistance to mineral deficiency and mineral toxicity have yet to be established.

Concerning biotic stresses, the main problem in sunflower cultivation is caused by fungal diseases. Genetic variability of cultivated sunflower is very low and deficient in disease-resistance genes. Due to wild sunflower species of the Helianthus genus, genes that confer resistance to certain diseases were discovered and incorporated into the genotypes of the cultivated sunflower. Based on the wild species, genes were found that confer resistance to Plasmopara *halstedii, Puccinia helianthi, Verticillium dahliae, V. albo-atrum,* and *Erysiphe cichoracearum.* Furthermore, wild sunflower species provide a high level of tolerance (field resistance) to *Phomopsis/Diaporthe helianthi, Macrophomina phaseolina, Albugo eragopognis,* and *Alternaria* ssp. Sources of resistance to other harmful diseases are sought after within wild sunflower species.

With the use of one wild species of H. *annuus* from Kansas (USA.), genes conferring resistance to a group of imidazolinone (IMI) or sulfonylurea herbicides were discovered. Moreover, similar genes were found through induced mutations. These sources of resistance provide successful control over a broad spectrum of weeds, which infest sunflower crops, including broomrape.

The growth of the parasitic weed sunflower broomrape (*Orobanche cumana* Wallr) is a major issue in sunflower production, especially in Central and Eastern Europe, as well as in Spain. Six races of broomrape have been detected (A, B, C, D, E, and F) and dominant resistance genes (Or_1, Or_2, Or_3, Or_4, Or_4, and Or_6) were found in wild sunflower species. During the last 4–10 years, new virulent races of broomrape emerged in several European countries. Geneticists and breeders work on finding the sources of resistance to the new broomrape races in wild sunflower species.

Numerous insect species cause economic damages during sunflower production, especially in North America (the homeland of sunflower). *Homoeosoma* species are the most widespread insects that infest sunflower. *Homoeosoma nebulella* infests sunflower in Europe and Asia, while infestation with *H. electellum* poses a major problem in USA, Canada, and Mexico. Based on the use of wild sunflower species H. *tuberosus*, genes conferring resistance to *Homoeosoma* species were incorporated. Sunflower has an armored layer in the hull, which provides resistance to this insect. Sources of resistance to other economically harmful insects are sought after.

New methods in biotechnology, particularly marker genes, have been frequently used in breeding for abiotic and biotic stresses.

Keywords: Abiotic and biotic stresses, breeding, interspecies hybridization, resistance, sunflower, wild species

1. Introduction

1.1. Sunflower breeding for resistance to abiotic stresses

Abiotic stresses not only determine the geographical and regional distribution of crops but also dictate if a potentially arable piece of land can actually be used for cultivation. According to an estimate, 24.2% of the world's geographic area is potentially arable. However, only 10.6% of the geographic area is under actual cultivation, while the rest is not available for cultivation due to one or more abiotic stresses [1]. According to the same author, drought is the main abiotic factor, as it affects 26% of the arable area. Mineral toxicities/deficiencies are second in importance, while frost stands third. Drought is the most limiting of all abiotic stresses, and it affects well over one-third of the soils worldwide. Plants that manage to survive the effects of drought stress show a decrease in fertility, yield, and product quality [2].

Characterization of drought tolerance is very complex and interrelated to many factors. Drought is a multidimensional stress affecting plants at various levels of their organization. Sunflower is grown in a number of countries on so-called marginal soils, often in semiarid conditions where almost every year an abiotic stress of one kind or another is present acting as a limiting factor on crop production. However, of all field crops, sunflower is best able to withstand drought conditions, primarily on account of the structure of its organs [3].

Drought is the main cause not only of differences between mean yield and potential yield but also of yield variations from year to year and therefore of yield instability [2].

Using the results of our own studies and those of other authors, the present chapter discusses the progress that has so far been made in sunflower breeding for resistance to abiotic stresses and indicates possible future directions in this area of sunflower research.

1.2. Sunflower breeding for resistance to drought

Previous experiences in sunflower cultivation have shown that drought can be a limiting factor in realizing the potential of a variety or a hybrid.

In sunflower breeding for resistance to drought, just like in the other crops, a number of physical and morphological parameters are at play. The accumulation of genes for these parameters in a single genotype makes it possible to increase resistance to drought [4].

Škorić [5] states that sunflowers must be resistant to both soil and air drought, that is, to high temperatures during flowering (pollination) and the oil synthesis stage. The ways to achieve this desired goal are as follows: a more efficient root system, a certain systemic composition of the main organs, and resistance to certain diseases (*Macrophomina phaseoli*). In addition to efficient water use, the root system must have the ability for efficient nutrient use under stress conditions.

On the one hand, resistance depends on the selection of genotypes whose flowering and maturity end before the occurrence of stress (early maturity).

On the other hand, mechanism of drought resistance incorporates the modification of certain physiological and morphological parameters, which enables a more efficient use of water reserves during the period of stress. The mechanism manifests itself through a more aggressive root system or water use reduction via a more efficient stomatal apparatus plus the interaction of these factors.

The inheritance of tolerance of drought based on high osmotic pressure was found to be controlled by partial dominance and overdominance. The inheritance of drought tolerance measured by temperature shock was found to be based on nonallelic interaction of genes contained in the system of partial dominance [6].

Soil drought limits water uptake and consumption by plants. Transpiration intensity decreases strongly, which, in combination with high air temperature, leads to overheating of plants. The protective reaction of plants against water shortage is the increased ability of cells to retain water. Respiration intensity typically increases under the influence of drought. Prolonged drought forces the plants to reduce the energy efficiency of respiration [22].

Fulda *et al.* [8] used their own results and those of other authors to conclude as follows. Obviously, water stress acclimation is a multigene acclimation, in which many different physiological processes and many drought stress-inducible genes are involved. Functionally, these gene products can be distinguished into osmolyte synthesis, protection factors for macromolecules (chaperons, LEA/dehyndrtype genes), proteases, membrane proteins (aquaporins, transporters, detoxification enzymes (glutathione-S-transferase (GST) and superoxide dismutase (SOD)), and genes of regulatory proteins such as transcription factors (TFs), protein kinases, and protein phosphatases. Although the alterations in all of these processes related to drought stress have been widely investigated in many model species and a few crop species, reports on sunflower are limited.

Studying the influence of water deficit and canopy senescence pattern on sunflower root functionality during the grain-filling phase, Lisanti *et al.* [9] have concluded that both water deficit and intrinsic canopy senescence dynamics can profoundly affect root functionality during grain-filling. The effects of these factors and their interactions, especially under drought, on yield merit focused attention in future research

According to Singh [1], drought seems rather difficult to define and more difficult to quantify. For example, the common criteria used in the various definitions are precipitation, air temperature, relative humidity, evaporation from free water surface, transpiration, wind, air flow, soil moisture, and plant conditions. A working definition of drought may be "the inadequacy of water availability, including precipitation and soil moisture storage capacity, in quantity and distribution during the life cycle of a crop to restrict the expression of its full genetic yield potential".

Therefore, under conditions of drought, water stress develops in the plants as the demand exceeds water supply; this may occur due to atmospheric or soil conditions and is reflected in a gradient of water potentials developed in the soil/soil–root interface and the leaf, the transpiring organ. Thus, moisture stress may be defined as the inability of plants to meet the evapotranspirational demand. Moisture stress is likely to develop to a different rate in different plant organs along this gradient [10].

Drought resistance may be defined as mechanism(s) causing minimal loss of yield in a drought environment relative to the maximum yield in a constraint-free, that is, optimal environment for the crop. However, it does not exist as a unique heritable plant attribute. The various mechanisms by which a crop can minimize yield loss due to drought are grouped into the following three categories:

1. drought escape
2. dehydration avoidance, and
3. dehydration tolerance [1]

Drought escape describes the situation where an otherwise drought-susceptible variety performs well in a drought environment simply by avoiding the period of drought. Early maturity is an important vehicle for drought escape, suitable for environments subjected to late-season drought stress [1].

Early sunflower hybrids generally have lower leaf area index (LAI), lower total evapotranspiration, and lower yield potential than the later ones. According to Škorić [11], early sunflower hybrids are most often susceptible to *Macrophomina,* and thus in cases where there is an early occurrence of drought such hybrids may become affected, thus nullifying any positive effect early maturity may bring.

Dehydration avoidance is the ability of a plant "to retain a relatively higher level of hydration under conditions of soil or atmospheric water stress." Therefore, the various physiological, biochemical, and metabolic processes involved in plant growth and yield production are not internally exposed to stress, but they are protected from water stress [10]. The common measure of dehydration avoidance is the tissue water status as expressed by water or turgor potential under conditions of water stress. This can be achieved by either reducing transpiration (such plants are often called water savers) or increasing water uptake (such plants are often termed as water spenders). Wild species are readily classifiable as water savers and water spenders, but crop plants ordinarily exhibit a combination of both features, probably as a result of selection by man.

Drought not only reduces the rate of photosynthesis but also directs the photosynthetic metabolism toward increased formation of low-molecular weight compounds such as alanine, hexoses, and malic acid [12]. When the drought ends, sunflower plants are capable of again having a high rate of photosynthesis, thus compensating for the negative effects of water deficiency.

As sunflower plants respond to drought, the free proline content of their leaves increases, because proline, due to its structure, increases the water retention capacity of the cell [13].

When breeding for dehydration avoidance, it is highly important that a considerable attention is paid to parameters such as reduced transpiration, osmotic adjustment, abscisic acid (ABA), cuticular wax, and leaf characteristics (leaf pubescence, altering the leaf angle, and leaf rolling). It is also especially important to find ways to increase water uptake by creating a more powerful, deeper, and well-branched root system [14].

1.2.1. Sources of drought resistance

Several types of germplasms are used in sunflower breeding for drought resistance:

1. landraces;
2. cultivated hybrids and varieties;
3. wild species of the genus *Helianthus;* [15]; and
4. genetically engineered germplasm.

Use of landraces and cultivated hybrids and varieties has produced some positive results, but not to the extent that would secure stable sunflower production under drought conditions. The best results in increasing the drought resistance of cultivated sunflower have been achieved using wild species of the genus *Helianthus.*

Over the last 10–14 years, highly drought-tolerant germplasms based on *H. argophyllus,* which have a commercial value, have been created in various breeding centers.

Research and characterization of physiological mechanisms in wild sunflower are just beginning. Škorić [16] suggests that in breeding for drought tolerance, there should be a greater effort to expand the use of other wild species such as *H. deserticola, H. hirsutus, H. maximiliani, H. Tuberosus,* and others.

1.2.2. Using different traits in sunflower breeding for drought resistance

Škorić [7] reported that over 30 different parameters were used in the study of drought resistance and breeding for drought resistance in sunflower. Among these, the most frequently used were physiological parameters.

Chimenti *et al.* [17] reported that high osmotic families extracted more water from the profile during the stress period and had greater grain yield and leaf area duration than families with a low degree of osmotic adjustment. The same authors concluded that osmotic adjustment can contribute to post-anthesis drought tolerance in sunflower through increased water uptake, reduced impact on grain number, grain size, and greater leaf area duration.

Andrei [18] concluded that high self-fertility (24–49%) in some hybrids ensured a greater stability in sunflower yield under stress conditions.

Studying the influence of drought stress on growth, protein expression, and osmolyte accumulation in sunflower, Fulda *et al.* [8] reported that osmolyte analysis revealed an accumulation of glucose (24–30-fold), inositol (20–30-fold), proline (10–20-fold), fructose (3–6-fold), and sucrose (4–4-fold) in extracts from leaves of drought-stressed plants. Changes in protein expression of drought-stressed versus control plants were detected in colloidal Coomasie-stained 2D-polyacrylamide gel electrophoresis (PAGE).

Sato *et al.* [19] studied the correlation between the responses of leaf expansion and hypocotyl elongation to water deficit in sunflower genotypes. Based on the results obtained, they reported that the response of hypocotyl growth to water deficit ranged between 31 and 48%, while that of leaf growth ranged between 40 and 63%. There was a significant positive correlation ($p < 0.01$ $R^2 = 0.61$) between both responses. The correlation was also significant using Pearson's correlation test ($p < 0.04$, $r = 0.78$).

Petcu *et al.* [20] studied physiological traits for the quantification of drought tolerance in sunflower and determined as follows. The reduction in leaf area, shoot size, and biomass accumulation of sunflower seedlings under water stress conditions determined the increase in root/shoot ratio. This suggests that for young plants the main sink was survival. In a late stage of vegetation, the root/shoot ratio decreased under drought stress in some hybrids but increased in others, suggesting that for mature plants the main sink was the yield. The physiology work has focused on morpho-physiological traits induced by drought and associated with drought tolerance of plants and the elaboration of screening methods for rapidly measuring drought tolerance using plants in an early stage of vegetation.

Based on the results of Škorić [7, 11], practical results in sunflower breeding for drought resistance have been achieved by using the stay-green phenomenon. Here, we should warn that in the selection of lines on the basis of stay-green criteria, only lines with a high degree of self-fertility should be looked for, otherwise a wrong choice of genotypes will be made.

The use of the stay-green criterion involves the selection of not only genotypes resistant to drought but also those resistant to *Macrophomina,* which tends to be a problem under stress conditions. Also, genotypes resistant to *Phomopsis* may be simultanously selected, as confirmed by the inbred lines Ha-48, Ha-22, CMS-1-40, PH-BC-2-91, PR-ST-3, RHA-SES, RHA-483, etc. as well as the hybrids made from these lines, which combine several resistance systems. Vrânceanu [21] confirmed the validity of using the stay-green criterion in the selection for drought resistance [22].

Petrović *et al.* [23] concluded that nitrate reductase activity and free-proline accumulation rate, which underwent large modifications in plants under water stress, may serve as parameters for the evaluation of sunflower genotypes for drought tolerance.

Working on the determination of water stress index in sunflower, Orta *et al.* [24] found statistically significant correlations between CWSI (crop water stress index) calculated from single leaf temperatures on the one hand and stomatal resistance, leaf area index, and available water in the root on the other.

Early sunflower hybrids generally have lower leaf area index, total evapotranspiration, and yield potential than the later hybrids. However, according to Škorić [11], early hybrids are typically sensitive to *Macrophomina,* so in the case of an early manifestation of drought they become infected and thus the advantage of earliness is nullified.

Some breeders believe that drought avoidance can be achieved by developing very early sunflower hybrids or by moving the sowing date (early or late sowing) in order to avoid the dry period. Dehydration avoidance can be achieved in several ways, for example, by selecting genotypes with reduced transpiration (water savers) or by increasing the uptake of available water from the soil by a powerful root system (water spenders).

Characteristics that appear to be correlated with drought tolerance include deeper rooting depth and more efficient root uptake of water, tolerance to high osmotic pressure, low transpiration rates, and plant ability to recover after wilting under heat stress.

The genetics of sunflower resistance to drought has not been studied sufficiently, despite numerous attempts and use of different plant characteristics. It appears safe to say that the drought resistance (tolerance) is controlled by a set of genes.

1.3. Sunflower breeding for resistance to salinity

Abiotic stress can be generated by mineral salts, which affect a considerable portion of the global arable land. Salinity ranks second after moisture stress. This stress may occur in the form of a specific mineral deficiency or toxicity, or as accumulation of an excess amount of soluble salts in the root zone [1].

Sunflowers are grown on low-to-medium-saline soils in many countries. These countries face soil salinity as a serious limiting factor in sunflower production. However, it should be remembered that there are several wild *Helianthus* species that naturally grow on saline soils. These species are important sources of genes for resistance to salinity. Breeders should apply effective screening methods in order to identify the wild species that possess genes useful in breeding for salinity resistance and equally effective breeding methods to transfer these genes into cultivated sunflower genotypes [22].

Seiler [25] stated that several wild species of *Helianthus* are native to salt-impacted habitats and may possess genes for salt tolerance. The same author reports that Chandler and Jan [26] evaluated three wild *Helianthus* species for salt tolerance, namely *H. paradoxus, H. Debilis,* and *H. annuus* population native to salty desert areas, and obtained the following results. *Helianthus debilis* tolerated a salt concentration about the same as cultivated sunflower, wilting at an NaCl concentration of 240–400 mM. The wild ecotype of *H. annuus* had a higher tolerance, with some plants surviving the NaCl concentration of 800 mM. *Helianthus paradoxus* was highly salt tolerant, with some plants surviving at 1300 mM of NaCl. Salt tolerance was a dominant trait in hybrids between *H. paradoxus* and cultivated *H. annuus,* which did as well as the wild parent.

The emergence percentage, emergence index, shoot length, and shoot fresh weight can be used as selection criteria for salt tolerance in sunflower at the seedling stage [27].

Tolerance of sunflower genotypes to salinity has been investigated by a number of researchers. Prakash *et al.* [28] found that turgor is not correlated with salt tolerance. The accumulation of proline shows a higher impact on tolerance to salinity. Since callus development, seed germination, and vigor are associated, the former could be a more reliable index of salt tolerance.

The involvement of turgor and proline in salt tolerance seems to be doubtful [29]. Prakash *et al.* [28] stated that turgor cannot be related to salt tolerance. However, proline accumulation seems to be more due to the effect of salinity.

Evidently, using *H. paradoxus* and possibly some other wild *Helianthus* species, sunflower breeders can successfully achieve high resistance to salinity. It is important to determine the selection criteria that can be applied in the breeding program, and these can be cell survival, seed germination, dry matter accumulation, leaf death or senescence, leaf ion content, leaf necrosis, root growth, osmoregulation, etc. [1].

1.4. Sunflower breeding for resistance to mineral deficiency and mineral toxicity

Sunflowers require only 10 macroelements (C, O, H, N, P, K, S, Ca, Fe, and Mg) and 6 microelements (B, Mn, Cu, Zn, Mo, and Co) for their growth and development. Air and water are the sources of carbon, oxygen, and hydrogen. The rest of the elements are taken up from the soil or fertilizers and are divided into primary elements, secondary elements, and microelements [14]. Sunflower nutrition has been the subject of many books and scientific papers, which have established optimum levels of each individual macro- and microelement needed for the normal growth and development of sunflower on different types of soil. There is also voluminous literature on the deficiencies or excess levels (toxicity) of individual elements and how they affect sunflower growth and development.

Studying the diversity of elements in sunflower inbred lines, Sarić *et al.* [30] came to the conclusion that the genetic specificity for mineral nutrition is manifested not only through different contents of mineral elements but also through their distribution into individual plant organs.

As there are unfortunately no major breeding programs anywhere in the world that deal specifically with sunflower resistance to mineral deficiency and mineral toxicity, sunflower breeders should consider a possibility of establishing one or more such programs. They would have to choose appropriate breeding methods and targets, define selection criteria, and select potential resistance sources (most likely wild *Helianthus* species) [16].

1.5. Sunflower breeding for heat resistance

Singh [1] made a very good definition of the heat and cold resistance, which reads: "Each plant species, more particularly genotype, has an *optimum range of temperatures* for its normal growth and development: the specific temperatures would depend not only on the genotype but also on the stage of growth and development of a given genotype. When temperature moves beyond this optimal range, it generates *temperature stress*, i.e., temperature interferes with the performance. Temperature stress may be grouped into the following three categories: (1) heat stress, (2) chilling stress and (3) freezing stress."

Sunflower is characterized by high adaptability to high temperatures. At high temperatures, sunflower intensifies the process of transpiration so that its leaves remain relatively cool. Transpiration rate can be increased only if sufficient water is supplied and this calls for a deep and well-developed root system. Therefore, the choice of genotypes with a deep and powerful root system is an important criterion in the selection for sunflower tolerance to high temperatures [22].

Another important criterion is the tolerance to intensive transpiration. For the environments in which high air temperatures frequently occur at the flowering stage, breeders should select genotypes capable of producing large quantities of pollen and maintain pollen viability under such conditions. It is also important for the pistil and its stigma, or for the disk flowers on the whole, to be tolerant to high temperatures, which ensures pollination and seed formation [22].

Yet another criterion for the selection of genotypes adapted to climates with high temperatures and air and soil drought is the capacity for high seed (formation) filling rate and rapid synthesis of oil in response to stress conditions.

In order for sunflower breeders to be able to determine the right breeding methods, targets, and selection criteria and to choose their breeding materials for selection for heat resistance, they must have a detailed knowledge of how sunflower organs respond to high temperatures. Sunflower is exposed to high temperatures in arid and semiarid conditions, which have been prevalent in much of Europe in 2007. High temperatures may be accompanied by high, but also low humidity levels.

The present knowledge on sunflower heat resistance allows sunflower breeders to define their selection criteria more easily and to search for sources of heat resistance in wild *Helianthus* species.

Breeding for resistance to high temperatures should be combined with selection for drought resistance. Intensive breeding programs on sunflower heat resistance should be organized in countries where excessive temperatures are a regular occurrence. Selection for heat resistance is an integral part of many breeding programs and is often combined with breeding for increased productivity and resistance to dominant diseases and drought [16].

1.6. Sunflower breeding for resistance to low temperatures (cold)

In many environments, crop productivity is limited by low temperatures. When temperatures remain above the freezing level, that is, >0°C, it is called chilling, while freezing describes temperatures below this level, that is, <0°C.

For sunflower, it is important to increase its resistance to cold in the early stages of growth and development, that is, at germination, emergence, and the stage of two to three leaf pairs, so as to enable successful early sowing. Cold resistance at maturation should be increased as well in order to enable sunflower growing at higher altitudes and in colder regions. Sources of cold resistance should be sought exclusively in the wild *Helianthus* species that are found growing wild in the mountains where winters are harsh and springs are cold [16].

Apart from wild *Helianthus* species, induced mutations can also be successfully used as sources of resistance to low temperatures.

Excellent results in the development of sunflower genotypes resistant to cold were achieved by Kalaydzhyan *et al.* [31, 32], who applied induced mutations by chemical mutagens, first of all DMS. Resistance to low temperatures was tested in 44,000 seeds of about 2.000 mutagenic progenies by planting them in late fall/early winter. Some 499 plants from 72 mutagenic progenies (0.91%) survived the harsh winter and low temperatures (down to −20°C). The following mutants showed highest resistance to low temperatures:

- in the case of M-1248 (progenies of 40–43), the overwintering rate was 63%;
- in the case of M-1976 (progenies of 14–20), the overwintering rate was 48%;
- in the case of M-2002 (progenies of 44–64), the overwintering rate was 42%;
- in the case of the cultivar Radnik (control), the freezing rate (death) was 100%.
- These mutants should be subjected to the cold test in the climatic chamber in order to obtain more reliable results.

In any case, Kalaydzhyan *et al.* [31, 32] evidently developed a unique germplasm, which can be used for the development of winter genotypes and genotypes tolerant to low temperatures. Unfortunately, sunflower geneticists and breeders around the world seem to be unaware of these outstanding results.

1.7. Sunflower breeding for tolerance to herbicides

In the past decade or so, significant results were achieved in sunflower breeding for resistance (tolerance) to herbicides from the class of imidazolinones and some herbicides from the class of sulfonylureas (SU).

Acetolactate synthase (ALS), also called acetohydroxyacid synthase (AHAS), is the first enzyme in the biosynthesis of three vital amino acids in plants: valine, leucine, and isoleucine. Four different classes of herbicides inhibit ALS, thus causing the herbicidal effect. The most common are imidazolinones and sulfonylureas. They have been widely used since their introduction in the early 1980s, and now they constitute one of the major weed control mode-of-action classes for many crops. Resistant (tolerant) plants rapidly metabolize the herbicide in herbicidally inactive form. Sensitivity is likewise due to the lack of metabolic detoxification (Stoenescu, personal communication).

Advantages of ALS-inhibiting herbicides are as follows: very low application rate, broad spectrum of weed control (broad leaf and grassy weed species), broad range of crop, selectivity, etc.

1.7.1. Development of IMI-resistant sunflower hybrids

A wild population of annual *H. annuus* from a soybean field in Kansas that had been repeatedly treated with imazethapyr for 7 consecutive years developed resistance to the imidazolinone and sulfonylurea herbicides [33]. Resistance to imazethapyr and imazamox herbicides has great potential for producers in all regions of the world for controlling several broad-leaved weeds.

Miller and Al-Khatib [34] reported that the USDA-ARS (NDSU) research team quickly transferred this genetic resistance into cultivated sunflowers and released public "IMISUN" lines in 1998. At the same time, Alonso *et al.* [35], IFVC research team, Novi Sad, and several private companies in Argentina incorporated IMI resistance from the wild population of *H. annuus* L. from Kansas into their elite lines and developed the first IMI-resistant hybrids [22]. Genetic stocks IMISUN-1 (oil maintainer), IMISUN-2 (oil restorer), and IMISUN-3 (confection maintainer) have been developed and released [36]. Miller and Al-Khatib [34] also released one oilseed maintainer and two fertility restorer breeding lines with imidazolinone herbicide resistance.

Malidža *et al.* [37] reported having transferred resistance to imidazolinones from the wild *H. annuus* L. from Kansas into the elite line HA-26 using three generations per year (one in the field and two in the greenhouse). They stated that the resistance was controlled by a single partially dominant gene. Alonso *et al.* [35] were among the first in the world to transfer genes from the wild *H. annuus* L population collected in Kansas into a cultivated sunflower genotypes resistant to the herbicide imazethapyr, which also 100% controlled (destroyed) broomrape in sunflowers.

Studying the mode of inheritance of resistance to imidazolinone herbicides by using F_2 and test-cross population, Bruniard and Miller [38] concluded that the resistance was controlled by two genes, a major gene having a semidominant type of gene action (*Imr1*) and a second gene (*Imr2*) with a modifier effect when the major gene is present.

Resistance in sunflower can only be achieved with homozygosity (*Imr1 Imr1, Imr2 Imr2*) of both resistance genes in inbred line or in a hybrid [38].

Sala *et al.* [39] reported having obtained a new source of IMI resistance, CLHA-PLUS, developed by means of induced mutations. The line was obtained through ethyl methanesulfonate mutagenesis and selection for the herbicide imazapyr. Also, the authors proved at the molecular level that CLHA-PLUS is different from *Imr1* and that both of them are allelic variants of the locus AHASL1 [40].

It has been shown experimentally that the gene CHLA-PLUS has a higher degree of IMI resistance than the gene *Imr 1 Imr 2*. Breeding centers wishing to use the CHLA-PLUS gene for breeding purposes have to sign a contract on its use with the company BASF. At the same time, BASF provides a protocol for screening for resistance at the molecular level (CLEARFIELD® Protocol SF30).

The recently established CLEARFIELD® (a BASF trademark) Production System for Sunflower provides growers with a new technology, which ensures broad-spectrum postemergence grass and broad-leaved weed control combined with high-performing sunflower hybrids from leading seed companies or public institutions.

BASF Corp. has also established two testing systems which serve to approve IMI-resistant sunflower hybrids as CLEARFIELD®, based mainly on relative tolerance compared with a standard resistant hybrid: Global and Country Qualification System.

Over the last 5 years, there has been a rapid spread of IMI (CLEARFIELD®)-resistant hybrids in the USA, Argentina, and especially central and eastern Europe, where new races of broomrape, which can be successfully controlled by this technology, have emerged.

1.7.2. Development of hybrids resistant to sulfonylurea (tribenuron-methyl)

Simultaneously with sunflower breeding for IMI resistance, work has been started on the development of hybrids resistant to herbicides from the tribenuron-methyl group of sulfonylureas. To date, two resistance sources have been discovered:

The first one was derived from SU-resistant wild *Helianthus annuus* plants collected from the same area in Kansas where IMI resistance was found. The USDA-ARS (NDSU) research group incorporated this genetic resistance into cultivated sunflower and released public lines SURES in 2001 [41].

At the same time, sunflower breeders in various breeding centers (public and private) in the world introduced the sulfonylurea resistance gene into their elite lines, and thus created resistant hybrids.

The second SU resistance was detected by DuPont within an artificial mutagenesis project conducted in the early 1990s. This material was reselected, purified, and tested by Pioneer/

DuPont during 1998–2000. Several mutation events were evaluated and selectivity to the sunflower mutation event SU7 was confirmed for a narrow range of SU herbicides.

Also, in SU-resistant hybrids, it is necessary that both parent lines possess resistance, because of the partial domination in inheritance of this trait.

1.7.3. The use of molecular techniques in sunflower breeding for resistance to abiotic stress

Molecular studies as part of sunflower breeding for resistance to abiotic stress should be focused on the recognition of chromosomal segments carrying genes that contribute to the determination of tolerance, provide the possibility to partition the character, and can be used as a tool for an efficient manipulation of the breeding material. For this purpose, genetic maps of neutral molecular markers, such as isozyme and restriction fragment length polymorphism loci, can be an efficient tool for the determination of useful genes [42].

Belhassen *et al.* [43] and Cellier *et al.* [44] were among the first to use molecular techniques in sunflower breeding for resistance to abiotic stress.

Belhassen *et al.* [43] started breeding for drought tolerance from an interspecific cross with *H. argophyllus*. Four cycles of divergent selection using the physiological criterion of leaf cuticular transpiration (relative water loss) allowed the production of two contrasting genotypes: T− (low level of leaf cuticular transpiration) and T+ (high level of leaf cuticular transpiration). Field experiments showed better yield tolerance index combined with good potential yield for T− hybrids in some locations. Physiological analyses conducted in the field and in controlled conditions allowed to distinguish the two genotypes for only one parameter – osmotic adjustment. Molecular comparison revealed the existence of a cDNA differentiating T− from T+. This cDNA has high homology with an amino acid transporter. A quantification of the amino acid concentrations during water deficit in T− and T+ lines showed that the T− plants accumulate significantly more proline than T+ ones. Using this cDNA, RFLP and STS analysis allowed the differentiation of the two lines.

Cellier *et al.* [44] studied a sunflower genotype showing drought tolerance in field conditions (R1 genotype) and another exhibiting drought sensitivity (S1 genotype). They found that R1 tolerance was characterized by a delay of both wilting and decrease of leaf water potential. To analyze R1 tolerance at a molecular level, they isolated different cDNAs (named SDI for Sunflower Drought Induced) corresponding to transcripts accumulated in water-stressed R1 leaves by subtractive hybridization. The analysis of transcript accumulation in both genotypes upon drought stress suggested a differential expression in the *sdi* genes. Abscisic acid-mediated induction in the tolerant genotype was observed for four of the *sdi* genes and was found to differ among them. Sequence analysis of SDI clones showed high identity with known proteins, including nonspecific lipid transfer proteins (nsLTPs), early light-inducible proteins (ELIPs), or dehydrin, predicted to be involved in various physiological processes.

Arce *et al.* [45] studied sunflower atypical transcription factors and miRNAs playing a key role in responses to abiotic stresses. In order to achieve the desired results, they used a series of molecular biology techniques. These techniques and strategies include database analysis, phylogenetic tree construction, screening of genomic DNA libraries, isolation of cDNA clones, expression studies using northern blots, western blots, and quantitative real-time reverse transcription polymerase chain reaction (qRT-PCR), functional analyses using plant transformation, both stable and transient, confocal microscopy, and microarrays.

Among their findings was the conclusion that transcription factors are proteins able to recognize and bind specific DNA sequences present in the regulatory regions of their target genes. Upon binding, entire signalization cascades are induced or repressed and the plant can adapt itself, at least temporarily, to the adverse conditions to which it is subjected.

Based on the copious results, Arce *et al.* [45] made the following conclusions.

The most amazing results obtained during these studies and other current studies are related to the divergence in structure and function of TFs and miRNAs found in sunflower, apparently conserved in some cases in other *Asteraceae* species but not in model plants. The release of the genomic sequence together with the advance in transformation techniques will certainly help to better understand how sunflower evolved to be adapted to abiotic stress factors and which novel regulating molecules are playing key roles in such an adaptation.

Alberdi *et al.* [46] studied the relationship between a set of molecular markers (amplified fragment length polymorphism (AFLP) and simple sequence repeat (SSR)) and leaf expansion parameters under water-deficit conditions in a cross of two public sunflower lines of contrasting response, in its F_2 and $F_{2:3}$ progenies, and in an independent F_8 recombinant inbred line (RIL) population.

Based on phenotypic trials (two in growth chambers – F_3 and $F_{2\text{–}3}$) and experiments in a greenhouse (RIL population), certain leaves collected during these experiments were used for DNA extraction. Using a set of 60 SSR and 41 AFLP markers, they achieved significant results, which may be useful for the development of molecular markers for assisted selection in breeding programs oriented to generate new cultivars with improved adaptation to water stress conditions.

Liu and Jan [47] closely studied the results of molecular studies about abiotic stresses in light of their own as well as other authors' research. They concluded that approaches using molecular biology, functional genomics, transcriptome, and proteomics have been used to identify genes or quantitative trait loci (QTLs) and proteins correlated with the network of the response to such stresses, which will provide knowledge for the development of hybrids with resistance or tolerance to them. Some wild species grow in locally extreme environments providing an opportunity to study species from these habitats.

Studying the phenomenon of salt tolerance in sunflower, Lexer *et al.* [48] identified an EST that codes for the Ca-dependent protein kinase with maps to a salt-tolerance QTL in sunflower.

1.8. Conclusions

Due to the basic structure of its main organs (root, stem, and leaves), sunflower is more resistant to abiotic stresses than other field crops. Therefore, it is usually grown on soils of lower quality ("marginal soils") and in semiarid and arid conditions, where it is often exposed to abiotic stresses.

When it comes to sunflower breeding for resistance to abiotic stresses, the greatest progress has been made in selection for drought resistance. The progress was achieved by using various criteria and parameters, but the most headway was made by using physiological parameters.

The best and the most affordable method for testing sunflower for drought resistance is the use of "stay-green" character. By using "stay-green" in sunflower selection for drought resistance, the selection for *Macrophomina* and *Phomopsis* resistance is made at the same time.

Wild sunflower species of *Helianthus* are successfully used in selection for drought resistance. *Helianthus argophyllus* is most commonly used in selection for drought resistance via interspecies hybridization. Thus, new germplasms have been developed in a number of breeding centers. Moreover, several more wild species deserve to be used in selection for drought resistance. The use of molecular breeding techniques enables faster and more efficient achievement of desired results in sunflower resistance to drought.

Significant results in sunflower selection for salinity resistance have been obtained by the use of *H. paradoxus* via interspecies hybridization.

Cold resistance can be increased by using certain wild species of sunflower, but especially induced mutations.

Wild species of sunflower are insufficiently used in selection for high temperature resistance, that is, heat resistance, as well as mineral deficiency and mineral toxicity resistance.

By using a population of wild *H. annuus* L. and induced mutations, great headway in sunflower selection for resistance to herbicides from the imidazolinones and sulfonylureas (tribenuron-methyl) group has been made. Sunflower resistance to broomrape (*Orobanche* spp.) has also been achieved.

2. Sunflower breeding for resistance to biotic stresses

Concerning biotic stresses in sunflowers, it can be safely concluded that diseases caused by different fungi present the most serious problem. Broomrape, the parasitic angiosperm, is in the second place, viruses and bacteria in third and fourth [22].

2.1. Sunflower diseases

The original variability of the cultivated sunflower is very narrow and different in genes applicable in selection for the improvement of different agronomic traits, especially those conferring resistance to diseases.

Diseases are a limiting factor in the production of sunflower in all continents where it is grown. Different diseases are dominant in different growing regions, depending on the prevailing environmental conditions. Some diseases cause economic damage to sunflower in all sunflower-growing regions of the world. More than 30 different pathogens that attack sunflowers and cause economic loss in production have been identified so far (Table 1). Sunflower breeders have achieved significant results in finding genes for resistance or high tolerance to certain diseases in wild species and incorporating them into cultivated sunflower genotypes possessing high combining ability [22].

Disease	**Pathogen**
Downy mildew	*Plasmopara halstedii*
Broomrape	*Orobanche cumana*
White rot	*Sclerotinia sclerotiorum*
Stem canker	*Diaporthe helianthi*
Alternaria blight	*Alternaria helianthi, A. helianthinficiens*
Rust	*Puccinia helianthi*
Phoma black stem	*Phoma macdonaldii*
Virus	*Sunflower chlorotic mottle virus*
Verticillium wilt	*Verticillium dahliae*
Charcoal rot	*Macrophomina phaseolina*
White blister rust	*Albugo tragopogonis*
Fusarium wilt	*Fusarium* spp.
Rhizopus head rot	*Rhizopus* spp.

Table 1. The most common sunflower diseases

Wild sunflower species have been a valuable source of resistance genes for many of the common pathogens of the cultivated sunflower. The relative severity of individual diseases varies widely, depending on climate and host cultivars. Breeding for resistance often is the most effective means of control. Sources of resistance or improved levels of tolerance for most diseases are available among the cultivated sunflower and the wild species of *Helianthus* [49].

Changes in the racial composition of certain pathogens have also been caused by the introduction of hybrids in commercial production, which are substantially more homogeneous with respect to the previous period when genetically heterogeneous open-pollinating varieties were grown.

Vear [50] recommended for efficient disease control in future breeding programs to combine vertical and horizontal resistance if available. If not, marker-assisted selection should be used to combine QTLs with different additive defense mechanisms [22].

Galina Pustovoit [51] evaluated new cultivars based on interspecific hybridization (*H. tuberosus* × cultivated sunflower) – Progress, October, Yubileyniy 60, and Novinka. Based on the results achieved in the field and by inoculation, the author concluded that the new cultivars possess group immunity, that is, resistance to downy mildew, rust, *Macrophomina*, *Phoma*, and broomrape.

To be successful in breeding for disease resistance, the sunflower breeder must be thoroughly acquainted with general principles of resistance breeding, major approaches to management of resistance genes, stability of sunflower resistance to certain pathogens, monitoring of interactions between the host (sunflower), pathogen and the environment, and resistance types (vertical and horizontal). Finally, he has to have an adequate germplasm at his disposal, select a method of breeding, and develop a strategy for achieving the desired goal [22].

The aim of this research is to review biotic stresses in sunflower, indicate their significance, and reveal the sources of resistance and methods of selection in order to achieve the desired goal.

2.1.1. Downy mildew [Plasmopara halstedii (Farl.) Berl. et de Toni)]

Downy mildew [*Plasmopara halstedii* (Farl.) Berl. et de Toni)] occurs in all regions around the world in which sunflower is grown as a major oil crop. Downy mildew occurs with light intensity in years with a wet spring.

Downy mildew control was successfully maintained with dominant genes for a long period. This period roughly corresponds to the presence of only two races of downy mildew, the European race, controlled by the dominant gene Pl_1, and North American, controlled by the Pl_2 gene. Unfortunately, changes took place in the past 14 years and there occurred a number of new races. These new races of downy mildew were registered in France, Hungary, USA, Argentina, and several other countries [22].

Viranyi [52] reported that the most detailed and up-to-date list of global distribution of *P. halstedii* pathogens has been compiled by Gulya [54] in a paper presented at the 2nd International Downy Mildew Symposium, Kostelec, Czech Republic. In the accurate overview, he comprised as many as 34 pathotypes (races), an unbelievably high number considering the fact that in most sunflower-producing countries from just a few to 12 well-distinguished virulence phenotypes exist. Europe, France, Germany, and Spain reported the highest numbers but the pathogen is rather diverse in the USA, Canada, and in South Africa as well. Furthermore, there are five *P. halstedii* pathotypes (300, 330, 710, 730, and 770) that are universally distributed globally, recorded from North and South America, Europe and Africa. Apart from the quantitative aspect of virulence, it is interesting to consider the dynamics of diversity as well, that is, the changes in a given region over time. In this respect, France leads with the highest number of new pathotypes arisen in the last 6–7 years [53].

Here, it should be mentioned that genes for resistance to the new races were quickly found in wild species and promptly transferred into genotypes of the cultivated sunflower [55]. An international set of differential lines has been made which makes it possible to determine which downy mildew races are present in a certain region. The set of differentials is supplemented with new lines as new downy mildew races occur [50].

The dynamism of changes in downy mildew races may be illustrated by the fact that, conclusive, with 2011, at least 18 downy mildew races have been determined in the world (100, 300, 304, 307, 314, 330, 700, 703, 704, 710, 711, 714, 717, 721, 730, 731, 770,...).

The testing of breeding materials by inoculation methods is in constant improvement and continuous progress. These issues have been dealt with by a large number of researchers, including Gulya *et al.* [56], Gulya *et al.* [57], Jouffret *et al.* [58], Tourvieille de Labrouhe *et al.* [59], Molinero-Demilly *et al.* [60], and others.

Tourvieille de Labrouhe *et al.* [61] reported that, in addition to major genes, nonrace-specific resistance contributes to the expression of resistance to downy mildew as well. The study also showed that the nonrace-specific resistance is inherited independently of major genes. Furthermore, Vear *et al.* [62] concluded that the inheritance of nonrace-specific resistance is under additive control. The authors reported that two QTLs may explain 42% variation in field reaction to downy mildew. This form of resistance was mapped as belonging to linkage groups 8 and 10. At the same time, they argued that this quantitative resistance is not related to any of the known major resistance gene clusters.

Also, Vear *et al.* [63] have developed a procedure for the development of new B-lines and parallel conversion into the *cms* form from source population. We should also mention here the procedure (scheme) of Vear *et al.* [64] for introgressing *Pl* genes into elite B-lines by backcrossing and simultaneous conversion into the *cms* form while performing resistance screening at the molecular level. This method significantly shortens the cycle of *Pl* gene introgression into elite lines.

According to Tourvieille de Labrouhe *et al.* [65] and Vear *et al.* [62], breeders should develop a strategy of simultaneous selection for nonspecific resistance and major gene resistance along with requisite use of molecular markers.

According to Seiler [49], complete resistance to the downy mildew pathogen was found in annual species *H. annuus, H. argophillus, H. debilis,* and *H. petiolaris* and perennial *H. decapetalus, H. divaricatus, H. eggertii, H. giganteus, H xlaetiflorus, H. mollis, H. nuttallii, H. scaberrimus, H. pauciflorus, H. salicifolius,* and *H. tuberosus* [66].

Diploid perennial species *H. divaricatus, H. giganteus, H. glaucophyllus, H. grosseserratus, H. mollis, H. nuttallii,* and *H. smithii* and their interspecific hybrids were resistant to downy mildew [67].

With the rapid improvement of molecular techniques and their use in plant pathology, new developments have opened new insights into research on fungal biology, detection technology, and genetics and host–pathogen interactions. For example, Hammer *et al.* [68] in Germany, using different approaches, were successful in detecting fungal structure from sunflower host tissues.

2.1.2. White rot [Sclerotinia sclerotiorum (Lib.) de Bary]

White rot is a major problem in countries with a humid climate or in years with an extremely wet summer. The fungus itself is polyphagous. It attacks over 360 plant species, which increases its variability and makes the selection for resistance difficult [69]. The major problem in the selection are the three types of the diseases (on the root, stem, and head) controlled by different mechanisms of resistance [11].

Sunflower stalk and head rot incident by *Sclerotinia sclerotiorum* (Lib.) de Bary is considered the most important disease of the crop in many parts of the world. Since cultural practices or fungicides are insufficient to control the disease, efforts are being made by breeders to develop resistant or tolerant cultivars. This may explain the dominance of publications dealing with various aspects of resistance [52].

When breeding sunflower for resistance to all three forms of *Sclerotinia* attack, it is necessary to combine two or three different tests [70].

Mancel and Shein [71] found that *Sclerotinia* isolates taken from different plant species differed in the degree of virulence. They also found that sunflower isolates that had been repeatedly subcultured in the laboratory were significantly less virulent than isolates recently obtained from sunflower.

When we discuss the three types of sunflower infection by *Sclerotinia,* it is easy to achieve high tolerance to the mid-stalk infection by selecting genotypes resistant to lice [11]. Young leaves of such genotypes are not injured by lice and therefore these plants avoid infections.

Using four different tests for the evaluation of *Sclerotinia* resistance (basal stem infection, ascospore, and oxalic acid injection into the back face of the head), Baldini *et al.* [72] found that the inbred line 28R was most tolerant to the basal stem and white head rot infections and it also showed the best performance in oxalate and culture filtrate tests, which indicated the presence of a specific resistance to oxalate.

Van Becelaere and Miller [73] tried different inoculation procedures for evaluation of resistance to *Sclerotinia* head rot. According to their results, the best method involved the spraying of heads at the beginning of flowering with 4 cm^3 of a suspension of ascospores, which contained 4000 ascospores per milliliter, and covering the heads with brown paper bags immediately after inoculation. Measurements of inoculation could begin as early as 34 days after the inoculation.

Vear *et al.* [50] studied the virulence of 10 *Sclerotinia* isolates. They found differences in in vitro growth rate and sclerotia production as well as some highly significant isolate and genotype effects. They concluded that the available resistance in sunflower genotypes has partial, nonrace-specific, and horizontal characteristics and that it should be durable.

Using an in vitro screening test based on callus induction to evaluate *Sclerotinia* resistance, Drumeva *et al.* [74] found that the test allowed the identification of the breeding material with high to moderate resistance to the pathogen.

When developing inbred lines, sunflower breeders should take note of the results of Van Becelaere [75], who found that the general combining ability (GCA) effects of female lines were relatively larger than the GCA effects of male lines, which indicated that, at least in that particular research, the female lines had a greater influence on the resistance of the hybrids.

When considering the methods of selection for white rot tolerance, recurrent selection and pedigree method were found to produce the best results.

Vear *et al.* [63] used the pedigree method to select sunflower heads resistant to *Sclerotinia*. They applied the ascospore test on F_2 and F_4 plants and the mycelium test on F_3 plants. Their results showed that in all cases there was a variation in the level of resistance among F_3 families. The gains in relation to their parents ranged from 24 to 61%.

Vear *et al.* [76] applied 14 cycles of recurrent selection to a sunflower restorer population developed in 1978 and they obtained significant results. The mycelium test was used in the first three cycles and a combined test with a suspension of ascospores in the subsequent cycles. About 80% reduction of the infected area was achieved in the fourth cycle. In the 12th cycle, the latency index (a measure of incubation period) in the ascospore test was doubled. Simple regression provided the best relation with this cycle, indicating that further increase in the degree of tolerance was possible.

Christov [66] and Christov *et al.* [77] reported that higher-ploidy perennial species (hexaploid and tetraploid species) exhibited greater susceptibility than the diploids, with *H. glaucophyllus, H. divaricatus, H. salicifolius,* and *H. mollis* having the highest frequency of healthy plants. Tolerance to *Sclerotinia* was observed in the perennials *H. eggertii, H. pauciflorus,* and *H. smithii* and annuals *H. annuus, H. argophillus, H. petiolaris,* and *H. praecox* [78].

Interspecific hybrids based on *H. nuttallii, H. giganteus,* and *H. maximiliani* were reported to show resistance against stem infection by Henn *et al.* [79]. Miller and Gulya [80] developed four maintainer and four restorer oilseed lines with improved tolerance to *Sclerotinia* stalk rot. The inbred line HA 410 released by Miller and Gulya [80] derived from a wild perennial hexaploid, *H. pauciflorus (=rigidus)*, had a moderate tolerance to stalk rot. *Sclerotinia* root rot tolerance was observed in perennials *H. mollis, H. nuttallii, H. resinosus,* and *H. tuberosus* [81].

Among the perennial species, resistance to *Sclerotinia* was observed in population of *H. tuberosus, H. divaricatus, H. hirsutus, H. maximiliani, H. mollis, H. nuttallii, H. occidentalis,* and *H. rigidus* (= *pauciflorus*) grown under natural infection conditions [82].

Sclerotinia head rot tolerance was observed in perennials *H. resinosus, H. tuberosus, H. decapetalus, H. grosseserratus, H. nuttallii,* and *H. pauciflorus* [83–85].

In the past decade, advances were made in the research of *Sclerotinia* resistance at the molecular level, particularly in the marker-assisted selection [86, 50, 62, and many others]. The new methods are expected to provide significant help to sunflower breeders [86].

2.1.3. *Sunflower rust (Puccinia helianthi Schw.)*

Rust is the second most important sunflower disease considering its global distribution. The disease causes economic loargophyllussses in sunflower production in North and South America, Australia and Africa. Based on our own observations, rust is present in several countries in Asia (China, India, Iran, Kazakhstan, and others), but its racial composition has not been determined yet. Fortunately for Europe, the local rust population is fairly stable. Rust races were studied most extensively in North America. Sackston [87] determined four North American races, 1, 2, 3, and 4. Race 4 was identified by Yang [88] and race 6 by Lambrides and Miller [89].

Antonelli [90] and Senetinner *et al.* [91] studied sunflower resistance to an Argentinean rust isolate, clone 340, and found that the lines MP 447, MP 444, and LC 74/74-20620 were resistant to it and that the resistance was controlled by a single dominant gene.

Hugues *et al.* [92] studied the occurrence and distribution of rust in Argentina in the period 1982–2008. Their results indicated that resistant cultivars were stable in terms of rust resistance. They also concluded that a single rust pathotype existed in central and southern sunflower-growing regions of Argentina, which was in contrast to previous studies.

In Africa, the determination of rust races in sunflower was done only in Mozambique. Using differential lines from Canada and USA, Vicente and Zazzerini [93] found that the rust race 4 was present in Mozambique.

In Europe, rust has been studied on a limited scale. Most of the work had been done at VNIIMK, Krasnodar. Studying various methods of inoculation by rust, Galina Pustovoit and Slyusar [94] concluded that growing a mixture of resistant genotypes in spatial isolation completed by selection of resistant plants was the most appropriate method.

Miller *et al.* [95] tested 343 genotypes for resistance to rust and found that 12 genotypes were resistant to race 4. The authors also found that the lines HA-R1, HA-R3, HA-R4, HA-R34, and 647-1 shared the same locus, R_4, while the line HA-R2 had a different one that was named R4.

Kochman and Goulter [96] proposed a system for identification of rust races in sunflower, and examined the slow-rusting phenomenon and resistance gene pyramiding to control sunflower rust.

Sendall *et al.* [97] studied the diversity of *Puccinia helianthi* pathosystem in sunflower in Australia at the molecular level and found a set of 24 lines and determined putative resistance genes.

Regarding the methods of artificial inoculation, Gulya and Maširević [98] provided a detailed description of inoculation techniques for evaluating sunflower resistance to rust under laboratory conditions (greenhouse experiments) as well as under field conditions. They also ranked the differential lines in three sets: set one (S-37-388, CM90RR, and MC29), set two (P-386, HA-R1, and HA-R2), and set three (R3-HA, HA-R4, and R4-HA).

Wild species of the genus *Helianthus* are a rich gene pool for further identification of resistance genes and their use to forestall the emergence of new races of *Puccinia helianthi*.

2.1.4. Stem canker (Phomopsis) Diaporthe helianthi

In the past three decades, *Phomopsis* has become the most destructive disease on the global scale. Its large-scale occurrence was first registered in the Vojvodina Province (Serbia) and Romania in 1980, when it caused large economic damage to sunflower production. Soon afterwards, it was registered in most sunflower-growing countries in Europe (France, Hungary, Slovakia, Bulgaria, Ukraine, Russia, and Italy). In the early 1980s, its presence was reported in the USA, Canada, Argentina, Uruguay, Australia, Iran, and some other countries [22].

The first significant results in sunflower breeding for resistance to *Phomopsis* were achieved in Serbia and Romania.

Škorić [99] reported that of 4000 inbred lines and 2000 experimental hybrids, only four lines exhibited field resistance to *Phomopsis*. Two of these lines had been derived from interspecific hybrids (cultivated sunflower × *H. tuberosus*): one was obtained from a cross of *H. argophyllus* × Armavirski 9344 and the restorer line SNRF-69 was derived from a local population from Hungary.

Based on extensive research, Vrânceanu *et al.* [100] found that the sunflower resistance to *Phomopsis* is of the horizontal type and that it is positively correlated with the stay-green phenomenon. The authors reported that, of all Romanian hybrids, Select has the highest degree of tolerance to *Phomopsis*.

Škorić [99] found that three female lines (Ha-22, Ha-74, and Ha-BCPL) and the restorer line SNRF-6 are field resistant to *Phomopsis*. Resistance was transferred to the hybrids NS-H-43, NS-H-44, and H-NS-44 developed from these lines. The same author also reported that *Phomopsis* resistance is positively correlated with *Macrophomina* and *Phoma* resistance as well as with drought tolerance.

Vrânceanu *et al.* [101] concluded that partial dominance is expressed in the inheritance of *Phomopsis* resistance in some cases, while additive inheritance is much more frequent. The same authors found that the stay-green stem at the ripening stage is positively correlated with *Phomopsis* resistance.

Much work has been done lately on the use of molecular markers in breeding for *Phomopsis* resistance.

Studying recombinant inbred lines derived by crossing LR4-17 (resistant) with HA89 (susceptible) at the molecular level, Langar *et al.* [102, 103] concluded that unlinked segments carried major QTLs for different components of resistance, and that the resistances of leaves and stems could be pyramided with a marker-assisted selection.

Molecular studies on the intraspecific diversity of this fungus using intergenic spacer sequence analysis revealed a high homology among French/Yugoslavian and among Italian isolates [104]. The phylogenetic tree obtained from the aligned data revealed three separate groups. The analysis also showed that all isolates originating from countries with regular and severe outbreaks of the disease (e.g., France, Yugoslavia, etc.) formed a well-defined taxon with relatively low variability compared with isolates from Italy where the disease seldom occurs. In another paper, Rekab *et al.* [105] pointed out a polyphyletic nature of this fungus.

Škorić [99] and Dozet [106] reported high levels of resistance to *Phomopsis* in *H. maximiliani*, *H. hirsutus*, *H. pauciflorus*, *H. mollis*, *H. resinosus*, and *H. tuberosus*.

Interspecific hybrids based on *H. eggertii* and *H. smithii* showed high tolerance to *Phomopsis* in Bulgaria [107].

Christov [78] identified annuals *H. annuus, H. argophyllus,* and *H. debilis* and perennials *H. pauciflorus, H. glaucophyllus,* and *H. eggertii* as potential sources of *Phomopsis* brown stem canker resistance, based on field screening in Bulgaria.

Nikolova *et al.* [108] reported resistance to stem canker in progenies of interspecific hybrids of perennial *H. pumilus*. Resistance to *Phomopsis* was reported in interspecific hybrids derived from *H. argophyllus, H. deserticola, H. tuberosus,* and *H. xlaetiflorus* [109].

Complete resistance to *Phomopsis* was reported in interspecific hybrids of *H. salicifolius* by Encheva *et al.* [110] and Škorić [22].

State research and private companies have developed a rich germplasm for *Phomopsis* resistance.

2.1.5. Verticillium wilt (Verticillium dahliae Kleb.)

In addition to *Verticillium dahliae* Kleb., sunflowers are attacked by *Verticillium albo-atrum* R. et B. and *Verticillium lateritium* Bertk. *Verticillium dahliae* Kleb. is the most harmful of these three fungi and it is also the most widespread globally. It causes economic damage to sunflower production in North and South America, Europe, North Africa, Australia, and some countries in Asia [22].

Sunflower breeding for *Verticillium* wilt resistance has been extensive in the USA, Canada, and Argentina. Putt [111] discovered the first sources of resistance to *Verticillium* wilt. His discovery was confirmed by Fick and Zimmer [112]. Resistance to the American race was found in the line HA-89 derived from the Russian cultivar VNIIMK 8931. It is controlled by a single dominant gene.

Bruniard *et al.* [113] and Bertero de Romano [114] found a *Verticillium* race in Argentina that could not be controlled by the gene V_1 (HA89).

Bruniard *et al.* [113] reported to have developed the lines V144, V99, V134, and V196 resistant to the Argentine race of *Verticillium*.

Gulya [54] reported that in 2002 he had found a new strain of *V. dahliae,* which was able to overcome the simple, V-1 dominant resistant gene used in oilseed and confection hybrids. The author tested a diverse germplasm and found that the Russian variety VNIIMK 8883 had genes for resistance to the new strain of *Verticillium dahliae*.

Several researchers used wild sunflower species in order to identify the source of resistance to *Verticillium* wilt.

Assessing the resistance of interspecific hybrids (cultivated sunflower × *H. tuberosus*) to *Verticillium* wilt, Galina Pustovoit and Krokhin [115] found a different mode of inheritance of resistance (two or three recessive genes or two complementary dominant genes), which hinders the development of resistant genotypes.

Putt [111] discovered a source of resistance in line CM144, which was derived from an interspecific hybrid of wild *H. annuus*. Škorić [116] determined high tolerance to *Verticillium dahliae* in *H. occidentalis, H. hirsutus,* and *H. tuberosus*.

2.1.6. Charcoal rot [Macrophomina phaseolina (Tassi) Goild]

Synonyms for this fungus are *Sclerotium bataticola* Taub., *Macrophomina phaseoli* (Maubl.) Ashby and *Rhizoctonia bataticola* (Taub.) Butler.

Charcoal rot causes economic damage to sunflower production in arid regions. It is widespread in most sunflower-growing countries.

Charcoal rot may cause premature death of sunflowers grown on light, sandy soil under hot and dry climate. The disease is well known in the southern part of Europe [52].

Manici *et al.* [117] concluded that the great variability in pathogenicity in all the climatic areas of Italy suggests good adaptation of *Macrophomina* to the host.

This pathogen has been studied by many authors. Iliescu [118] and Ionita and Iliescu [119] published a detailed review of charcoal rot symptomatology, taxonomy, epidemiology, pathogenesis, and control of *Macrophomina* in sunflowers. To our knowledge, a most detailed description of charcoal rot has been provided by Aćimović [120].

Walcz and Piszkev [121] have developed an inoculation method for screening sunflower lines for resistance to this pathogen.

Mihaljčević [122, 123] conducted the most detailed studies on the effectiveness of inoculation methods with *Macrophomina*. According to his results, the method of Hsi (1961) was the best of the four inoculation methods tested. Hsi developed this method for sorghum testing and Mihaljčević [122] adapted it for sunflower testing.

Ahmad *et al.* [124] examined 13 exotic sunflower inbred lines and eight *Macrophomina* isolates. The tested inbred lines differed significantly in agronomic characteristics (head diameter, head weight, number of seeds per head, 1000-seed weight, and yield per unit area). The inbred lines HAR 1 and HAR 2 were resistant/tolerant across all charcoal rot isolates, while HA 822 was susceptible to the disease development and two charcoal rot isolates (MP9 and MP21) were virulent in affecting the head weight.

Mihaljčević [122] also found high resistance levels in lines derived from the Argentine cultivars Pehuan INTA, Ciro, and Klein as well as in the lines GVP-1 and GVP-2, derived from varietal populations (VNIIMK, Krasnodar) developed by interspecific hybridization with *H. tuberosus*.

Galina Pustovoit and Gubin [83] found the sources of resistance to *Macrophomina* in the F_{14} progenies of the interspecific hybrid VNIIMK8931 × *H. tuberosus*. A radical inoculation method (injecting fungus suspension into the head tissue) confirmed a complete resistance in 62 lines.

Studies of wild sunflower species have been insufficient to enable the identification of resistance genes as the sources of resistance against charcoal rot. Seiler [49] concluded that interspecific hybrids based on *H. tuberosus* have resistance to charcoal rot. Wild species *H. mollis*, *H. maximiliani*, *H. resinosus*, *H. tuberosus*, and *H. pauciflorus* have also shown resistance.

2.1.7. *Phoma black stem (Phoma macdonaldii Boerema)*

According to Aćimović [120], the synonym for this fungus is *Phoma oleracea* var. *helianthi-tuberosi* Sacc.

Phoma black stem is in large expansion in several countries in the world. It causes premature drying of plants (forced ripening) resulting in economic damage that increases from 1 year to another [22].

Viranyi [52] points out that *Phoma* black stem is extremely severe in France where basal stem lesions often result in lodging.

The inoculation method described by Maširević [125] is recommended to sunflower breeders. For efficiency, molecular markers should be used when screening breeding material.

Fayzalla [126] examined in detail the resistance to *Phoma macdonaldii* in a large set of Novi Sad genotypes of cultivated sunflower and several wild species. Using an inoculation method, he found that there was no satisfactory tolerance to *Phoma macdonaldii* in the genotypes of the cultivated sunflower. Among the wild species, however, high tolerance was registered in *H. maximiliani, H. argophillus, H. tuberosus,* and *H. pauciflorus.*

Phoma black stem resistance has been reported in several perennial species: *H. eggertii, H. hirsutus, H. resinosus,* and *H. tuberosus* [99].

Encheva *et al.* [110] stated that interspecies of hybrids based on *H. salicifolius* are highly resistant to *Phoma black stem.*

Christov [78] also confirms that interspecies hybrids based on *H. eggertii, H. debilis, and H. argophillus* exhibit *high* levels of *resistance* to *Phoma.*

Darwishzadeh *et al.* [127] undertook experiments to determine the partial resistance of sunflower genotypes to seven isolates and highly significant differences were observed among genotypes, isolates, and their interactions. Two genotypes exhibited specific resistance with a wide range of isolate-nonspecific partial resistance appearing as well. In addition, QTLs were also found associated with isolate-specific and nonspecific resistance [128]. Alignan *et al.* [129] developed a 1000-element cDNA microarray-containing genes putatively involved in primary metabolic pathways in order to identify genes responsible for partial resistance. They were successful in identifying 38 genes differently expressed among genotypes, treatments, and times.

According to Škorić [99], resistance to *Phoma* black stem is positively correlated with resistance to *Phomopsis* stem canker and charcoal rot.

2.1.8. *Alternaria blight (Alternaria helianthi Tub. et Nish.)*

Aćimović [120] cited the following synonyms for *Alternaria* blight: *Helminthosporium helianthi* Hansf., *Alternaria leucanthemum* Nelena et Vas. and *Embellisia helianthi* (Hansf.). The same author stated that sunflowers are also attacked by *Alternaria zinniae* Pape, *Alternaria alternata* (Fr.) Keiss (synonym *Alternaria tenuis* Ness.) and *Alternaria helianthinficiens* Simmons, Walcz, and Roberts. Of these species, *Alternaria helianthi* is the most common on sunflowers and the best studied from the point of view of sunflower resistance. It was found to attack sunflowers in all continents where this oilseed crop is grown. In the previous decade, it caused the most extensive economic damage on sunflowers in India and Brazil. According to Aćimović [120], most of the cultivated sunflower genotypes are sensitive to *Alternaria* blight.

Regina *et al.* [130] concluded that the occurrence of *Alternaria helianthi* in southern Brazil depended on the pathogen race and sunflower cultivar to a large extent. Attack is most intensive on crops sown in December and least intensive in October-sown crops. Dudienas *et al.* [131] claimed that *Alternaria* causes economic damage in Brazil, especially in humid conditions.

Aćimović [132] tested 1389 inbred lines for 4 years under field conditions and found that only six lines possessed satisfactory tolerance to *Alternaria* blight.

Madhavi *et al.* [133] found the sources of resistance to *Alternaria* blight in *H. tuberosus* and *H. occidentalis.*

Lipps and Herr [134] examined 496 sunflower genotypes for resistance to *Alternaria* for 3 years and found tolerance in eight genotypes only. A different situation was encountered when the *H. tuberosus* population was inoculated in the greenhouse. Based on the obtained results, the authors concluded that *H. tuberosus* can be used as a source of resistance to *Alternaria helianthi.*

Morris *et al.* [135] confirmed that all 21 annual taxa and 18 of 21 perennial species evaluated were susceptible to *A. helianthi* using applied spore suspensions, while perennial species *H. hirsutus, H. pauciflorus* ssp. *subrhomboideus,* and *H. tuberosus* appear to resist infection by *A. helianthi.*

Sujatha *et al.* [136] determined that nine perennial *Helianthus* species, *H. maximiliani, H. mollis, H. divaricatus, H. simulans, H. occidentalis, H. pauciflorus* and *H. decapetalus, H. resinosus,* and *H. tuberosus* were highly resistant to *Alternaria* leaf spot; all annuals were susceptible.

Christov [78] reported that perennial *H. decapetalus, H. laevigatus, H. glaucophyllus,* and *H. ciliaris* were potential sources of genes for *Alternaria* resistance.

Complete resistance to *Alternaria* leaf spot was reported in interspecific hybrids of *H. salicifolius* by Encheva *et al.* [110]. Škorić [81] obtained similar results.

2.2. Other fungal diseases

There is a large number of other fungal diseases of sunflower that cause economic damage to sunflower production in some regions and in some years. Unfortunately, most of them have not been included in breeding programs yet [22].

2.2.1. Fusarium wilt (Fusarium spp.)

According to Aćimović [120], several species of the genus Fusarium attack sunflowers: *Fusarium solani, Fusarium solani* var. *minus, Fusarium oxysporum, Fusarium oxysporum, F.*

helianthi, Fusarium moniliforme (syn. *Gibberella fujikuroi*), *Fusarium equiseti, Fusarium tabacum, Fusarium culmorum, Fusarium* sp. and *Fusarium* spp.

Viranyi [52] states that *Fusarium wilt (Fusarium* spp.) has been reported as a pathogen of concern only from Russia [137] where it appeared to be harmful for sunflower production. Based on the extent of necrosis incited by the fungus on the main root and the root–hypocotyl transition zone of sunflower seedlings, some tolerance to pathogen attack could be detected among the genotypes [138]. In a breeding program, a number of new breeding lines were developed exhibiting relatively good field tolerance [139].

There are few research papers dealing with sunflower resistance to *Fusarium*. In one of these earlier papers, Orellana [140] reported that out of 49 inbred lines tested, 23 were resistant to *Fusarium moniliforme*. In recent years, Goncharov [139] produced plants tolerant to *Fusarium* on the basis of laboratory tests and individual selection of plants from three double-cross combinations and an F_3 cross (UV.680 × o.p. *cv.* Leader).

2.2.2. *Rhizopus head rot (Rhizopus spp.)*

Dry rot of sunflower is caused by the following fungi from the genus *Rhizopus: Rhizopus arrhizus* Fisch. (syn. *Rhizopus nodosus* Namysl.), *Rhizopus nigricans* Ehr. (syn. *Rhizopus stolonifer* Eh. et Fr.) and *Rhizopus oryzae* Nent et Geer [120].

Dry rot occurs typically in regions with dry climate and high temperatures. It often causes a significant yield reduction and it particularly reduces the oil content in seeds [22].

It has become an important disease of sunflower in the USA [141]. The disease reduced oil quality and quantity in oilseed sunflower [142]. Infection of sunflower with *Rhizopus* head rot is enhanced by larval feeding of sunflower moth, *Homoeosoma electellum* (Hulst), which contributes to a secondary infection [141].

Yang *et al.* [143] reported that 4 out of 32 wild species and subspecies were resistant when inoculated with *R. arrhizus* and *R. oryzae* Went. The resistant sources were perennial *H. divaricatus, H. hirsutus, H. xlaetiflorus,* and *H. resinosus.*

One of the pioneer works was that of Agrawat *et al.* [144], who studied the resistance to *Rhizopus nodosus* in 91 sunflower cultivars. Their results based on an inoculation test indicated that resistance existed only in cultivars – Armavirec, Armavirskiy 3497, EC 40277, and K-2217, all from Krasnodar.

Rhizopus head rot brings great economic damage in many countries, by decreasing seed yield, seed oil content, and seed development. Unfortunately, few researchers in the world work on the examination of this pathogen.

2.2.3. *Powdery mildew (Erysiphe cichoracearum DC)*

Sunflower is a host to three fungal genera that cause powdery mildew [120, 145]: *Erysiphe cichoracearum* DC, which is widespread in all continents where sunflower are grown, *Leveillula compositarum* Golow., *Leveillula taurica* (Lev.) Arn., and *Sphaerotheca fuliginea* (Schlecht. ex Fr.) Poll.

Since *Erysiphe cichoracearum* DC is common on sunflowers around the world, resistance to this pathogen has been studied most extensively. Saliman *et al.* [146] was among the first to identify wild species from the genus *Helianthus* resistant to *Erysiphe* [120].

Jan and Chandler [147] transferred the resistance from *H. debilis* Nutt. into the line P21. The mode of inheritance in this resistance source was partially dominant. According to unpublished results of Škorić, resistance to powdery mildew exists in several inbred lines, especially those that incorporate genes from *H. tuberosus.*

Breeding programs conducted in Argentina, Australia, and the Republic of South Africa have been targeted on Albugo resistance and several highly tolerant hybrids were obtained.

Seiler [49] indicates that *Helianthus debilis* ssp. *praecox*, and *H. bolanderi*, and 14 perennial species were tolerant of powdery mildew in both field and greenhouse tests [146]. Not all population of perennial species are resistant: population of *H. grosseserratus* and *H. maximiliani* showed differential reactions. Škorić [116] reported that interspecific hybrids with *H. giganteus, H. hirsutus, H. divaricatus,* and *H. salicifolius* had no powdery mildew symptoms.

Jan and Chandler [147] transferred the resistance from *H. debilis* Nutt. into the line P21. The mode of inheritance in this resistance source was partially dominant. According to unpublished results of Škorić, resistance to powdery mildew exists in several inbred lines, especially those that incorporate genes from *H. tuberosus.*

2.2.4. *Botrytis cinerea Pers.*

Sunflower geneticists and breeders have unjustly neglected the polyphagous fungus *Botrytis cinerea* Pers., although it causes economic damage in sunflower production in some regions.

Prats [148] was the first to discover a source of resistance to *Botrytis cinerea* in the cultivar INRA 64-01.

Burlov and Artemenko [149] found the line Od-2624 to be resistant to *Botrytis.*

Kostyuk [150] studied some 1400 sunflower genotypes and found that none of them were resistant and only some were tolerant to *Botrytis* under natural and inoculation conditions.

2.2.5. *White rust (Albugo tragopogonis Schr.)*

According to Aćimović [120] the synonym for this fungus is *Albugo tragopogonis* (Pers.) Schr. White rust has been registered on sunflowers in several countries and is particularly aggressive in South America (Argentina), Africa (Republic of South Africa), some Asian countries, several countries from the former Soviet Union, and Australia.

Breeding programs conducted in Argentina, Australia, and the Republic of South Africa have been targeted on *Albugo* resistance and several highly tolerant hybrids were obtained [22].

An established breeding centre, which focuses its research on identifying sources of resistance to white rust in wild species of genus *Helianthus*, unfortunately does not exist anywhere in the world.

2.3. Viruses and Bacteria

Some viruses are capable of causing disease in sunflowers. The number of viruses that are specific and attack only sunflowers is very limited. In most cases, the main host is another agricultural crop and sunflower is only a secondary host [120]. According to Gulya *et al.* [145], several viruses attack sunflower: aster yellows virus, cucumber mosaic virus, sunflower mosaic virus, sunflower ringspot virus, sunflower yellow blotch virus, leaf crinkle virus, tobacco ringspot virus, tobacco streak virus, tomato spotted wilt virus, potyvirus, etc.

Viruses are typically transmitted by vectors, the most important among which are aphids.

Srechari *et al.* [151] reported three aphid species, *Aphis gossypii* Glove., *Aphis craccivora* Koch, and *Rhopalosiphum maidis* (Fitch), as virus vectors. Among them, *A. gossypii* is best known as the vector that transmits the sunflower mosaic disease.

Lenardon *et al.* [152] detected the sunflower chlorotic mottle virus (SuCMoV) in several regions of Argentina.

Lenardon *et al.* [153] tested 232 lines in the greenhouse using an inoculation method. Only three lines exhibited partial resistance (L33, L74, and L42) to the sunflower chlorotic mottle virus. Of these three lines, L33 was the most resistant. Screening F_2 population of crosses between resistant and sensitive lines in the greenhouse and in field, the authors concluded that the resistance is controlled by a single dominant gene (Remo-1).

In recent years, the sunflower chlorotic mottle virus has been studied intensively at the molecular level.

Dujovny *et al.* [153] conducted a molecular characterization of a new potyvirus (SuCMoV). Arias *et al.* [155] described the effect of SuCMoV on some aspects of carbon metabolism in sunflower plants.

Mailo *et al.* [156] mechanically inoculated one sensitive (20 016) and one tolerant line (B-133) with SuCMoV. Total RNA was isolated from infected leaf tissue for study at the molecular level. The achieved results indicated that the gene expression profiles in the inoculated plants (of the sensitive and the tolerant line) were statistically significant compared with leaves of plants that were not inoculated. Eighty-eight genes were differentially expressed in the tolerant line.

2.3.1. Bacterial diseases

Bacterial diseases of sunflower are caused by pathogenic bacteria. They can be found on sunflowers in most countries where this oil crop is grown. In addition to sunflower, most of these bacteria also attack other crops. The sunflower is typically a secondary host and quite rarely the main host [120].

The most widespread bacteria on sunflowers are *Agrobacterium tumefaciens* (E. F. Sm. and Town.) Conn, *Pseudomonas syringae* pv. *tabaci* (Wolf and Foster 1917) Young, Dye and Wilkie 1978 (synonyms *Pseudomonas tabaci* and *Bacterium tabacum* Wolf and Foster), *Xanthomonas campestris* pv. *phaseoli* (Smith) Due., *Pseudomonas syringae* pv. *helianthi* (Kawamura) Dye, Wilkie and Young, *Pseudomonas solanacearum* (Smith), *Erwinia carotovora* pv. *carotovora* (Jones) Bergey *et al.*, etc. [120].

There are few papers in the literature dealing with sunflower selection for resistance to bacterial diseases.

Among these few, Nemeth and Walcz [157] reported the occurrence of *Erwinia carotovora* on sunflowers in Hungary in the period 1984–1986. Testes of inbred lines and commercial hybrids conducted under natural conditions indicated that there existed significant differences in resistance to this bacterial disease. However, the tests showed that the breeding material can be tested by inoculation methods under field conditions.

2.4. Sunflower breeding for resistance to broomrape (*Orobanche cumana* Wallr.)

The parasitic angiosperm broomrape (*Orobanche cumana* Wallr. = *Orobanche cernua* Loelf.) is the cause of many economic losses in sunflower production in a number of countries in the world, especially in central and eastern Europe, Spain, Turkey, Israel, Kazakhstan, and China. Its presence has also been established in Australia.

Sunflower breeders have been fighting *Orobanche cumana* Wallr. for almost a century [22].

According to past researches, there have been different mechanisms of sunflower resistance to *Orobanche*. Most often these are genetic mechanisms, but there are also physiological, biochemical, mechanical, and others.

According to Morozov [158], the first reports of broomrape in sunflower came from Saratov Oblast in Russia and date back to the 1890s. The same author mentions that the first sunflower varieties resistant to race A of *Orobanche* were developed by Plachek at the Saratov breeding station.

At the beginning of the 20th century, broomrape spread across Russia significantly and endangered the mass production of sunflower. The first cultivar resistant to race A, Saratovskij 169, was created by Plachek. In the years that followed, other cultivars resistant to race A were also produced (Kruglik A/41, Zelenka, and Fuksinka). As the mass production of sunflower spread quickly, it was followed by a relatively fast production of a new race called B. Zhdanov in Rostov on the Don announced that he had produced several cultivars resistant to a new race (B). During the period 1924–1960, Pustovoit in VNIIMK, Krasnodar created highly productive cultivars, which were resistant to race B [22, 158].

In order to attain their breeding goals and identify sources of broomrape resistance, sunflower breeders must develop a breeding strategy, decide on a breeding method, secure the necessary germplasm and differential lines for broomrape race identification, and choose the appropriate inoculation method and molecular marker technique (marker-assisted selection (MAS)) – Škorić [22].

Vrânceanu *et al.* [159] defined a set of differential lines for the evaluation of the composition of broomrape races. Among them was the AD-66 line, which represented a tester line susceptible to all broomrape races. On the other hand, the differential line (cultivar) Kruglik A41 was used for race A, the Jdanov-8281 cultivar for race B, the Romanian cultivar Record for race C, line S-1348 for race D, and line P-1380-2 for race E. Pâcureanu-Joița *et al.* [160] included the line LC-1093 as a tester for race F. Unfortunately, there have been no tester lines for the latest broomrape races [160].

Genes for resistance to broomrape races A, B, C, and D are present in varietal populations of sunflower developed in breeding programs from Krasnodar, Armavir, Odessa, Fundulea, and several other places [149]. Genes that confer resistance to races E, F, G, and the latest ones, on the other hand, have been identified in certain wild species of the genus *Helianthus* and have been incorporated into cultivated sunflower genotypes by interspecific hybridization [161, 162].

Galina Pustovoit [163] and her team made a great contribution in this area by developing sunflower varieties through interspecific hybridization in which *H. tuberosus* was used as the donor of *Or* genes. These varieties were used in the identification of Or_4 and Or_6 genes.

Fernandez-Martinez *et al.* [164] tested 44 wild sunflower accessions (representing 27 perennial and 4 annual species) and 44 cultivated sunflower accessions, which they raised in a growth chamber and then transplanted to a greenhouse. The material was inoculated with the virulent race F (population SE 296). Most of the perennial species proved fully resistant to race F.

Among the wild annual species, *H. anomalus* and *H. agrestis* were completely resistant, while *H. debilis* ssp. *cucumerifolius* and *H. exilis* segregated with regard to *Orobanche* resistance [164].

Interspecific hybrids based on *H. eggertii* and *H. smithii* showed total resistance to broomrape in Bulgaria [107]. Broomrape resistance to the local race in Bulgaria was reported in *H. divaricatus, H. eggertii, H. giganteus, H. grosseserratus, H. glaucophyllus, H. mollis, H. nuttallii, H. pauciflorus* (=*rigidus*), *H. resinosus*, and *H. tuberosus* [107, 165].

Diploid perennial species *H. divaricatus, H. giganteus, H. glaucophyllus, H. grosseserratus, H. mollis, H. nuttallii*, and *H. smithii* and their interspecific hybrids were resistant to broomrape [67]. Christov [78] reported that several perennial *Helianthus* species showed 100% resistance including *H. tuberosus, H. eggertii, H. smithii, H. pauciflorus*, and *H. strumosus*.

Jan *et al.* [166] crossed the wild sunflower species *H. maximiliani* Schrad, *H. grosseserratus* Mart., and *H. divaricatus* L. with cultivated sunflower and developed four populations (BR1-BR4) resistant to race F in Spain.

Numerous authors in public institutions and private companies use wild sunflower species as donors of genes for resistance to broomrape.

The sources of resistance to broomrape, which have been discovered so far, mostly use the gene of resistance taken from the wild species of the genus *Helianthus*. According to the results obtained so far, there are over 20 wild species of the genus *Helianthus*, which contain the gene of resistance to broomrape [22].

When broomrape occurred, breeders used infested fields for testing selection materials for broomrape resistance in many countries. This method is not reliable enough since in natural conditions infestation with broomrape on those fields is not equally spread, which causes large experimental errors. A much more reliable method, which is applied on testing resistance, is mixing of broomrape seed (a certain amount) with the selection material in the process of cultivation. In order to accelerate the process of testing the resistance of selection material, there are certain containers used in a greenhouse during the period autumn/winter, which are filled with a mixture of soil and sand filled with broomrape seeds and the seed of sunflower genotypes that are being tested [22].

Panchenko [167] developed a method of testing the selection materials in the greenhouse, which enables simultaneous testing of a great number of lines that are being created. The purpose of this method is preparation of a medium (sterilized soil + sand or perlite) on the tables in the medium. Following that, the appropriate amount of broomrape seed is added and the selection material is cultivated. Within 3–4 weeks after germination, it is possible to make the evaluation of resistance.

In Rustica Prograin Genetique, Grezes-Besset [168] developed a fast method of testing the selection material of sunflower in plastic tubes (a mixture of sand and perlite), which enables a reliable testing of a large number of lines (hybrids) in small space and the cycle lasts about 3 weeks.

However, the most reliable and the most easily applied method of screening breeding materials for broomrape resistance is the use of molecular markers, QTL, RFLP, RAPD, TRAOP, and SSR markers which have been used for this purpose.

Increased use of marker-assisted selection, which gives quick and reliable results, is very positive for sunflower breeding.

The best example of that is the production of hybrids resistant to the imidazolinone group of herbicides, which has proven itself in mass production by cultivating IMI-resistant hybrids followed by controlling broomrape. IMI-resistant hybrids are very important in regions where new races of broomrape have occurred [22].

Dominant genes for resistance to races A, B, C, D, E, and F have been found and incorporated into cultivated sunflower genotypes. In the last 2–3 years, new broomrape populations have been discovered in several countries. None of the existing commercial hybrids resistant to races A, B, C, D, E, and F have proven resistant to these new populations.

2.5. Sunflower breeding for insect resistance

Several hundred different species of insects cause infestations in sunflower. However, economic losses are caused only by a few insect species [169]. Some insects transmit several sunflower diseases [170]. *Homoeosoma* spp. are a significant problem in cultivated sunflower on four continents. *Homoeosoma nebulella* (Hubner) infests sunflower in Europe and Asia. In South America, sunflower is infested by *H. heinrichi* (Pastrana), whereas *H. electellum* (Hulst) causes damage to sunflower in Mexico, USA, and Canada.

Cultivar resistance to European sunflower moth (*H. nebulella*) was incorporated in USSR 60-70 years ago through interspecies hybridization of cultivated sunflower and *H. tuberosus* spp. *Purpurellus,* Cockerell [171, 172].

The mechanism of resistance to sunflower moth exists due to the phytomelanin (carbon) layer in the husk. Black hull colour is positively correlated with phytomelanin content in the husk.

North-American species of sunflower moth (*H. electellum*) is far more virulent on cultivated sunflower than *H. nebulella*; hence, resistance breeding to this insect is of great importance in North America.

According to the results of Rogers and Kreitner [173], the presence of phytomelanin in sunflower seed pericarp prevents seed infestation with *H. electellum*. By monitoring the formation of pericarp (husk), it was determined that phytomelanin starts to accumulate between the hypodermis and sclerenchyma 3 days after fertilization, whereas its formation is clearly visible 13 days after fertilization.

In the major production area of North America, there are about 14 principal insect pests of cultivated sunflower, and of this total about six are considered potential economic pests [174].

"Two breeding procedures are recommended for identifying lines with improved resistance to insects attacking cultivated sunflower. These procedures are based on the initial evidence that the resistance to the insects is quantitatively inherited, that is, controlled by several genes. Both are based on recurrent selection and random mating, with the main objective to combine as many of the alleles controlling resistance as possible" [175].

"The recurrent phenotypic selection breeding procedure could be utilized for selection against stem and/or foliage infesting insects. An original (C_0) source population may be created by random mating cultivars or lines (e.g., Plant introductions, open-pollinated populations), which are then screened for resistance to a particular insect attack sunflower" [175].

"Recurrent phenotypic selection with S_1 line progeny evaluation could be utilized for selection for head and/or seed infesting insects. An original (C_0) source population is created similarly as in the recurrent phenotypic selection procedure. The C_0 population is planted in a normal breeding nursery with the most vigorous plants selected for bagging and self-pollination" [175].

According to Seiler [49], the insects causing most economic damage in North America are: sunflower beetle [*Zygogramma exclamationis* (Fabritius)], the sunflower stem weevil [*Cylindrocopturus adspersus* (LeConte)], the red and gray seed weevils [*Smicronyx fulvus* (LeConte) and *S. sordidus* (LeConte)], the banded sunflower moth (*Cochylis hospes* Walsingham), the sunflower moth [*Homoeosoma electellum* (Hulst)], and the sunflower midge *Contarinia schulzi* Gagne. The sunflower head moth, *Homoeosoma electellum* is the most widespread and damaging sunflower insect pest in North America, while in Europe and Asia it is *Homoeosoma nebulella* (Hubner).

According to the results obtained by Rogers [176, 177], the following sunflower species exhibit significant levels of resistance to sunflower moth: *H. arizonensis, H. ciliaris, H. pumilus, H. resinosus, H. rigidus × laetiforus, H. silphiodes* and *H. smithii*.

Among the insects that cause economic losses to sunflower production, the biggest success was achieved in breeding for resistance to sunflower head moth above all in Europe by the development of cultivars in which an armored layer was induced in the husk from some specific wild species. Similar results were obtained in North America.

High level of resistance to *Bothynus gibbosus* was exhibited by the following species: *H. tuberoses, H. maximiliani, H. niveus, H. xlaetiflorus, H. salicifolius, H. mollis, H. grosseseratus, H. Argophyllus,* and *H. ciliaris* [178].

The results of Rogers [176, 179, 180], as well as Rogers and Thompson [178, 181, 182], confirm high levels of resistance to *Zygogramma exclamationis, Bothynus gibbosus, Masonaphis masoni* and *Empoasca abrupta* in wild sunflower species *H. tuberosus* and *H. maximiliani,* and recommend the use of these species in breeding programs.

Results of Rogers and Thompson [183, 184] confirm significant levels of resistance in two annual and 10 perennial wild species (*Masonaphis masoni*). The highest resistance to aphids was seen in *H. carnosus, H. exilis, H. floridanus* and *H. radula.*

Weak point in sunflower breeding for resistance to insects is that only few sunflower researchers deal with this issue. Insecticides can be used to some extent, more or less successfully, as a solution to this problem in some species.

2.6. Conclusions

Biotic stresses cause great economic damages and act as a limiting factor for the production of sunflower.

Diseases are the main problem among biotic stresses. Using wild sunflower species of the *Helianthus* genus, genes conferring resistance to most dominant diseases were discovered and incorporated to the genotypes of the cultivated sunflower.

Regarding the achievements in sunflower breeding for disease resistance, the results can be divided into four different groups.

The first group consists of work that resulted in the discovery of genetic resistance to certain causative agents of sunflower diseases (*Plasmopara halstedii, Puccinia helianthi, Verticillium dahliae, Verticillium albo-atrum,* and *Erysiphe cichoracearum*).

The second group comprises work in which a high level of tolerance (field resistance) was achieved. This group includes the results achieved in breeding for resistance to *Phomopsis/ Diaporthe helianthi, Macrophomina phaseolina, Albugo tragopogonis,* and *Alternaria* ssp.

The third group consists of results in which a satisfactory level of tolerance was achieved (*Phoma macdonaldii* and to some extent *Sclerotinia sclerotiorum*).

The fourth group consists of results that were partly achieved, where the level of favorable tolerance, that is, resistance, was not reached (*Rhizopus* ssp., *Botrytis cinerea* and other fungal pathogens).

Viruses and bacteria only pose a minor problem in comparison with diseases. Breeding for resistance to viruses and bacteria also includes wild sunflower species.

Broomrape (*Orobanche cumana* Walr.) is a major global issue in sunflower production, particularly in Central and Eastern Europe. Genes conferring resistance to six races of broomrape have been discovered in some wild sunflower species and incorporated into genotypes of the cultivated sunflower. Research work, which aims at finding genes of resistance to the newest races within specific wild sunflower species, is in progress.

Sources of resistance to imidazolines and sulfonylurea herbicides (tribenuron-methyl) have been found in a population of wild *H. annuus* from Kansas and incorporated in cultivated sunflower genotypes. Moreover, genes conferring resistance to these herbicides were discovered using induced mutations. The newly developed hybrids, resistant to the abovementioned herbicides, provide successful weed and broomrape control through imidazolines.

Insects are a major issue in sunflower production, especially in North America. Significant results have been obtained through breeding for resistance to sunflower head moth. Wild sunflower species are used in research work aimed at finding the sources of resistance to other economically harmful insects.

Different biotechnological methods (tissue culture, embryo culture, protoplast fusion, molecular markers, in vitro screening, and other methods) have been included in breeding for resistance to biotic stresses.

Author details

Dragan Škorić*

Address all correspondence to: draganskoric@sbb.rs

Serbian Academy of Sciences and Arts, SASA – Novi Sad Branch, Novi Sad, Serbia

References

[1] Singh BD. Plant Breeding: Principles and Methods. Kalyani Publishers, Ludhiana, New Delhi, Noida, India; 2006. 1018 p. ISBN: 8127220744.

[2] Monti LM. Breeding plants for drought resistance: The problem and its relevance. *In:* Monti LM and Porceddu E, editors. Drought Resistance in Plants: Physiological and Genetic Aspects. Series Agriculture. Office for Official Publications of the European Communities, Luxembourg; 1987. 388 p. ISBN: 9282573230.

[3] Škorić D. Sunflower Breeding. Uljarstvo, Belgrade. 1988;1:1–90.

[4] Rengel D, Arribat S, Maury P, Martin-Magniette ML, Hourlier T, Laporte M, Vares D, Carrere S, Grieu P, Balzerque S, Gouzy J, Vincourt P and Langlade N. Joint study of gene regulation and physiological responses to drought reveals a constitutive care network and genetic variability. *In:* Proc. 18th Intl. Sunflower Conf., Mar del Plata & Balcarce, Argentina, February 27–March 1, 2012. Intl. Sunflower Assoc., Paris, France; 2012. pp. 1–17.

[5] Škorić D. Desired model of sunflower hybrids and the newly developed NS-hybrids. Helia. 1980;3:19–24.

[6] Kamali V and Miller JF. The inheritance of drought tolerance in sunflower. *In:* Proc. 10th Intl. Sunflower Conf., Surfers Paradise, Australia, March 14–18, 1982. Intl. Sunflower Assoc., Paris, France; 1982. pp. 228–233.

[7] Škorić D. Achievements and future directions of sunflower breeding. Field Crop. Res. 1992;30:231–270. http://dx.doi.org/10.1016/0378-4290(92)90003-R.

[8] Fulda S, Stegmann H and Horn R. Influence of drought stress on growth protein expression and osmolyte accumulation in sunflower *Helianthus annuus* L. c.v. Peredovik L008. *In:* Velasco L, editor. Proc. 17th Intl. Sunflower Conf., Cordoba, Spain, 8–12 June 2008. Intl. Sunflower Assoc., Paris, France; 2008;1:347–362.

[9] Lisanti S, Carbacos JM, Hall AJ and Chimenti C. Influence of water deficit and canopy senescene pattern on sunflower root functionality during the grain-filling phase. *In:* Proc. 18th Intl. Sunflower Conf., Mar del Plata & Balcarce, Argentina, February 27–March 1, 2012. Intl. Sunflower Assoc., Paris, France; 2012. p. 120.

[10] Blum A. Plant Breeding for Stress Environments. Boca Raton, Florida USA, CRC Press, Inc.; 1988. 232 p. ISBN: 0-8493-6388-8.

[11] Škorić D. Sunflower breeding. *In:* Polak V. editor. Sunflower-Monograph, Nolit, Beograd; 1989. pp. 284–393. (In Serbian) ISBN:

[12] Turčevskij IA. Fotosintez i zasuha. Izd. Kazanskogo gosudarstvenogo Univerziteta, Kazan; 1964. 197 p.(In Russian)

[13] Gusev NA. Fiziologija vodoobmena i ustojčivosti rastenij. Izd. Kazanskogo gosudarstvenogo Univerziteta, Kazan; 1968. 211 p.(In Russian) UDC: 581.13

[14] Ćupina T, Sakač Z. Fiziološki aspekti formiranja prinosa suncokreta. Poljoprivredni fakultet, Novi Sad; 1989. 224 p. ISBN:

[15] Seiler GJ. Evaluation of wild sunflower species for potential drought tolerance. *In:* Proc. 5th Sunflower Research Workshop, Fargo, ND, USA, January 26 1983. Natl. Sunflower Assoc., Bismarck, ND, USA; 1983. 24 p. http://www.sunflowernsa.com:80/research/research-workshop/documents/1983_seiler_evaluation.wild.drought.tolerance.pdf.

[16] Škorić D. Sunflower breeding for resistance to abiotic stresses. Helia. 2009;32(40):1–16.

[17] Chimenti C, Giuliano J, Hall A. Osmotic adjustment, its effects on yield maintance under drought in sunflower. *In:* Seiler GJ, editor. Proc. 16th Intl. Sunflower Conf., Fargo, ND, USA, 29 August–4 September 2004. Intl. Sunflower Assoc., Paris, France. 2004; I:261–267.

[18] Andrei E. Drought effects on growth stages and yield components of sunflower hybrids. *In:* Seiler GJ, editor. Proc. 16th Intl. Sunflower Conf., Fargo, ND, USA, 29 August–4 September 2004. Intl. Sunflower Assoc., Paris, France. 2004; I:251–255.

[19] Sato NE, Pereyra-Irujo G, Creus CM and Aguirrezabal AN. Correlation between the responses of leaf expansions and hypocotyl elongation to water deficit in sunflower (*Helianthus annuus* L.) genotypes. *In:* ISA, edition. Proc. 18th Intl. Sunflower Conf., Mar del Plata & Balcarce, Argentina, February 27–March 1, 2012. Intl. Sunflower Assoc., Paris, France; 2012. p. 127.

[20] Petcu E, Stanciu M, Stanciu D and Raducanu F. Physiological traits for quantification of drought tolerance in sunflower. *In:* Velasco L, editor. Proc. 17th Intl. Sunflower Conf., Cordoba, Spain, 8–12 June 2008. Intl. Sunflower Assoc., Paris, France; 2008;1:345–349.

[21] Vrânceanu AV. Floarea-soarelui hibrida. Editura Ceres., Bucuresti, Romania; 2000. 1147 p. (In Romanian). ISBN: 9734004530.

[22] Škorić D. 2012. Sunflower breeding. *In:* Škorić D and Sakač Z, editors. Sunflower Genetics and Breeding. (International monography). Serbian Academy of Sciences (SASA), Branch in Novi Sad, Novi Sad, Republic of Serbia; 2012. pp. 164–344. ISBN: 978-88-81125-82-3.

[23] Petrović M, Kastori R, Škorić D and Petrović N. Sunflower lines and hybrids response to water stress. Helia. 1992;14(17):47–64.

[24] Orta AH, Erdem T and Erdem Y. Determination of water stress index in sunflower. Helia. 2002;24(37):27–38. http://dx.doi.org/10.2298/hel0237027o.

[25] Seiler GJ. Utilization of wild sunflower species for the improvement of cultivated sunflower. Field Crop. Res. 1992;30:194–230. http://dx.doi.org/10.1016/0378-4290(92)90002-Q.

[26] Chandler JM and Jan CC. Identification of salt-tolerant germplasm sources in the *Helianthus* species. Agron. Abstr. Am. Soc. Agron., Madison, WI, USA. 1984; p. 61.

[27] Hussain MK, Rehman OU. Breeding sunflower for salt tolerance: Physiological basis for salt tolerance in sunflower (*Helianthus annuus* L.). Helia. 1993;16(18):77–84.

[28] Prakash AH, Vajranabhaiah SN and Reddy P.C. Effect of salt stress on callus development from hypocotyl segments of sunflower (*Helianthus annuus* L.) genotypes. Helia. 1993;16(18):71–76.

[29] Prakash AH, Vajranabhaiah SN, Reddy PC and Purushothama MG. Differences in growth, water relations and solute accumulation in the selected calluses of sunflower (*Helianthus annuus* L.) under sodium chloride stresses. Helia. 1996;19(14):149–146.

[30] Sarić M, Krstić B and Škorić D. Element diversity in sunflower inbred lines. Helia. 1991;14(14):41–48.

[31] Kalaydzhyan AA, Khlevnoy LV, Neshchadim NN, Golovin VP, Vartanyan VV, Burdun AM. Rossiyskiy solnechnyy tsvetok. Krasnodar, Sovet. Kuban. Russia. 2007; 342 p. (In Russian).

[32] Kalaydzhyan AA, Neshchadim NN, Osipyan VO and Škorić D. Kuban sunflower-gift to the world. Monograph. Ministry of Russian Agriculture - Russian Academy of Agriculture-Kuban State Agrarian University, Krasnodar. Russia. 2009; 498 p. (In Russian).

[33] Al-Khatib K, Beungurtner JR, Peterson DE and Currie RS. Imazethapyr resistance in common sunflower (*Helianthus annuus* L.). Weed Science. 1998;46:403–407. http://www.jstor.org/stable/4046055.

[34] Miller JF and Al-Khatib K. Development of herbicide resistant germplasm in sunflower. *In:* ISA, edition. Proc. 15th Intl. Sunflower Conf., Toulouse, France, 12–15 June 2000. Intl. Sunflower Assoc., Paris, France. 2000;2:419–423.

[35] Alonso LC, Rodriguez-Ojeda MI, Fernandez-Escobar J and Lopez-Ruiz-Calero G. Chemical control of broomrape (*Orobanche cernua* Loefl.) in sunflower (*Helianthus annuus* L.) resistant to imazethapyr herbicide. Helia. 1998;21(29):45–54.

[36] Al-Khatib K and Miller JF. Registration of four genetic stocks of sunflower resistant to imidazolinone herbicides. Crop Sci. 2000;40:869–870. doi: 10.2135/cropsci2000.403869x.

[37] Malidža G, Škorić D and Jocić S. Imidazolinone resistant sunflower (*Helianthus annuus* L.): Inheritance of resistance and response towards selected sulfonyl urea herbicides. *In:* ISA, edition. Proc. 15th Intl. Sunflower Conf., Toulouse, France, 12–15 June 2000. Intl. Sunflower Assoc., Paris, France. 2000;2:42–47.

[38] Bruniard JM and Miller JF. Inheritance of imidazolinone-herbicide resistance in sunflower. Helia. 2001;24(35):11–16. http://dx.doi.org/10.1515/helia.2001.24.35.11.

[39] Sala C, Bulos M, Echarte M, Whitt S, Budziszewski G, Howie W, Singh B and Weston B. Development of CLHA-PLUS: a novel herbicide tolerance trait in sunflower conferring superior imidazolinone tolerance and ease of breeding. *In:* Velasco L, editor. Proc. 17th Intl. Sunflower Conf., Cordoba, Spain, 8–12 June 2008. Intl. Sunflower Assoc., Paris, France; 2008;1:489–494.

[40] Miller JF and Seiler GJ, 2005. Tribenuron resistance in accessions of wild sunflower collected in Canada. *In:* Proc. 27th Sunflower Research Workshop, Fargo, ND, January 2005. Natl. Sunflower Assoc., Bismarck, ND, USA; 2005. 5 p. http://www.sunflowernsa.com/research/research-workshop/documents/Miller_Tribenuron_05.PDF.

[41] Miller JF and Al-Khatib K. Registration of two oilseed sunflower genetic stock. SURES-1 and SURES-2, resistant to tribenuron herbicide. Crop Sci. 2004;44(3):301–302. doi: 10.2135/cropsci2004.1037.

[42] Ottaviano E. Conclusive remarks. pp. 377–379. *In:* Monti LM and Porceddu E, editors. Drought Resistance in Plants: Physiological and Genetic Aspects. Series Agricul-

ture. Office for Official Publications of the European Communities, Luxembourg; 1987. 388 p. ISBN: 9282573230.

[43] Belhassen E, Castiglioni VPR, Chimenti C, Griveau Y, Jamaux I, Steinmetz A. Looking for physiological and molecular markers of leaf cuticular transpiration using interspecific crosses between *Helianthus argophyllus* and *Helianthus annuus*. *In:* ISA, edition. Proc. of the ISA-Symposium II: Drought Tolerance in Sunflower. Beijing P.R. China, June 14, 1996. Intl. Sunflower Assoc., Paris, France; 1996. pp. 39–44.

[44] Cellier F, Ouvrard O, Ferrare K, Tousch D, Lamaze T, Dupuis JM, Casse-Delbart F. Differential expression of water stress-regulated genes in drought tolerant or sensitive sunflower genotypes. *In:* ISA, edition. Proc. of the ISA-Symposium II: Drought Tolerance in Sunflower. Beijing P.R. China, June 14, 1996. Intl. Sunflower Assoc., Paris, France; 1996. pp. 36–39.

[45] Arce AL, Giacomelli JI, Ribichich KF and Chan RL. Sunflower atypical transcription factors and miRNAs playing a key role in responses to abiotic stresses. *In:* ISA, edition. Proc. 18th Intl. Sunflower Conf., Mar del Plata & Balcarce, Argentina, February 27–March 1, 2012. Intl. Sunflower Assoc., Paris, France; 2012. p. 47.

[46] Alberdi I, Pereyra-Irujo G, Zambelli A, Kaspar M, Feingold S, Alvarez D and Aguirrezabal L. Molecular markers associated with leaf expansion response to water deficit conditions. *In:* ISA, edition. Proc. 18th Intl. Sunflower Conf., Mar del Plata & Balcarce, Argentina, February 27–March 1, 2012. Intl. Sunflower Assoc., Paris, France; 2012. p. 103.

[47] Liu Z and Jan CC. Molecular techniques for sunflower breeding. *In:* Škorić D and Sakač Z, editors. Sunflower Genetics and Breeding. (International monography). Serbian Academy of Sciences (SASA), Branch in Novi Sad, Novi Sad, Republic of Serbia; 2012. pp. 431–476. ISBN: 978-88-81125-82-3.

[48] Lexer C, Lai Z and Rieseberg LH. Candidate gene polymorphisms associated with salt tolerance in wild sunflower hybrids: Implications for the origin of *Helianthus paradoxus*, a diploid hybrid species. New Phytol. 161: 225–233. doi: 10.1046/j.1469-8137.2003.00925.x.

[49] Seiler G. Utilization of wild *Helianthus* species in sunflower breeding. *In:* Škorić D and Sakač Z, editors. Sunflower Genetics and Breeding. (International monography). Serbian Academy of Sciences (SASA), Branch in Novi Sad, Novi Sad, Republic of Serbia; 2012. pp. 344–430. ISBN: 978-88-81125-82-3.

[50] Vear F. Breeding for durable resistance to the main diseases of sunflower. *In:* Seiler GJ, editor. Proc. 16th Intl. Sunflower Conf., Fargo, ND, USA, 29 August–4 September 2004. Intl. Sunflower Assoc., Paris, France. 2004; I:14–28.

[51] Pustovoit GV. 1978. Main results of sunflower breeding for group resistance in VNIIMK. *In:* VNIIMK, edition. Zbornik VNIIMK: Pests and diseases of oil crops. Krasnodar. 1978. pp. 42–48. (In Russian)

[52] Viranyi F. Research progress in sunflower diseases and their managment. *In:* Velasco L, editor. Proc. 17th Intl. Sunflower Conf., Cordoba, Spain, 8–12 June 2008. Intl. Sunflower Assoc., Paris, France; 2008;1:1–12.

[53] Vear F, Serre F, Roche S, Walser P and Tourvieille de Labrouhe D. Recent research on downy mildew resistance useful for breeding industrial-use sunflowers. *In*: 1st Symposium on Sunflower Industrial Uses, September 10–14, 2006. Lignano Sabbia d'Oro, Udine, Italy; 2006. pp. 1–8. http://isasunflower.org/publications/isc-symposia/single-view/article/1st-symposium-on-sunflower-industrial-uses-lignano-sabbia-doro-udine-italy.html.

[54] Gulya TJ. Distribution of *Plasmopara halstedii* races from sunflower around the world. *In*: Lebeda A and Spencesr-Phillips PTN, editors. Advances in Downy MIldew Research. Proc. 2nd Int. Downy Mildew Symposium, Olomouc, Czech Republic, Palacky Univeristy and JOLA, Kostelec na Hané, Czech Republic; 2007;3:121–134.

[55] Miller JF and Gulya TJ. 1988. Registration of six downy mildew resistant sunflower germplasm lines. Crop Sci. 1988;28(6):1040–1041. doi: 10.2135/cropsci1988.0011183X002800060073x.

[56] Gulya TJ, Sackston WE, Viranyi F, Maširević S and Rashid KY. New races of the sunflower downy mildew pathogen (*Plasmopara halstedii*) in Europe and North and South America. J. Phytopath. 1991;132(4):303–311. doi: 10.1111/j.1439-0434.1991.tb00125.x.

[57] Gulya TJ, Tourvieille de Labrouhe D, Maširević S, Penaud A, Rashid K and Virany F. Proposal for standardized nomenclature and identification of races *Plasmopara halstedii* (sunflower downy mildew). ISA Symposium II. Sunflower Downy Mildew, January 13–14, 1998, Fargo, ND, USA; 1998. pp. 130–136. http://isasunflower.org/fileadmin/documents/Symposia/3rd_Symposium_on_Sunflower_Mildew__USA__1998/T03-130_-_Proposal_for_Standardized_nomenclature_and_Identification_of_races_of_Plasmopara_halstedii.pdf.

[58] Jouffret P, Pilorge E and Pinochet X. Nuisibilite et anjeux economiques du mildiou du tournesol en France. *In:* INRA, edition. Le mildou du tournesol. RD 10, 78026 Versailles Cedex, CETIOM- Centre de Grignon, Bp-78840 Thiverval Grignon, France; 2000. pp. 31–34.

[59] Tourvieille de Labrouche D, Ducher M, Philippon J, Meliala C and Walser P. Les methods d'analyse du mildion. *In:* INRA, edition. Le mildou du tournesol. RD 10, 78026 Versailles Cedex, CETIOM- Centre de Grignon, Bp-78840 Thiverval Grignon, France; 2000. pp. 44–67.

[60] Molinero-Demilly V, Guenard M, Giroult C, Herbert O, Tourvieille de Labrouhe D, Philippon J, Penaud A, Tardin MC, Argeles G, Costes M, Mezzaroba A, Masse P, Pachet I, Cellier V and Gregoire S. 2004. Improving the sunflower downy mildew re-

sistance test. *In:* Seiler GJ, editor. Proc. 16th Intl. Sunflower Conf., Fargo, ND, USA, 29 August–4 September 2004. Intl. Sunflower Assoc., Paris, France. 2004; I:99–104.

[61] Tourvieille de Labrouhe, D., 2000. Disease control concerns every body: the example of sunflower downy mildew. *In:* ISA, edition. Proc. 14th Intl. Sunflower Conf., Toulouse, France. June 12–14. 2000. Intl. Sunflower Assoc., Paris, France. 2000; PLIC:1–8.

[62] Vear F, Jouan-Dufournet I, Bert PF, Serre F, Cambon F, Pont C, Walser P, Roche S, Tourvieille de Labrouhe D and Vincourt P. 2008. QTL for capitulum resistance to *Sclerotinia sclerotiorum* in sunflower. *In:* Velasco L, editor. Proc. 17th Intl. Sunflower Conf., Cordoba, Spain, 8–12 June 2008. Intl. Sunflower Assoc., Paris, France; 2008;2:604–610.

[63] Vear F, Serre F, Walser P, Bony H, Joubert G and Torvieille de Labrouhe D. 2000. Pedigree selection for sunflower capitulum resistance to *Sclerotinia sclerotiorum*. *In:* ISA, edition. Proc. 14th Intl. Sunflower Conf., Toulouse, France. June 12–14. 2000. Intl. Sunflower Assoc., Paris, France. 2000; IIK:42–47.

[64] Vear F, Pinochet X, Cambon F and Philippon J., 2000. La lutte génétique, une methode efficace à privilégier. *In:* Tourvieille de Labrouhe D, Pilorge E, Nicolas P and Vear F, editors. Le mildiou du tournesol. Paris: INRA editions. CETIOM-INRA. Paris, France; 2000. pp. 124–134.

[65] Tourvieille de Labrouche D, Serre F, Walser P, Philippon J, Vear F, Tardin MC, Andre T, Castellanet P, Chatre S, Costes M, Jouve Ph, Madeuf JL, Mezzaroba A, Plegades J, Pauchet I, Mestries E, Penaud A, Pinochet X, Serieys H and Griveau Y. Partial, non-race-specific resistance to downy mildew in cultivated sunflower lines. *In:* Seiler GJ, editor. Proc. 16th Intl. Sunflower Conf., Fargo, ND, USA, 29 August–4 September 2004. Intl. Sunflower Assoc., Paris, France. 2004; I:104–110.

[66] Christov M. Hybridization of cultivated sunflower and wild *Helianthus* species. *In*: Caligari PDS and Hind DJN, editors. Compositae: Biology and Utilization, Proc. Intl. Compositae Conf., The Royal Botanic Gardens, 24 July–4 August, 1994. Kew, London. 1996;2:602–614.

[67] Nikolova LM, Christov M and Shindrova P. 1998. New sunflower forms resistant to *Orobanche cumana* Wallr. Originating from interspecific hybridization. *In*: Wegmann K, Muselman LJ and Joel DM, editors. Current Problems of *Orobanche* Researchers. Proc. 4th Intl. Workshop on Orobanche, 23–26 September, 1998. Albena, Bulgaria; 1988. pp. 294–299.

[68] Hammer TR, Thines M and Spring O. Transient expression of *gfp* in the obligate biotrophic ooomycete *Plasmopara halstedii* using electroporation and a mechanoperforation method. Plant Pathol. 2007;56(1):177–182. doi: 10.1111/j.1365-3059.2006.01494.x.

[69] Škorić D and Jocić S. Achievements of sunflower breeding at the IFVC in Novi Sad. *In:* Seiler GJ, editor. Proc. 16th Intl. Sunflower Conf., Fargo, ND, USA, 29 August–4 September 2004. Intl. Sunflower Assoc., Paris, France. 2004; II:441–448.

[70] Castaño F, Vear F and Tourvieille de Labrouhe D. Resistance of sunflower inbred lines to various forms of attack by *Sclerotinia sclerotiorum* and relations with some morphological characters. Euphytica. 1993;68(1–2):84–98. doi: 10.1007/BF00024158.

[71] Mancel MK and Shein SE. 1982. Field inoculation of sunflower for *Sclerotinia sclerotiorum* basal stalk rot and virulence of isolates from various hosts. *In:* ISA, edition. Proc. 10th Intl. Sunflower Conf., Surfers Paradise, Australia, 14–18 March 1982. Intl. Sunflower Assoc., Toowoomba, Australia. 1982. pp. 167–170.

[72] Baldini M, Turi M, Vischi M, Vannozzi GP and Olivieri AM. Evaluation of genetic variability for *Sclerotinia sclerotiorum* (Lib.) de Bary resistance in sunflower and utilization of associated molecular markers. Helia. 2002;25(36):177–190. http://dx.doi.org/10.2298/hel0236177b.

[73] Van Becelaere G and Miller JF. Methods of inoculation of sunflower heads with *Sclerotinia sclerotiorum* (Lib.) de Bary. Helia. 2004;27(41):137–142. http://dx.doi.org/10.2298/hel0441137v.

[74] Drumeva M, Nenov N and Kiryakov I. Study on an *in vitro* screening test for resistance to *Sclerotinia sclerotiorum* in sunflower. *In:* Velasco L, editor. Proc. 17th Intl. Sunflower Conf., Cordoba, Spain, 8–12 June 2008. Intl. Sunflower Assoc., Paris, France; 2008;1:181–186.

[75] Van Becelaere G. Methods of inoculation and inheritance of resistance to *Sclerotinia* head rot in sunflower. *Ph.D.* Thesis. North Dakota State University of Agriculture and Applied Sci. Fargo, ND, USA. 2003. 81 p.

[76] Vear F, Serre F, Roche S, Walser P and Tourvieille de Labrouhe D. Improvement of *Sclerotinia sclerotiorum* head rot resistance in sunflower by recurrent selection of a restorer population. Helia. 2007;30(46):1–12. http://dx.doi.org/10.2298/hel0746001v.

[77] Christov M, Shindrova P and Encheva V. Transfer of new gene material from *Helianthus* species to sunflower. *In:* ISA, edition. Proc. 14th Intl. Sunflower Conf., Beijing/Shenyang, China, 12–20 June 1996. Intl. Sunflower Assoc., Paris, France. 1996; II: 1039–1046.

[78] Christov M. *Helianthus* species in breeding research in sunflower. *In*: Velasco L, editor. Proc. 17th Intl. Sunflower Conf., Cordoba, Spain, 8–12 June, 2008. Intl. Sunflower Assoc., Paris, France. 2008. pp. 709–714.

[79] Henn HJ, Steiner U, Wingender R and Schnabl H. Wild type sunflower clones: Source of resistance against *Sclerotinia sclerotiorum* (Lib.) de Bary stem infection. J. Appl. Bot. (Angewandt Botanie). 1997;71(1–2):5–9. ISSN 0949-5460.

[80] Miller JF and Gulya TJ. Registration of eight Sclerotinia-tolerant sunflower germplasm lines. Crop Sci. 1999;39(1):301–302. doi: 10.2135/cropsci1999.0011183X003900010075x.

[81] Škorić D. 1987. FAO sunflower sub-network report 1984–1986. *In*: Škorić D, editor. Genetic Evaluation and Use of *Helianthus* Wild Species and their Use in Breeding Programs. FAO, Rome, Italy. 1987. pp. 1–17.

[82] Tikhomirov VT and Chiryaev PV. Sources of resistance to diseases in original material of sunflower. Helia. 2005;28(42):101–106. http://dx.doi.org/10.2298/hel0542101t.

[83] Pustovoit GV and Gubin IA. 1974. Results and prospects in sunflower breeding for group immunity by using the interspecific hybridization method. *In:* Proc. 6th Intl. Sunflower Conf., Bucharest, Romania, 22–24 July 1974. Intl. Sunflower Assoc., Toowoomba, Australia. 1974. pp. 373–382.

[84] Mondolot-Cosson L and Andary C. Resistance factors of a wild species of sunflower, *Helianthus resinosus,* to *Sclerotinia sclerotiorum.* Acta Hortic. 1984;381:642–645. http://dx.doi.org/10.17660/ActaHortic.1994.381.88.

[85] Ronicke S, Hahn V, Horn R, Grone I, Brahn H and Friedt W. Interspecific hybrids of sunflower as sources of Sclerotinia resistance. Plant Breed. 2004;123(2):152–157. doi: 10.1046/j.1439-0523.2003.00925.x.

[86] Burke JM, Gardner KA and Rieseberg LH. The potential for gene flow between cultivated and wild sunflower (*Helianthus annuus*) in the United States. Am. J. Bot. 2002;89(9):1550–1552. doi: 10.3732/ajb.89.9.1550.

[87] Sackston WE. Studies of sunflower rust III. Occurrence, distribution and significance of races of *Puccinia helianthi* Schw. Can. J. Bot. 1962;40(11):1449–1458. doi: 10.1139/b62-139.

[88] Yang SM. 1986. A new race of *Puccinia helianthi* of sunflower. Ann Phytopathol. Soc. Japan. 1986;52(2):248–252. http://ci.nii.ac.jp/lognavi?name=crossref&id=info:doi/10.3186/jjphytopath.52.248.

[89] Lambrides CJ and Miller JF. 1994. Inheritance of rust resistance in a source of MC 29 sunflower germplasm. Crop Sci. 1994;34:1224–1230. doi: 10.2135/cropsci1994.0011183X003400050015x.

[90] Antonelli EF. Variabilidad de la población patógena de *Puccinia helianthi* Schw. en la Argentina. *In:* ISA, edition. Proc. 11th Intl. Sunflower Conf., Mar del Plata, Argentina, 10–13 March 1985. Intl. Sunflower Assoc., Toowoomba, Australia. 1984;2:591–596.

[91] Senetinner AC, Atonelli EF and Luduena PM. Analisis genetico de la resistancia a *Puccinia helianthi* Schw. de cuatro lineas de girasol. *In:* ISA, edition. Proc. 11th Intl. Sunflower Conf. Mer del Plata, Argentina, March 10–13. Intl. Sunflower Assoc., Toowoomba, Australia. 1984;2:597–602.

[92] Hugues N, Perez-Fernandez J and Quiroz F. *Puccinia helianthi* Schw. infecciones en hibridos comerciales en Argentina y su evolucion durante dos decadas. *In*: Velasco L, editor. Proc. 17th Intl. Sunflower Conf., Cordoba, Spain, 8–12 June, 2008. Intl. Sunflower Assoc., Paris, France. 2008;1:214–218.

[93] Vicente PA and Zazzerini A. Identification of sunflower rust (*Puccinia helianthi*) physiological races in Mozambique. Helia. 1997;20(27):25–30. http://isasunflower.org/fileadmin/documents/HELIA_issues/Helia27/HELIA_27_-_PG_25.pdf.

[94] Pustovoit GV and Slysar EL. 1978. Sunflower selection to resistance to rust. *In:* VNIIMK, edition. Zbornik VNIIMK: Pests and diseases of oil crops. Krasnodar, USSR. 1978. pp. 42–47. (In Russian)

[95] Miller JF, Rodriguez RH and Gulya TJ. Evaluation of genetic materials for inheritance to race 4 rust in sunflower. *In:* ISA, edition. Proc 12th Intl. Sunflower Conf., Novi Sad, Yugoslavia, 25–29 July 1988. Intl. Sunflower Assoc., Paris, France. 1998;2:361–364.

[96] Kochman JK and Goulter KC. A proposed system for identifying races of sunflower rust. *In:* ISA, edition. Proc. 11th Intl. Sunflower Conf. Mer del Plata, Argentina, March 10–13. Intl. Sunflower Assoc., Toowoomba, Australia. 1984;2:391–396.

[97] Sendall BC, Kong GA, Goulter KC, Aitken EAB, Thompson SM, Mitchell JHM, Kochman JK, Lawson W, Shatte T and Gulya TJ. Diversity in the sunflower: *Puccinia helianthi* pathosystem in Australia. Australas. Plant Pathol. 2006;35(6):657–670. doi: 10.1071/AP06071.

[98] Gulya TJ and Maširević S. Proposed methodologies for inoculation of sunflower with *Puccinia helianthi* and for disease assessment. Proposed methodologies for inoculation of sunflower with different pathogens and for disease assessment. FAO-European Research Network on Sunflower (Rome, Italy). Bucharest, Romania. 1994:31–47.

[99] Škorić D. 1984. Sunflower breeding for resistance to *Diaporthe/Phomopsis helianthi*. Munt.-Cvet. *et al.* Helia. 1984;8:21–24.

[100] Vrânceanu AV, Csep N, Pirvu N and Stoenescu FM. 1983. Genetic variability of sunflower reaction to the attack of *Phomopsis helianthi*. Helia. 1983;6:23–24.

[101] Vrânceanu AV, Craiciu DS, Soare G, Păcureanu-Joiţa M, Voinescu G and Sandu I. Sunflower genetic resistance to Phomopsis attack. *In:* ISA, edition. Proc. 13th Intl. Sunflower Conf. Pisa. Italy. September 7–11. 1992. Intl. Sunflower Assoc. Paris. France. 1992;2:1301–1306.

[102] Langar K, Griveau Y, Serieys H and Berville A. Genetic analysis of *Phomopsis (Diaporthe helianthi* Munt.-Cvet. *et al.)* disease resistance in cultivated sunflower (*Helianthus annuus* L.). *In:* ISA, edition. Proc. 14th Intl. Sunflower Conf., Toulouse, France. June 12–14. 2000. Intl. Sunflower Assoc., Paris, France. 2000; IIK:90–94.

[103] Langar K, Griveau Y, Serieys H, Kaan F and Berville A. Mapping components of resistance to *Phomopsis* (*Diaporthe helianthi*) in a population of sunflower recombinant inbred lines. *In:* Seiler GJ, editor. Proc. 16th Intl. Sunflower Conf., Fargo, ND, USA, 29 August–4 September 2004. Intl. Sunflower Assoc., Paris, France. 2004; II:643–649.

[104] Pecchia S, Mercatelli E and Vannacci G. Intraspecific diversity within *Diaporthe helianthi*: evidence from rDNA intergenic spacer (IGS) sequence analysis. Mycopathologia. 2004;157(3):317–326. doi: 10.1023/B:MYCO.0000024185.66158.7e.

[105] Rekab D, Del Sorbo G, Reggio C, Zoina A and Firrao G. Polymorphisms in nuclear rDNA and mtDNA reveal the polyphyletic nature of isolates of *Phomopsis* to sunflower and a tight monophyletic clade of defined geographic origin. Mycol. Res. 2004;108(4):393–402. http://dx.doi.org/10.1017/S0953756204009372.

[106] Dozet BM. Resistance to *Diaporthe/Phomopsis helianthi* Munt.-Cvet. *et al.* in wild sunflower species. *In*: Proc. 12th Sunflower Research Workshop, Fargo, ND, 9–10 January, 1990. Natl. Sunflower Assoc., Bismarck, ND. USA. 1990:86–88.

[107] Christov M, Shindrova P, Encheva V, Bacharova R and Christova M. New sunflower forms resistant to broomrape. *In:* Wegmann, K., Musselman, L.J. and Joel, D.M. (Eds.), Current Problems of *Orobanche* Research. *In*: Wegmann K, Muselman LJ and Joel DM, editors. Current Problems of *Orobanche* Researchers. Proc. 4th Intl. Workshop on Orobanche, 23–26 September, 1998. Albena, Bulgaria; 1988. pp. 317–319.

[108] Nikolova LM, Christov M and Seiler G. Interspecific hybridization between *H. pumilis* Nutt. and *H. Annuus* L. and their potential for cultivated sunflower improvement. Helia. 2004;27(41):151–162. http://dx.doi.org/10.2298/hel0441151n.

[109] Degener, J., Melchinger, A.E. and Hahn, V., 1999. Interspecific hybrids as sources of resistance to *Sclerotinia* and *Phomopsis* in sunflower breeding. Helia. 1999;22(30):49–60.

[110] Encheva V, Valkova D and Christov M. Reaction of some annual and perennial sunflower species of the genus *Helianthus* to the casual agent of gray spot of sunflower. DAI, General Toshevo, Bulgaria. Field Crop. Stud. 2006;3(1):141–148. ISSN 1312-3882.

[111] Putt ED. Breeding behaviour of resistance to leaf mottle or *Verticillium* in sunflower. Crop Sci. 1964;4(2):177–179. doi: 10.2135/cropsci1964.0011183X000400020016x.

[112] Fick GN and Zimmer DE. Monogenic resistance to *Verticillium* wilt in sunflowers. Crop Sci. 1974;14:603–604. doi: 10.2135/cropsci1974.0011183X001400040036x.

[113] Bruniard JM, Luduena PM and Ivanovich A. Selection of sunflower lines with resistance to *Verticillium dahliae* Kleb./IDIA 413/416. Estac. Exp. Reg. Agropec. INTA, Pergamino, Argentina. 1984. pp. 96–98.

[114] Bertero de Romano A and Vazquez AN. *Verticillium dahliae* Kleb. Estimacion de perdidas de rendimiento pare distintas intensidades de ataque. *In:* ISA, edition.Proc. 11th Intl. Sunflower Conf.Mar del Plata, Argentina, March 10–13. Intl. Sunflower Conf. Toowoomba, Australia. 1984;2:379–384.

[115] Pustovoit GV and Krokhin EJ. The inheritance of resistance to major pathogens in interspecific sunflower hybrids. *In*: VNIIMK Rewiev: Pests and diseaseas of oil crops, Krasnodar. 1978. pp. 40–44. (In Russian).

[116] Škorić D. Genetic resources in the *Helianthus* genus. *In*: Proc. of the Int. Symp. on Science and Biotechnology for an Integral Sunflower Utilization. October 24, 1984. Bary, Italy. 1984. pp. 37–73.

[117] Manici, L.M., Cerato, C. and Caputo, F. Pathogenic and biological variability of *Macrophomina phaseolina* (Tassi) Goid. isolates in different areas of sunflower cultivation in Italy. *In:* ISA, edition. Proc. 13th Intl. Sunflower Conf. Pisa, Italy, September 7–11 1992. Intl. Sunflower Assoc. Paris, France. 1992;1:779–784.

[118] Iliescu H. FAO-Progress report–Working subgroup on *Macrophomina phaseolina.* FAO–Rome, Italy. Helia. 1999;22 (Special issue):334–342.

[119] Ionita A and Iliescu H. Proposed methodologies for inoculation of sunflower with *Macrophominaphaseolina* and for disease assessment. FAO (Rome) – European Research Network on Sunflower: Proposed methodologies for inoculation of sunflower with different pathogens and for disease assessment, Bucharest. 1994. pp. 17–19.

[120] Aćimović M. Diseases of Sunflower. Institute of Field and Vegetable Crops, Novi Sad. Feljton, Novi Sad, Republic of Serbia. 1998. 717 p. (In Serbian).

[121] Walcz I and Piszker Z. Artificiaal inoculation technique for selecting resistance of sunflower to *Macrophomina phaseolina* (Tassi) Goidanich. *In*: Book of Abstr. 10th national Plant Breeding Conference, Budapest, Hungary. 2004. p. 171. (In Hungarian).

[122] Mihaljčević M. Choosing a suitable technique for screening sunflower for resistance to charcoal rot in climatic conditions of Yugoslavia. *In:* ISA, edition. Proc. 8th Intl. Sunflower Conf. Minneapolis, Minnesota, USA, July 23–27. Intl. Sunflower Assoc. Toowoomba, Australia. 1978. pp. 248–263.

[123] Mihaljčević M. Finding the sources of resistance to *Sclerotium bataticola* Gaiv. for inbred lines and sunflower hybrids. M.Sc. Thesis, University of Novi Sad, Faculty of Agriculture.1980. 49 p. (In Serbian).

[124] Ahmad M, Hussain T and Mehdi SS. Effect of various charcoal rot isolates on agronomic traits of exotic sunflower inbred lines. Helia. 1991;14(14):79–84.

[125] Maširević S. Proposed methodologies for inoculation of sunflower with *Phomopsis* spp./*Diaporthe* spp. and for disease assessment. FAO (Rome)–European Research Network on Sunflower. Proposed methodologies for inoculation of sunflower with different pathogens and for disease assessment. Bucharest. 1994. pp. 11–13.

[126] Fayzalla ES. Studies on biology, epidemiology and control of *Phoma macdonaldi* Boerema of sunflower. MSc. Thesis, University of Novi Sad, Faculty of Agriculture. 1978. 118 p.

[127] Darwishzadeh R, Dechamp-Guillaume G, Hewezi T and Sarrafi A. Genotype-isolate interaction for resistance to black stem in sunflower (*Helianthus annuus*). Plant Pathol. 2007;56(4):654–660. doi: 10.1111/j.1365-3059.2007.01603.x.

[128] Darwishzadeh R, Kiani SP, Dechamp-Guillaume G, Gentzbittel L and Sarrafi A. Quantitative trait loci associated with isolate specific and isolate non-specific partial resistance to *Phoma macdonaldii* in sunflower. Plant Pathol. 2007;56(5):855–861. doi: 10.1111/j.1365-3059.2007.01620.x.

[129] Alignan M, Hewezi T, Petitprez M, Dechamp-Guillaume G and Gentzbittel L. A cDNA microarray approach to decipher sunflower (*H. annuus* L.) responses to the necrotrophic funges *Phoma macdonaldii*. New Phytol. 2006;170(3):523–536. doi: 10.1111/j.1469-8137.2006.01696.x.

[130] Regina MV, Leite BC, Amorins L, Filho ABN, De Oliveira MC and De Castro C. Effect of sowing date and initial inoculums of *Alternaria heliathi* on sunflower in the south region of Brazil. *In*: Velasco L, editor. Proc. 17th Intl. Sunflower Conf., Cordoba, Spain, 8–12 June, 2008. Intl. Sunflower Assoc., Paris, France. 2008;1. pp. 193–198.

[131] Dudienas C, Ungaro MRG and Moraes SA. *Alternaria* disease development under tropical conditions. Helia. 1998;21(29):63–72.http://isasunflower.org/publications/helia-scientific-review.html.

[132] Aćimović M. Inbred sunflower lines as a source of resistance to economically important diseases. Agric. Sci. Rev. 1976;39(49):79–88. (In Serbian)

[133] Madhavi KJ, Sujatha M, Raja Ram Reddy D and Chander Rao S. Biochemical characterization of resistance against *Alternaria helianthi* in cultivated and wild sunflowers. Helia. 2005;28(43):13–24. http://dx.doi.org/10.2298/hel0543013m.

[134] Lipps PE and Herr LJ. Reactions of *Helianthus annuus* and *H. tuberosus* plant introductions to *Alternaria helianthi*. Plant Dis. 1986;70(9):831–834. doi: 10.1094/PD-70-831.

[135] Morris JB, Yang SM and Wilson I. Reaction of *Helianthus* species to *Alternaria helianthi*. Plant Dis. 1983;67(5):539–540. doi: 10.1094/PD-67-539.

[136] Sujatha M, Prakabaran AJ and Chattopadhyay C. Reaction of wild sunflowers and certain interspecific hybrids to *Alternaria helianthi*. Helia. 1997;20(27):14–24. http://isasunflower.org/publications/helia-scientific-review.html.

[137] Antonova TS. Sunflower disease caused by the fungi of *Fusarium* sp. Nauka Kubani. 2004;3:33.

[138] Antonova TS, Saukova SL and Klippert YV. The reaction of sunflower roots and stems on infection by fungi of *Fusarium* species. Nauka Kubani. 2004;4:92.

[139] Goncharov SV, Antonova TS and Saukova SL. Sunflower breeding for resistance to *Fusarium*. Helia. 2006;29(45):49–54. http://dx.doi.org/10.2298/hel0645049g.

[140] Orellana RG. Sources of resistance to soilborne fungal disease complex of sunflower. Plant Dis. Reptr. 1973;47(4):318–320.

[141] Rogers CE, Thompson TE and Zimmer DE. *Rhizopus* head rot of sunflower: Etiology and severity in the Southern Plains. Plant Dis. Reptr. 1978;62:769–771.

[142] Thompson TE, Rogers CE and Zimmerman DC. Sunflower oil quality and quantity as affected by rhizopus head rot. J. Am. Oil Chem. Soc. 1980;57(3):106–108. doi: 10.1007/BF02678814.

[143] Yang SM, Morris JB and Thompson TE. 1980. Evaluation of *Helianthus* spp. for resistance to *Rhizopus* head rot. *In*: ISA, edition. Proc. 9th Intl. Sunflower Conf., Torremolinos, Spain, 8–13 July, 1980. Intl. Sunflower Assoc., Paris, France. 1980. pp. 147–151.

[144] Agrawat JM, Vaish OP, Mathur SJ and Shhipa HP. Some observations on *Rhizopus* head rot of sunflower in Fajasthan, India. *In:* ISA, edition. Proc. 8th Intl. Sunflower Conf. Minneapolis, Minnesota, USA, 23–27 July 1978. Intl. Sunflower Assoc. Toowoomba, Australia. 1978. pp. 264–270

[145] Gulya T, Rashid KY and Maširević SM. Sunflower Diseases. *In:* Schneiter AA, editor. Sunflower Technology and Production. Agronomy Monograph 35. American Society of Agronomy, Crop Science Society of America, Soil Science Society of America, Madison, WI, USA. 1997. pp. 263–379. http://dx.doi.org/10.2134/agronmonogr35.c6.

[146] Saliman M, Yang SM and Wilson L. Reaction of *Helianthus* species to *Erysiphe cichoracearum*. Plant Dis. 1982;66(7):572–573. doi: 10.1094/PD-66-572.

[147] Jan CC and Chandler JM. Transfer of powdery mildew resistance from *Helianthus debilis* Nutt. to cultivated sunflower. Crop Sci. 1984;25(4):664–666. doi: 10.2135/cropsci1985.0011183X002500040020x.

[148] Prats J. Avenir du tournesol (The future of the sunflower). Bull. tech. Inf. Ingrs Servs agric. Paris. France. 1970: pp. 615–622. (CABI Record Number 19711608593).

[149] Burlov VV, Artemenko YP. Prospects of sunflower breeding for immunity to head grey mold *Botrytis cinerea.* (Title translated from Russian). Nauchno-tekhnickeskii biulleten Vsesoiuznogo selektsionno-geneticheskogo instituta. Odessa, USSR. 1980 (3): pp. 62–65.

[150] Kostyuk SV. Results of on ecological study of the sunflower gene pool for resistance to grey mould. *In*: Proceeding, Problemy selektsii i semenovodstva podsolnechnika dlya zasushlivykh usloviy stepi. VSGI Odessa, USSR. 1985: pp. 67–74. (P.B. Abstr.).

[151] Sreehari M, Govinda-Rao N and Sastry KS,. Studies on insect transmission of sunflower mosaic disease. Helia. 1992;14(17):65–68.

[152] Lenardon SL, Giolitti F, Leon A, Bazzalo ME and Grondona M. Effect of sunflower chlorotic mottle virus infection on sunflower yield paremeters. Helia. 2001;24(35):55–66. http://dx.doi.org/10.1515/helia.2001.24.35.55.

[153] Lenardon SL, Bazzalo ME, Abratti G, Cimmino CJ, Galella MT, Grondona M, Giolitti F and Leon AJ. Screening sunflower for resistance to sunflower chlorotic mottle virus and mapping the Rcmo-1 resistance gene. Crop Sci. 2004;45(2):735–739. doi: 10.2135/cropsci2005.0735.

[154] Dujovny G, Usugi T, Shohara K and Lenardon S. Characterization of a potyvirus infecting sunflower in Argentina. Plant Dis. 1998;82(5):470–474. http://dx.doi.org/10.1094/PDIS.1998.82.5.470.

[155] Arias MC, Lenardon S and Taleisnik E, 2003. Carbon metabolism alterations in sunflower plants infected with the sunflower chlorotic mottle virus. J. Phytopatology, 2003;151(5):267–273. doi: 10.1046/j.1439-0434.2003.00718.x.

[156] Mailo D, Poverene M, Giolitti F and Lenardon S. Differential gene expression in SuCMoV–tolerant and susceptible sunflower lines. *In:* Velasco L, editor. Proc. 17th Intl. Sunflower Conf. Cordoba, Spain, June 8–12, 2008. Intl. Sunflower Assoc. Paris, France. 2008;2:634–640.

[157] Nemeth J and Walcz I. Bacterial disease of sunflower in Hungary caused by *Erwinia carotovora*. Helia. 1992;15(16):79–84.

[158] Morozov VK, 1947. Sunflower breeding in USSR. Pishchepromizdat. Moscow. USSR. 1947: 274 p. (In Russian).

[159] Vrânceanu AV, Tudor VA, Stoenescu FM and Parvu N. Virulence groups of *Orobanche cumana* Wallr., differential hosts and resistance sources and genes in sunflower. *In*: ISA, edition. Proc. 9th Intl. Sunflower Conf. Torremolinos, Spain. June 8–13. 1980. Intl. Sunflower Assoc., Paris, France. 1980;2:74–81.

[160] Pâcureanu-Joița M, Vrânceanu AV, Scare G, Marinescu A and Sandu I. The evaluation of the parasite-host interaction in the system *Helianthus annuus* L. *Orobanche cumana* Wallr. in Romania. *In:* Stamenković S, editor. Proc. 2nd Balkan Symposium of field crops, Novi Sad, Yugoslavia, June 16–20, 1998, Institute of Field and Vegetable Crops, Novi Sad, Yugoslavia. 1: pp. 143–147.

[161] Pâcureanu-Joița M, Raranciue S, Procopovici E, Sava E and Nastase D. The impact of the new races of broomrape (*Orobanche cumana* Wallr.) parasite in sunflower crop in Romania. *In:* Velasco L, editor. Proc. 17th Intl. Sunflower Conf. Cordoba, Spain, June 8–12, 2008. Intl. Sunflower Assoc. Paris, France. 2008;1:224–231.

[162] Pâcureanu-Joița M, Veronesi C, Raranciue S and Stancin D. Parasite-host interaction of *Orobanche cumana* Wallr. with *Helianthus annuus* L. *In:* Seiler G, editor. Proc. 16th Intl. Sunflower Conf. Fargo, ND, USA. August 29–September 2. 2004. Intl. Sunflower Assoc. Paris. 2004;1:171–177.

[163] Pustovoit GV. Sunflower breeding for group resistance by interspecific hybridization. Kolos, Moscow. USSR. 1974: pp. 164–210. (In Russian)

[164] Fernandez-Martinez JM, Melero-Vara J, Munoz-Ruiz J, Ruso J and Dominguez J. Selection of wild and cultivated sunflower for resistance to a new broomrape race that overcomes resistance of the Or_5 gene. Crop Sci. 2000;40(2):550–555. doi: 10.2135/cropsci2000.402550x.

[165] Christov M. Characterization of wild *Helianthus* species as sources of new features for sunflower breeding. *In*: Hind DJN, Caligari P and Smith SAL, editors. *Compositae*:

Biology and Utilization. Proc. Intl. Composite Conf., The Royal Botanic Gardens, 24 July–4 August, 1994. Kew, London.1996;2:447–470. ISBN: 1900347008, 9781900347006

[166] Jan CC, Ruso JA, Munoz-Ruz J and Fernandez-Martinez JM. Resistance of sunflower (*Helianthus*) perennial species, interspecific amphiploides, and backcross progeny to broomrape (*Orobanche cumana* Wallr.) races. *In:* ISA, edition. Proc 14th Intl. Sunflower Conf. Toulouse, France, June 12–14, 2000. Intl. Sunflower Assoc. Paris, France. 2000;2:14–19.

[167] Panchenko AY. Early diagnosis of broomrape resistance in breeding and improving seed production. Vestnik-S.H.N. 1974: (2):107–114. (In Russian)

[168] Grezes-Besset B, 1994. Evaluation de la resistance du tournesol a l'*Orobanche*. Rustica Prograin Genetique. Protocoleno. 1994: E-16. Version (1):1–7.

[169] Schulz JT. Insects pests. *In:* Carter JF, editor. Sunflower Science and Technology. Agronomy Monograph 19. CSSA, ASA and SSSA, Madison, WI. USA. 1978: Agron. Soc. Am. Agron. Monog. 1978;19(6):169–223. doi: 10.2134/agronmonogr19.c6.

[170] Zimmer DE and Hoes JA. Diseases. *In*: Carter JF, editor. Sunflower Science and Technology. Agronomy Monograph 19. CSSA, ASA and SSSA, Madison, WI. USA. 1978;19(7):225–262. doi: 10.2134/agronmonogr19.c7.

[171] Pustovoit GV. Distant interspecific hybridization of sunflowers. *In:* Proc. 2nd Int. Sunflower Conf., Morden, Manitoba, Canada, 17–18 Aug. 1966. Int. Sunflower Assoc. Toowoomba, Australia. 1966: pp. 82–101.

[172] Gundaev AI, Basic principles of sunflower selection. *In*: Genetic Principles of Plant Selection, Nauka, Moscow. USSR. 1971: pp. 417–464. (Transl. Dep. of the Secretary of State, Ottawa, Canada, 1972)

[173] Rogers CE and Kreitner GL. Phytomelanin of sunflower achenes: a mechanism for pericarp resistance to abrasion by larvae of the sunflower moth (*Lepidoptere*: *Phyralidae*). Environ. Entomol. 1983;12(2):277–285.

[174] Charlet LD and Brewer GJ. Management strategies for insect pests of sunflower in North America. *In:* Recent research developments in entomology. Research Signpost, Trivandrum, India. 1997;1:215–229.

[175] Miller JF and Charlet LD. Breeding for insect resistance. *In:* Škorić D and Sakač Z, editors. Sunflower Genetics and Breeding. (International monography). Serbian Academy of Sciences (SASA), Branch in Novi Sad, Novi Sad, Republic of Serbia; 2012. pp. 447–499. ISBN: 978-88-81125-82-3.

[176] Rogers CE. Biology and breeding for insect and disease resistance in oilseed crops. *In:* Harris MK, editor. Biology and Breeding for resistance to Arthropods and Pathogenes in Agricultural Plants. Proceedings of the International Short Course in Host Plant Resistance. Texas. Texas A&M University. College Station. USA. 1980. 605 p.

[177] Rogers CE, Thompson TE and Seiler GJ. Sunflower Species of the United States. Published by the National Sunflower Association, P.O. Box, 433, Bismarck, ND 48402. USA. 1982. 75 p.

[178] Rogers CE and Thompson TE. *Helianthus* resistance to the carrot beetle. J. Econ. Entomol. 1978;71(5):760–761. http://dx.doi.org/10.1093/jee/71.5.760.

[179] Rogers CE. Cerambycid pests of sunflower: Distribution and behaviour in the Southern Plains. Environ. Entomol. 1978;6(6):833–838. http://dx.doi.org/10.1093/ee/6.6.833.

[180] Rogers CE. Resistance of sunflower species to the western potato leafhopper. Environ. Entomol. 1981;10(5):697–700. http://dx.doi.org/10.1093/ee/10.5.697.

[181] Rogers CE and Thompson TE. *Helianthus* resistance to the sunflower beetle (*Coleoptera: Chrysomelidae*). J. Kansas Entomol. Soc. 1980;53(4):727–730. http://www.jstor.org/stable/25084103.

[182] Rogers CE, Thompson TE and Wellik MJ. Survival of *Bothynus gibbosus* (Coleoptera: Scarabaeidae) on Helianthus Sp. J. Kansas Entomol. Soc. 1980;53(3):490–494. http://www.jstor.org/stable/25084064.

[183] Rogers CE and Thompson TE. Resistance of wild Helianthus species to an Aphid, Masonaphis masoni. J. Econ. Entomol. 1978;71(2):221–222. http://dx.doi.org/10.1093/jee/71.2.221.

[184] Rogers CE, Thompson TE and Gagne RJ. Cecidomyiidae of Helianthus: Taxonomy, hosts and distribution. Ann. Entomol. Soc. Am. 1979;72(1):109–113. http://dx.doi.org/10.1093/aesa/72.1.109.

Permissions

All chapters in this book were first published in ABSPRAFP, by InTech Open; hereby published with permission under the Creative Commons Attribution License or equivalent. Every chapter published in this book has been scrutinized by our experts. Their significance has been extensively debated. The topics covered herein carry significant findings which will fuel the growth of the discipline. They may even be implemented as practical applications or may be referred to as a beginning point for another development.The contributors of this book come from diverse backgrounds, making this book a truly international effort. This book will bring forth new frontiers with its revolutionizing research information and detailed analysis of the nascent developments around the world.

We would like to thank all the contributing authors for lending their expertise to make the book truly unique. They have played a crucial role in the development of this book. Without their invaluable contributions this book wouldn't have been possible. They have made vital efforts to compile up to date information on the varied aspects of this subject to make this book a valuable addition to the collection of many professionals and students.

This book was conceptualized with the vision of imparting up-to-date information and advanced data in this field. To ensure the same, a matchless editorial board was set up. Every individual on the board went through rigorous rounds of assessment to prove their worth. After which they invested a large part of their time researching and compiling the most relevant data for our readers.

The editorial board has been involved in producing this book since its inception. They have spent rigorous hours researching and exploring the diverse topics which have resulted in the successful publishing of this book. They have passed on their knowledge of decades through this book. To expedite this challenging task, the publisher supported the team at every step. A small team of assistant editors was also appointed to further simplify the editing procedure and attain best results for the readers.

Apart from the editorial board, the designing team has also invested a significant amount of their time in understanding the subject and creating the most relevant covers. They scrutinized every image to scout for the most suitable representation of the subject and create an appropriate cover for the book.

The publishing team has been an ardent support to the editorial, designing and production team. Their endless efforts to recruit the best for this project, has resulted in the accomplishment of this book. They are a veteran in the field of academics and their pool of knowledge is as vast as their experience in printing. Their expertise and guidance has proved useful at every step. Their uncompromising quality standards have made this book an exceptional effort. Their encouragement from time to time has been an inspiration for everyone.

The publisher and the editorial board hope that this book will prove to be a valuable piece of knowledge for researchers, students, practitioners and scholars across the globe.

List of Contributors

Paula Andrea Castrillón-Arbeláez
Escuela Nacional de Estudios Superiores (ENES)-UNAM, Unidad León, Predio El SaucilloyEl Potrero. C.P.. León, Gto., México

John Paul Délano Frier
Centro de Investigación y de Estudios Avanzados del I. P. N. (Cinvestav), Unidad Irapuato, Gto., México

Roberto Bobadilla and Alexandre Berr
Institut de Biologie Moléculaire des Plantes, CNRS UPR2357, Université de Strasbourg, Strasbourg, France

Ludger Wessjohann
Leibniz Institute of Plant Biochemistry, Department of Bioorganic Chemistry, Halle/Saale, Germany

Uta Greifenhagen, Dominic Brauch and Claudia Birkemeyer
Universität Leipzig, Faculty of Chemistry and Mineralogy, Leipzig, Germany

Gagan Paudel and Andrej Frolov
Leibniz Institute of Plant Biochemistry, Department of Bioorganic Chemistry, Halle/Saale, Germany
Universität Leipzig, Faculty of Chemistry and Mineralogy, Leipzig, Germany

Elena Lukasheva, Natalia Osmolovskaya and Elena Tarakhovskaya
St. Petersburg State University, Department of Plant Physiology and Biochemistry, Saint Petersburg, Russia

Tatiana Bilova
Leibniz Institute of Plant Biochemistry, Department of Bioorganic Chemistry, Halle/Saale, Germany
Universität Leipzig, Faculty of Chemistry and Mineralogy, Leipzig, Germany
St. Petersburg State University, Department of Plant Physiology and Biochemistry, Saint Petersburg, Russia

Gerd Ulrich Balcke, Alain Tissier and Thomas Vogt
Leibniz Institute of Plant Biochemistry, Department of Cell and Metabolic Biology, Halle/Saale, Germany

Carsten Milkowski
Martin Luther University Halle-Wittenberg, Interdisciplinary Center for Crop Plant Research (IZN), Halle/Saale, Germany

Shirin Mirlohi and Yuke He
National Key Laboratory of Plant Molecular Genetics, Shanghai Institute of Plant Physiology and Ecology, Shanghai Institutes for Biological Sciences, Chinese Academy of Sciences, Shanghai, China

Zbynek Heger
Department of Chemistry and Biochemistry, Laboratory of Metallomics and Nanotechnologies, Mendel University in Brno, Brno, Czech Republic

Miguel Angel Merlos Rodrigo, Ondrej Zitka, Adam Vojtech and Rene Kizek
Department of Chemistry and Biochemistry, Laboratory of Metallomics and Nanotechnologies, Mendel University in Brno, Brno, Czech Republic
Central European Institute of Technology, Brno University of Technology, Brno, Czech Republic

Naser A.Anjum and Eduarda Pereira
CESAM-Centre for Environmental and Marine Studies & Department of Chemistry, University of Aveiro, Aveiro, Portugal

Maria Antonia Machado Barbosa and Allan Klynger da Silva Lobato
Núcleo de Pesquisa Vegetal Básica e Aplicada, Universidade Federal Rural da Amazônia, Paragominas, Brazil

Milton Hélio Lima da Silva
Coordenação de Botânica, Museu Paraense Emílio Goeldi, Belém, Brazil

Gabriel Mascarenhas Maciel
Instituto de Ciências Agrárias, Universidade Federal de Uberlândia, Monte Carmelo, Brazil

Douglas José Marques
Setor de Olericultura e Experimentação, Universidade José do Rosário Vellano (UNIFENAS), Alfenas, Brazil

Allan Klynger da Silva Lobato, Emily Juliane Alvino Lima and Elaine Maria Silva Guedes Lobato
Núcleo de Pesquisa Vegetal Básica e Aplicada, Universidade Federal Rural da Amazônia, Paragominas, Brazil

Gabriel Mascarenhas Maciel
Instituto de Ciências Agrárias, Universidade Federal de Uberlândia, Monte Carmelo, Brazil

Douglas José Marques
Setor de Olericultura e Experimentação, Universidade José do Rosário Vellano (UNIFENAS), Alfenas, Brazil

Shekhar S. Mudagandur, Gopikrishna Gopalapillay and Koyadan K. Vijayan
Central Institute of Brackishwater Aquaculture, Chennai, India

Dragan Škorić
Serbian Academy of Sciences and Arts, SASA - Novi Sad Branch, Novi Sad, Serbia

Index

www.ingramcontent.com/pod-product-compliance
Lightning Source LLC
LaVergne TN
LVHW080919210926
848485LV00004B/682
9781639891344